Emerging Technologies
for Integrated Pest Management

Concepts, Research, and Implementation

Edited by

George G. Kennedy and Turner B. Sutton

Proceedings of a Conference
March 8-10, 1999
Raleigh, North Carolina
U.S.A.

APS PRESS
The American Phytopathological Society
St. Paul, Minnesota

This book has been reproduced directly from computer-generated
copy submitted in final form to APS Press by the editors
of the volume. No editing or proofreading has been done by the Press.

Reference in this publication to a trademark, proprietary product,
or company name by personnel of the U.S. Department of Agriculture
or anyone else is intended for explicit description only and does not
imply approval or recommendation to the exclusion of others that
may be suitable.

Library of Congress Catalog Card Number: 00-100194
International Standard Book Number: 0-89054-246-5

Printed in the United States of America on acid-free paper

The American Phytopathological Society
3340 Pilot Knob Road
St. Paul, Minnesota 55121-2097, USA

We dedicate this book to

Robert L. Rabb

whose intellectual probings and intolerance of "fuzzy thinking" have
been an inspiration for students and colleagues

Preface

In the decades since the concepts of IPM were first articulated, there have been dramatic advances in both the theory and practice of IPM. Today, throughout much of the world, IPM is the prevailing paradigm for crop protection. Recently, however, the U.S. National Research Council (NRC), in a report titled *Ecologically Based Pest Management: New Solutions for a New Century,* expressed concern that a focus on pesticides and individual pests continues to dominate IPM. The NRC has called for a shift in the IPM paradigm away from managing components or individual organisms to an approach that examines processes, flows, and relationships among organisms, an approach that is based on new knowledge and employs new methods to study, monitor, and evaluate new pest management tools. In essence, the NRC has called for a renewal of the original vision for IPM as articulated by Ray F. Smith, Carl Huffaker, Robert L. Rabb and others.

Achieving this vision requires new knowledge across the continuum from molecular to ecosystem. It also requires new technologies and an agricultural and institutional infrastructure that can both accommodate and facilitate the development and implementation of new IPM approaches and foster progress toward a realization of the vision.

During the last decade, there have been numerous scientific and technological advances that have tremendous potential for re-focusing and advancing IPM. These advances have the potential to alter the trajectory of IPM and to dramatically advance the practice of IPM. The development and synthesis of new information, the formulation of new concepts, and increased interdisciplinary collaboration will be required to accomplish this. These Proceedings are the results of a conference intended to facilitate the development and adoption of key emerging pest management technologies in a way that will optimize their contribution to IPM. The specific objectives of the conference were to:

- Stimulate the development of new concepts and approaches to the implementation and integration of emerging technologies in IPM;
- Provide a forum for interdisciplinary education and the sharing of perspectives among disciplinary specialists that will be necessary to sustain the success of IPM.

- Identify constraints, problems, and needs with respect to the development, implementation, and adoption of emerging technologies in IPM.

In selecting the major areas of focus for the conference, the organizing committee was frustrated by the constraints imposed by the 2.5-day format of the conference and the resulting inability to include presentations on all of the important areas in which significant and relevant technological advances are being made. The decision to emphasize diagnostics, genetic engineering, biological control, pesticide technology, global positioning and geographical information systems, and information processing and delivery technology reflects our view that advances in these areas have the potential to dramatically advance IPM. In restricting the focus of the conference to these areas, we do not wish to imply that advances in other areas do not have similar potential. Important areas that were not covered in the conference include semiochemicals, area-wide IPM, and artificial intelligence and expert systems.

The conference was well attended by participants from academia, government, and the private sector, representing research, extension, business, and public policy, as well as entomology, plant pathology, weed science, economics, and sociology. The conference was international, with representatives from 14 countries.

While it would be gratifying to state that all of the objectives of the conference were achieved, it is not possible to do so at this time. Clearly, the conference provided an assessment of the current status and potential of many emerging technologies for IPM, as well as an interdisciplinary forum for the presentation of new information and the exchange of ideas. Many of the presentations included discussions of constraints, problems, and needs with respect to the development, implementation, and adoption of emerging technologies, and these were a major topic of discussion. It became clear during the conference, however, that the program would have been significantly enhanced by formal presentations on the various technologies by economists.

If and how the information presented at the conference and in these Proceedings contributes to the development of new concepts and approaches to the integration of new technologies for crop protection and the implementation of IPM will only be revealed over time. At this point, we feel confident that the conference was successful in providing an interdisciplinary forum and in focusing attention on the status and the potential of important emerging technologies for IPM.

The conference was made possible by financial support from the following organizations/institutions, to whom we are most grateful: the USDA/CSREES Pest Management Program, the North Carolina Agricultural Research Service, the North Carolina Cooperative Extension Service, the NSF Center for Pest Management, the U.S. Environmental Protection Agency, the USDA Animal and Plant Health Inspection Service, and the USDA/CSREES NRI Competitive Grants Program.

G. G. Kennedy
T. B. Sutton
November 3, 1999
Raleigh, NC

Acknowledgments

We would like to acknowledge the American Phytopathological Society and the Entomological Society of America for their co-sponsorship of the conference and the advertising they provided on their Internet sites. We also extend our appreciation to the following organizations and institutions for their financial support: the USDA/CSREES Pest Management Program, the North Carolina Agricultural Research Service, the North Carolina Cooperative Extension Service, the NSF Center for Pest Management, the U.S. Environmental Protection Agency, the USDA Animal and Plant Health Inspection Service, and the USDA/CSREES NRI Competitive Grants Program (Award number 9802443).

We would like to acknowledge the support of our Department Heads, O. W. Barnett, Plant Pathology, and James Harper, Entomology, for their encouragement and support throughout the planning and execution of the conference. We also thank the members of the organizing committee for the many hours that they spent discussing, planning, and helping with the various aspects of the conference. The members were: T. E. Anderson, BASF Corporation, J. L. Apple, Professor Emeritus, Plant Pathology, O. W. Barnett, Plant Pathology, D. M. Benson, Plant Pathology, R. L. Brandenburg, Entomology, F. L. Gould, Entomology, J. D. Harper, Entomology, H. M. Linker, Crop Science, D. W. Monks, Horticultural Science, D. F. Ritchie, Plant Pathology, R. M. Roe, Entomology, and R. E. Stinner, Entomology. We want to especially thank Ron Stinner and the personnel at the NSF Center for IPM for creating and maintaining our World Wide Web site. Additionally, we would like to thank Kitty Kershaw and Catherine Phillips for help with preparation of the grants and mailings associated with the conference, Jennifer Ballard for design of the conference announcement and program, Clyde Sorenson for the conference cover design, Pat Robertson for coordinating the registration, and Mary Sue Harris for help with the financial accounting. Finally, we are particularly indebted to

Leslie Kennedy for copy editing the Proceedings.

In addition, we would like to recognize the student registration scholarship winners. Registration scholarships were established for graduate students through funding from the NSF IPM Center and the Director of Academic Programs in the College of Agriculture and Life Sciences at North Carolina State University. Students competed for these scholarships by writing letters indicating how attendance at the conference would help them in their research and professional development. Additionally, their major professors wrote letters supporting their application. The awardees were: Doug Anspaugh, Virginia Aris, Woody Bailey, Juan Cabrera, Kieran Clements, Eugenia Gonzalez, Nada Gudderra, Eric Honeycutt, Jaesoon Huang, Heike Meissner, Leah Millar, Godfrey Nalyana, Ana Romero, Nick Storer (all from North Carolina State University), Mahmood Ahmad (University of Adelaide, Australia), Alyssa Collins (Rutgers University), Soraya Franca (Brazil), Jennifer Grant (Cornell University), Esther Kioko (National Museum of Kenya), Marlo Lascano (Equador), and Larry Osburne (Oklahoma State University).

Table of Contents

Emerging Technologies for Integrated Pest Management

Concepts, Research, and Implementation

Section 1

Background

The authors contributing to this section were charged with providing the context within which to assess the potential roles of emerging technologies in IPM and to determine the needs with respect to research on and implementation of these technologies. In the first chapter, George Kennedy provides a historical perspective from which to assess the current status of IPM implementation. He identifies 10 key points that have emerged from our experiences with IPM and that determine the parameters within which IPM must operate. In the second chapter, Paul Zorner examines the changing agricultural and ecological context in which IPM operates, and focuses attention on the critically important issue of sustainability for pest management and for agriculture in general. In the final chapter in this section, Ann Sorensen and Patrick Stewart discuss the factors that influence the adoption of new technologies, and suggest that traditional innovation diffusion theory does not adequately describe the adoption of new technologies by IPM practitioners. They argue that pest management systems are being modified to accommodate new technologies, and caution that we must guard against "allowing new IPM technologies to drive producers' pest management systems instead of having the systems determine the technologies."

Perspectives on Progress in IPM

George G. Kennedy
Department of Entomology, North Carolina State University,
Raleigh, NC 27695-7630 USA

Twenty-nine years ago this month, an international conference on "Concepts of Pest Management" was held at North Carolina State University. That conference, organized by R. L. Rabb and F. E. Guthrie (13), focused primarily on the conceptual underpinnings for insect pest management. It stressed the ecological and economic principles in selecting and integrating methods of pest control. The conference was an important contribution to the debate that refined the IPM concept and ultimately led to the widespread, philosophical embrace of IPM by farmers, crop consultants, government regulatory bodies, research and extension scientists in the crop protection disciplines, and the agricultural chemical industry.

The IPM concept was formulated in response to growing concerns over adverse health and environmental effects of pesticides and limitations of the singular reliance on chemical pest control that had developed in much of agriculture. Those limitations included: outbreaks of secondary pests and the resurgence of target pest populations following destruction of their natural enemies, problems with pesticide resistance, hazards from residues in the harvested crop, direct hazards to pesticide applicators, and hazards to non-target organisms, including wildlife (15).

The developing IPM concept emphasized that the occurrence of pest problems was under the influence of the total agroecosystem (16), and that not all levels of pest abundance were damaging (17,18). It stressed that multiple methods should be used to control single pests as well as pest complexes. And it emphasized that pest management is a multidisciplinary endeavor that must be based on decision rules built on ecological principles and economic, social and environmental considerations (8).

2

In the case of plant pathology, effective, inexpensive chemical controls were not available to control many plant diseases. Much of the ongoing research on disease management and many of the recommended disease management practices were clearly consistent with the developing concept of IPM. However, as pointed out by Jacobsen (7), the early definitions and philosophical underpinnings of IPM reflected strongly the entomological perspectives of those who contributed most to the early, formal development of the concept and who were instrumental in its promotion. The emphasis of IPM on pest populations and economic thresholds did not readily accommodate the nature of the pest problems addressed by plant pathologists and weed scientists. Nor did the early language of IPM foster embrace of the concept by these disciplines, whose jargon was distinctly different. Thus, despite the strong emphasis of IPM on understanding the ecological context for pest problems, these disciplines were slow to formally embrace IPM.

The economic threshold concept as a decision rule for implementing pest control actions was fundamental to the development of IPM. As originally formulated, the economic threshold concept addressed pest populations that: 1) increased over time within the crop; 2) could be readily censused; 3) could be related in a predictable way to a reduction in crop yield or quality (i.e., damage); and 4) could be readily controlled by an insecticide to prevent yield loss (17,18). Because most plant pathogens and weeds do not share these attributes, the economic threshold concept was not readily embraced by plant pathologists and weed scientists. In the case of plant pathogens, populations are difficult to census directly in a meaningful way; the relationship of population size to reductions in yield is not always predictable; and for most pathogens, effective remedial controls, namely therapeutic fungicides, were not available (7,22). Similarly, in the case of weeds, damaging weed infestations almost always develop in the absence of control measures; the typical weed infestation consists of a great diversity of weed species; and most weed species produce large numbers of seeds that are extremely long-lived in the soil. Further, until relatively recently, therapeutic agents in the form of effective, post-emergence herbicides with a high degree of selectivity were not available to control weed infestations that reach economic thresholds (9).

Even within entomology, the economic threshold concept did not apply readily to complexes of pests or to insect pests that suddenly invade a crop in damaging numbers from sources outside of the crop. Further, the dynamic nature and uncertainty of the variables that determine an economic threshold made the practical application of the formal concept extremely difficult.

Despite these shortcomings, the concept provided a basis for eliminating unnecessary pesticide applications. In the broader context of IPM, modifications of the threshold concept (6,9,22) have led to the development of decision rules to assess the need to apply preventive tactics for pests that have extremely low thresholds and for complexes of pests that cause similar types of injury to the crop (e.g., 3). Implementation of the threshold concept also has been fostered by technological advances in a number of areas, including sampling methodology, sex pheromone-based monitoring systems for insect pests, temperature-driven pest development models, the development of computerized environmental monitoring systems, therapeutic fungicides, and selective post-emergence fungicides.

The impact of the economic threshold concept on the development and application of IPM has been dramatic, so much so that the use of scouting and the application of pesticides based on thresholds has been considered a minimum criterion for adoption of IPM in a recent assessment of IPM adoption on U.S. crop land by the USDA Economic Research Service (20). In this assessment, four levels of IPM implementation were recognized:

1. No IPM: chemically based pest control without the use of thresholds.

2. Low level IPM: use of scouting and thresholds to determine the need for and timing of pesticide applications, but few if any preventive measures used. In the case of corn, crop rotation without the use of a soil insecticide qualified as IPM implementation regardless of whether scouting was used.

3. Medium level IPM: use of scouting and thresholds plus one or two additional practices indicative of pest management (e.g., crop rotation, biological control, use of pest-resistant cultivar).

4. High level IPM or Bio-intensive IPM: use of scouting and thresholds plus three or more additional practices indicative of IPM.

4

Based on this scheme, the ERS report estimated that IPM is being used on 50% or more of the U.S. land planted to fruit, nut, vegetable and field crops. Thus, scouting and thresholds are being applied on at least 50% of land area devoted to these crops.

The importance of this achievement should not be depreciated, nor should it be dismissed as a sell-out by IPM to the agrichemical industry, as some have implied (1,2,10). Scouting and thresholds have contributed to significant reductions in pesticide use in many cropping systems and have enhanced the effectiveness of pesticide applications by improving their timing. Further, they have served to heighten farmers' awareness of the role that other pest management tactics can play in preventing pest populations from reaching damaging levels. In short, the threshold concept and its application in the broadest sense have provided the means by which much of U.S. crop production has been moved away from the routine, prophylactic use of insecticides and, to a lesser extent, fungicides and herbicides.

Critics of IPM are correct in pointing out that agriculture remains overly dependent on pesticides (1,11,12). Clearly, there is much more to IPM than the use of scouting and thresholds, and much remains to be done in the development and implementation of IPM if the agricultural, economic, social, and environmental goals of IPM are to be realized.

In an effort to promote progress in pest management and reduce our reliance on chemical pesticides, some have chosen to emphasize our continued dependence on pesticides and the extent to which higher levels of IPM have not been adopted on U.S. crop land (1). Although it is important that we recognize how much remains to be accomplished, it is equally important that we recognize the significant progress that has been made in the adoption of higher levels of IPM.

Benbrook et al. (1) developed a system for categorizing the level of IPM adoption that is more sophisticated and potentially more informative than that used by the Economic Research Service. The system of Benbrook et al. (1) distinguishes between practices that are designed to manage pesticide use cost effectively and practices that are designed to reduce pest pressure. In addition, their system weights practices according to their potential to decrease pest pressure. Using this system, Benbrook et al. (1) estimated that IPM is not practiced on 28% to 34% of U.S. crop land, and that low level IPM (scouting and thresholds; few if any preventive measures) is practiced on 34% to 42% of U.S. crop land. Significantly, they estimated that medium-level IPM (scouting and thresholds plus some preventive measures and efforts to reduce the use of broad-spectrum pesticides) is practiced on 22% to 28% of U.S. crop land, and that high level, or Bio-intensive, IPM (integration of multiple preventive practices resulting in control of pests without routine reliance on pesticides) is practiced on 4% to 8% of the U.S. crop land. Thus, under their more stringent criteria, medium to high levels of IPM are being practiced on 26% to 34% of U.S. crop land.

Does this level of IPM adoption constitute a significant accomplishment or not? Is the glass half full or is it half empty? The answer depends on one's perspective. At the time the federal government began to invest significantly in IPM during the early 1970s, agricultural pest control emphasized the prophylactic use of broad-spectrum pesticides where such were available. Reliance on them had been promoted for two decades and had become ingrained. Thus, pesticide use patterns that had developed over two decades had to be changed, and a new, more complex approach had to be accepted by farmers. The knowledge base for many pests and on many crops in the early 1970s was inadequate to develop and implement cost-effective and reliable IPM systems. The public research and extension infrastructure was not focused on IPM and required re-direction, and the infrastructure of private crop consultants necessary to implement IPM on a large scale was virtually nonexistent. The technology needed to effect a large-scale reduction in reliance on the routine use of broad-spectrum pesticides was inadequate or unavailable, and the private-sector research and development infrastructure was not focused on meeting IPM needs. In addition, an array of government policies and regulations impeded the development and use of new technologies and the adoption of many IPM practices (1,19,4). In view of these constraints, and given that adoption of complex new technologies is typically a slow process, it would seem that progress has been significant.

Over the years, public concern over the health and enviromental risks associated with agricultural pesticides has increased dramatically, and IPM has been looked to increasingly to address these concerns (8,14). The success or failure of IPM is being judged increasingly on the basis of reductions in the amounts of pesticides used (1,8,10,14). When viewed from this perspective, progress in IPM appears less dramatic (1,11,12). However, as pointed out by Kogan (8), reductions in pesticide use are a desirable consequence of

6

IPM, but they are neither the only measure of success, nor are they necessarily the best measure of success. The real issue should be whether pesticide use is in accordance with the principles of IPM and whether IPM is advancing us toward the multiple goals of maintaining a safe, reliable, and economic food supply with minimum risk to humans and other non-target organisms and to the environment.

Regardless of how one views the progress that has been made, it is clear that much remains to be done. The path before us is long and difficult. What have we learned from our experiences with IPM to date that can help guide its development and implementation in the future? I offer the following list of significant points based on my own experiences, my readings, and discussions with farmers, extension agents, crop consultants, and colleagues:

1. Pest management must be "conceptually and operationally linked to the total crop production system" (21). Yet, pest management is never the only priority in crop production, and is rarely the highest. It rarely determines the structure of an agricultural production system, and it never does so to the exclusion of economic factors. Pest management systems must be cost effective and logistically compatible with the overall farming operation in which they are to be implemented.

2. Agricultural production systems are dynamic. Change can occur rapidly. Some important agents of change include: economic factors, market opportunities, government policies and regulations, technological advances (which may or may not be related to pest management), the introduction of exotic pest species, and pest control crises.

3. The concepts of IPM are general but their application is site-specific. The array of pests that must be managed, as well as the context in which they must be managed, varies from site to site and year to year.

4. Biological knowledge alone is not adequate to develop and implement IPM programs. Such programs must be not only biologically sound, but also technically feasible and compatible with the goals and objectives of the farmer and the institutions (e.g., banks, landowners, governments) that influence what the

farmer does (19).

5. IPM is an interdisciplinary endeavor that must address entire pest complexes to maximize adoption and efficiency, and to minimize adverse interactions (e.g., 5). Implementation requires both extensive disciplinary knowledge and interdisciplinary collaboration and communication.

6. IPM implementation is information intensive and requires the ability to acquire, process, and act on a variety of categories of information, in real time. Training for IPM practitioners is critical and must be ongoing.

7. IPM technologies and tactics must be cost competitive with, or otherwise offer a clear advantage over, alternative technologies and tactics already in use if they are to be adopted.

8. Therapeutic measures that can be implemented easily and quickly and that are highly effective against pest infestations that have exceeded threshold levels will continue to be necessary.

9. IPM tactics are not risk free. In the face of increasing concern over environmental and health risks of all types, it is important to recognize the risks associated with the use patterns of each IPM tactic. Risks associated with IPM tactics and combinations of tactics must be weighed against risks associated with alternative tactics and balanced against the benefits to be derived from their use.

10. No single technology or pest management tactic acting alone will remain durable. The development of resistance to pest management tactics by pests will remain a constant threat.

Advances in technology have played a crucial role in the development and implementation of IPM, and they will continue to do so. During the last decade, there have been numerous scientific and technological advances that have tremendous potential for advancing IPM. Advances in immunology, recombinant DNA technology, geographic information systems, information processing and communications technology, pesticide science, and biological control are providing new tools for and approaches to pest

8

management. These advances have the potential to alter the trajectory of IPM and to dramatically advance its implementation. Accomplishing this will require the development and synthesis of new information, the formulation of new concepts, and an increase in interdisciplinary collaboration. This conference was conceived with the goal of facilitating the development and adoption of key, emerging pest management technologies in a way that will optimize their contribution to IPM. The specific objectives for the conference are to:

1. Stimulate the development of new concepts and approaches to the implementation and integration of emerging technologies in IPM;

2. Provide a forum for interdisciplinary education and the sharing of perspectives among disciplinary specialists that will be necessary to achieve integration in pest management;

3. Identify constraints, problems, and needs with respect to the development, implementation, and integration of emerging technologies in IPM.

In his introduction to the conference on *Concepts of Pest Management* (13), Bob Rabb stated, "As each new method or agent of pest control is developed . . . a decision as to its ecological and practical acceptability must be made." That statement is even more true today than it was 29 years ago. As we explore the various emerging technologies, I urge you to keep this point in mind and to consider the context in which these technologies must function. Each of these technologies holds great promise for IPM. Realizing this promise is the formidable challenge before us.

Literature Cited

1. Benbrook, C. M., Groth, E. III, Halloran, J. M., Hansen, M. K., and Marquardt, S. 1996. Pest Management at the Crossroads. Consumers Union, Yonkers, NY.
2. Cate, J., and Hinkle, M. 1993. Integrated Pest Management: the Path of a Paradigm. National Audubon Society, Washington, D.C.

3. Cartwright, B., Edelson, J. V., and Chambers, C. 1987. Composite action thresholds for the control of lepidopterous pests on fresh-market cabbage in the Lower Rio Grande Valley of Texas. J. Econ. Entomol. 80:185-181.
4. Cartwright, B., Collins, J. K., and Cuperus G. W. 1993. Consumer influences on pest control strategies for fruits and vegetables. Pages 151-170 in: Successful Implementation of Integrated Pest Management for Agricultural Crops. A. R. Leslie, and G. W. Cuperus, eds. Lewis Publishers, Boca Raton.
5. Gadoury, D. M. 1993. Integrating management decisions for several pests in fruit production. Plant Disease 77:299-302.
6. Higley, L. G., and Pedigo, L. P. eds. 1996. Economic Thresholds for Integrated Pest Management. Univ. Nebraska Press, Lincoln, NB.
7. Jacobsen, B. J. 1997. Role of plant pathology in integrated pest management. Annu. Rev. Phytopathol. 35:373-391.
8. Kogan, M. 1998. Integrated pest management: historical perspectives and contemporary developments. Annu. Rev. Entomol. 43:243-270.
9. Mortensen, D. A., and Coble, H. D. 1996. Economic thresholds for weed management. Pages 89-113 in: Economic thresholds for integrated pest management. L. G. Higley, and L. P. Pedigo, eds. Univ. Nebraska Press, Lincoln, NB.
10. National Research Council. 1996. Ecologically based pest management: new solutions for a new century. National Academy Press, Washington, D.C.
11. Pimentel, D. 1997. Pest management in agriculture. Pages 1-11 in: Techniques for Reducing Pesticide Use. D. Pimentel, ed. J. Wiley & Sons, NY.
12. Perkins, J. H., and Patterson B. R. 1997. Pests, pesticides, and the environment: a historical perspective on the prospects for pesticide reduction. Pages 13-33 in: Techniques for Reducing Pesticide Use. D. Pimentel, ed. J. Wiley & Sons, NY.
13. Rabb, R. L., and Guthrie, F. E. eds. 1970. Concpets of Pest Management. North Caorlina State University, Raleigh.
14. Smith, R.F. 1970. Pesticides: their use and limitations in pest management. Pages 103-118 in: Concepts of Pest Management. R. L Rabb, and F. E. Guthrie, eds. North Carolina State University, Raleigh.
15. Rajotte, E. G. 1993. From profitability to food safety and the

environment: shifting the objectives of IPM. Plant Disease 77: 296-299.

16. Smith, R. F., and van den Bosch, R. 1967. Integrated control. Pages 295-340 in: Pest Control: Biological, Physical and Selected Chemical Methods. W. W. Kilgore and R. L. Doutt, eds. Academic Press, New York.

17. Stern, V. M. 1966. Significance of the economic threshold in integrated pest control. Pages 41-56 in: Food and Agriculture Organization. Proceedings of. FAO Symposium on Integrated Pest Control. Rome, Oct. 11-15, 1965. Rome Italy. FAO-UN.

18. Stern, V. M., Smith, R. F., van den Bosch, R., and Hagen K. S. 1959. The integrated control concept. Hilgardia 29:81-101.

19. Stoner, K. A., Sawyer, A. J., and Shelton, A. M. 1986. Constraints to the implementation of IPM programs in the U.S.A.: a course outline. Agric. Ecosys. & Environ. 17:253-268.

20. Vandeman, A. J., Fernandes-Cornejo, Jans, S. and Lin, B. H. 1994. Adoption of integrated pest management in U.S. Agriculture. Agricultural Information Bulletin No. 707, Economic Research Service, Washington, D.C.

21. Waters, W. E., Huffaker, C. B., and Newsom, L. D. 1980. Summary. Pages 471-473 in: New Technology of Pest Control. C. B. Huffaker ed. J. Wiley & Sons, New York.

22. Zadoks, J. C. 1985. On the conceptual basis of crop loss assessment: the threshold theory. Annu. Rev. Phytopathol. 23:455-473.

Factors Affecting the Adoption of New Technologies

Ann Sorensen
American Farmland Trust, Center for Agriculture in the Environment
DeKalb, IL USA

Patrick A. Stewart,
Arkansas State University, State University, AR USA

Integrated pest management (IPM) relies primarily on biological and ecological interventions to manage the pests that threaten agriculture. The cost-effectiveness and sustainability of an IPM system depend on its capacity to maintain pest populations below economic damage thresholds across varying, often unpredictable, circumstances. Systems must be resilient across a wide range of weather, pest pressure, and agronomic conditions. In addition, they must perform well over time by continuously undermining the ability of pests to adapt to the tactics used to manage them.

Ecological theory suggests, and field experience confirms, that IPM systems based on a diversity of control tactics are the most resilient because of redundancy in the system. In a system where a single tactic is relied upon for control, growers are vulnerable when pests adapt to the tactic. However, regardless of the tactic used, there are combinations of weather, agronomic conditions, and pest adaptations that will likely lead to problems. The best way to assure cost-effective management across all conditions is to build redundant mechanisms of control into production systems, in other words, bio-intensive, prevention-based IPM systems (3). This means that growers first need access to information concerning the variety of pest control tactics available. Second, producers must be willing to adopt these practices, or failing that, be convinced to do so.

While traditional IPM practices, such as crop rotations, have been used since the dawn of agriculture, new IPM technologies are emerging on a regular basis. The question remains as to whether new IPM

12

practices and technologies are being adopted and, if not, what barriers are preventing their adoption. We fear that not all aspects of traditional innovation diffusion theory necessarily apply to the adoption of new technologies by producers who are already adopters of IPM practices. In essence, diffusion theory states that only a few people adopt a particular technology at the onset (the "early adopters"). Each adopter will generally move through five stages in the process of adopting before the technology/innovation is fully integrated into the production system: awareness, interest, evaluation, trial, and adoption (2). However, our hypothesis is that producers are adapting their IPM systems to fit these new IPM technologies rather than vice versa (as diffusion theory emphasizes). As a result, this paper is organized to analyze and test diffusion theory using a small-scale sample of IPM farmers from three sites. First, we review diffusion theory, then relate it to IPM technologies and practices. We next examine determinants of adoption of traditional IPM practices before considering the new information technologies of geographic information systems (GIS) and geographic positioning systems (GPS) and barriers to their adoption by IPM producers. Other new technologies that are soon to reach farmers and affect management systems are considered before conclusions are drawn.

Profile of an Adopter

Early adopters are not necessarily "progressive," while non-adopters are not necessarily "laggards" or "traditional." According to Nowak (8) producers are either *unwilling* or *unable* to adopt a new technology. While these two notions are not necessarily mutually exclusive, ability is a precondition for willingness to adopt. A producer may be willing to try a new technology but may be unable to do so. Or the producer may be able to do so but is unwilling. Certainly, the distinction between willingness and ability has a major influence on who will adopt and on the way in which a project manager should go about recruiting producers as potential adoptors for a new technology. If ability is a barrier, the project manager has only to worry about providing the technology or teaching the practice. If willingness is a barrier, the potential adopters must be convinced of the benefits, either through more information or through pilot projects.

All adopters of new technologies and practices go through stages in the adoption process (7). In the early stages, producers become aware of a specific technique, product, or process. According to Nowak (7), this awareness may come from two distinct sources. First, either the producer has a problem and seeks a solution, or a third party calls attention to a problem previously unrecognized by the producer. The producer then seeks information about potential solutions, evaluating both financial cost and effort involved. This information may come from the scientific community, neighboring producers (the most likely and trusted source of information), dealer information brochures, and/or technical journals. The problem is that certain producers are not even aware of a technology, or if they are, do not have sufficient information to stimulate their interest in learning more about it and, hence, believe that the technology would not benefit their operation.

A second source of awareness comes from what is termed "technology push." This occurs when the research community provides producers with the latest findings on technology and new innovations. However, the producers may fail to understand the information that the research community is trying to diffuse, and, as a result, may fail to integrate these technologies into production systems. This problem is exacerbated as technologies become more complex and may lead either to inappropriate utilization or non-utilization.

In the literature about innovation diffusion and technology adoption, many factors have been identified that influence the adoption or non-adoption decision-making process. These include the complexity of a new technology or practice, the extent that the innovation is part of a set of other technologies or ideas (its divisibility), and the cost and compatibility with the current management style and production practices being used. It can be expected that simple and inexpensive technologies will be adopted before complex and expensive ones, and that "stand-alone" technologies will be used before technologies that are part of a "bundle" of practices and technologies, unless the producer has already invested in that "bundle." In this case, the ability to observe a system plays a part by demonstrating the degree to which the system works. In other words, the producer will have been "sold" on an approach by its success, and will be more likely to invest in more and similar technologies/practices, as will witnesses (2).

Social constraints to adoption may include growers' perception of the technology, communication channels among farmers, and the demographic attributes of the producer. Being unable to adopt a new technology implies that some kind of obstacle is in place, and therefore the reason not to adopt is logical and rational. The producer is simply making the decision not to adopt based on this identified obstacle, even though the producer may be willing to adopt. There are several reasons why a producer may be unable to adopt. He/she may perceive that the necessary information is lacking or that the information is not sufficient for an informed decision about the agronomic and economic attributes of this new technology. Likewise, the information may be too costly to obtain. Even in the "information age" we live in, the time, expense and difficulty of obtaining the information may be a real obstacle. Generally, information is not free to the agricultural community (or even the general public), even if the cost of obtaining information is just the time it takes to become fully informed.

Specific management system characteristics influence adoption. Size is important, as larger operations are more likely to adopt new technologies and practices than are smaller ones. In what may be considered to be a related factor, higher-income producers have greater ability to take risks and thus are more likely to adopt new technologies and/or practices. Finally, specialized operations, by dealing with fewer production variables, are more likely to adopt than are more diversified operations (8).

Therefore, adoption is determined by interactions among producer and farm-level characteristics and aspects of the technology itself. The question remains as to the role one or the other plays in the innovation diffusion process. Specifically, while awareness is a required prerequisite for adoption, it does not assure it. However, if there is an overlap in groups who are aware and most prone to adopt, both information and promotional strategies may be focused on the same group.

Integrated Pest Management: Adoption of Traditional and New Technologies

Although IPM traditionally utilizes fairly simple techniques, many of the new technologies require conceptual and technical skills for which producers may not be trained. Studies show, in fact, that the complexity of a new IPM technology is inversely related to the rate and degree of adoption (7). Producers may see new IPM technologies and practices as having excessive labor requirements, both in terms of people and required skills. Related to this particular barrier are the managerial skills (or lack thereof) needed to decipher and apply the information collected, and the capabilities of people in the field collecting data and implementing IPM strategies. Financial costs, investments, and the influence on net returns may be the most important concern for producers, and may impede the adoption of new IPM technologies.

Producers practicing IPM must collect, analyze, and use all necessary information as the very basis of their pest management decision-making process. As Nowak (7) states, "the requirement for the analytical use of quality information occurs in a context where the producer is often overwhelmed" by the diversity of data. The information base from which the producer can make the best possible decisions regarding pest control is generated from on-farm, site-specific observations. The question is, then, what is the value of information such as scouting records? If the producer puts little value on accurate record-keeping and information-gathering, then this very foundation of IPM can be the most important barrier to IPM adoption.

Many of the new technologies used in IPM appear to require conceptual and technical skills for which the producer may not be trained. Studies show, in fact, that the complexity of a new technology is inversely related to the rate and degree of adoption (7). In line with financial concerns is the notion that producers may see new IPM technologies as having excessive labor requirements in terms of both people and skills, which may be a financial barrier to adoption. Related to this particular barrier are the managerial skills on the part of the producer, which may impede the adoption of new IPM technologies. Lack of understanding may result in inadequate application or even use of unnecessary new IPM techniques. Too often, new IPM practices are designed for the more-educated producers, rather than for the majority of producers. Because very few producers adopt new technologies without significant support from the outside, the number and attitudes of support people may limit availability and accessibility. USDA Extension has been downsized to the point that it cannot possibly be informed about every detail of new technologies, especially since technologies are generally site-specific. Extension personnel cannot possibly know every operation that may or may not benefit from a particular new technology. They may choose to focus on more successful and educated farmers, creating a situation where less-than-average managers receive inadequate assistance.

A producer may not want to adopt new IPM technologies or practices because they do not "fit" easily into existing production patterns. Producers may not even need a particular new technology but feel "pressured" to adopt by neighbors or consultants. Some of the new IPM technologies are so sophisticated that it is not apparent whether or not a particular technology is needed, useful, or economical for a specific production site. In other words, producers may be pushed to adapt their systems to "fit" the new technologies rather than integrating the new technologies into their systems, as it should be. Producers may see a new technology as a "silver bullet" to their pest control problems and readily adopt it. On the other hand, producers may not want to adopt a new IPM technology or practice because they prefer the status quo of traditional IPM practices currently being used. In other words, a producer may believe that the current production practices are not consistent with the new IPM technology. Whether this represents risk aversion, skepticism, or intrinsic barriers remains to be tested.

Small Survey Results

To better understand the current dynamics of new technology adoption, we developed a survey instrument and mailed it to three groups of IPM producers in Washington (apple and pear growers), California (wine grape growers), and Massachusetts (sweet corn growers). A total of 148 surveys were mailed in January 1999, with a response rate of 42.6% (20/40 in Washington; 22/40 in California; 21/68 in Massachusetts and New Jersey[1]). The producers were not randomly sampled from an overall population of IPM producers; rather, the surveys were sent to IPM coordinators and project directors, who then forwarded them to their producers. This limits conclusions that may be drawn. For the sake of statistical needs, the samples were aggregated with a variable for location entered into the equation to control for geographic and cultural effects that may occur in these far-flung places. The 63 responses, while ample for statistical analysis, represent a small sample. As a result, in addition to the aforementioned warning, any inferences from this study may not conform with the general population of IPM producers, and should be accepted with caution.

The dependent variable in our analysis is an additive index based on 10 well-established, "traditional" IPM practices. The practices include: scouting for pests, pheromone traps, action thresholds, spot treatments, plowing at night, soil nutrient testing, row and plant spacing, release of

---

1/ New Jersey's responses were too few in number to include the state as a separate region. Including New Jersey with Massachusetts as one region enabled us to use the New Jersey data.

beneficial insects, calibrating sprayers, and using "no glug" jugs. While these practices tend to vary across survey sites, they are baseline practices used in most IPM systems. When summed, the measure suggests that all producers responding to the survey use at least three of these traditional practices (as can be expected from a group identified as IPM producers), with only one individual using all 10 practices. The average respondent uses 5.9 (SD = 1.51) of these practices, which are distributed normally.

Determinants of traditional IPM practice adoption include attributes of the farm operation, demographic characteristics of the farmer, and the farmer's behavior. Farm operation attributes include the location (as stated above) and size of the farm (ranging from 2.8 to 3237 ha and averaging 201 ha [SD = 498.12]), and the percentage of the land farmed by a producer that is owned by the producer (average ownership = 69.97% [SD=34.09%]).

Demographic characteristics measured and tested in this study included the producer's age, education, and gross income range from agriculture. The mean income category of 9.93 (SD = 4.29) represents an income of $90,000 to $100,000, although the modal income category (22 respondents) is more than $200,000 a year. The age of respondents ranges from 27 to 82 years, with an average age of 50.3 years (SD = 11.72).

IPM producers in this study tended to be highly educated, a point suggested by the literature, as 30% had some college, 49% had a college degree, and 8% had a graduate degree. On the other hand, only 11% had only a high school diploma, and 2% had a vocational/technical education. While higher levels of education do not necessarily assure adoption of IPM techniques, they are expected to be positively related to a greater use of traditional IPM techniques.

Use of computers and the ongoing education it implies are also expected to have a positive influence on the number of IPM practices adopted. Computer use, which assumes a greater appreciation for and use of information, is a key component of IPM. We measured this through three Yes/No questions concerning the use of computerized record-keeping (46%), decision-making software for pest control (12.7%), and use of the Internet to obtain information on the latest IPM technologies (31.1%). These three questions were then summed to provide a 4-point scale (0-4), with an average score of less than one (0.89; SD 0.97).

Likewise, educating oneself through a variety of information sources is expected to have a positive effect on the number of IPM measures employed. We asked what sources respondents rely on for IPM information, of which there are six options: trade journals, neighbors, extension personnel, chemical company representatives, crop consultants, and other options defined by the respondent. This information was then summed to create a scale from 0-6 and had a mean score of 3.5 (SD = 1.5).

Finally, it is expected that actively promoting IPM options will have a positive effect on the use of IPM tools as farmers preach what they practice. To measure this, an additive scale was created based on whether the respondent gave talks at the local farm bureau (6.3%), wrote articles in newsletters (19%), talked with neighbors about IPM (79.4%) or used other approaches (30.2%). This scale, which ranges from 0-4, had a mean score of 1.35 and a standard deviation of .92.

TRADITIONAL IPM PRACTICE ADOPTION

Analysis of the variables was carried out through Ordinary Least Squares Regression, which assumes a linear relationship between the dependent variable (number of traditional IPM practices) and the independent variables, or determinants. The dependent variable was regressed on 10 variables. The resulting model is highly significant (F value=3.627) and explains nearly 30% of the variance when controlling for the number of variables in the model. The truncated model, which removed non-significant variables in a stepwise manner (removing the least significant first), explains nearly 34% of the variance, a slight improvement. More meaningful is that the model is statistically more significant (F-value=6.252).

Analysis of the effects of the variables themselves suggested that six of the 10 are significant, at least at P=0.10 in a one-tailed test of significance (see Table 1). Of the farm characteristic variables, both Washington and California are significantly different from the reference category of Massachusetts, with each location leading to the use of a little more than one less IPM practice. Interestingly, both hectares farmed and percent land owned, variables expected to have a good deal of explanatory power, are not significant and have been omitted from Table 1.

While the age of the respondent and the education of the respondent do not have a significant effect on the number of IPM practices used, for every increase in income range ($10,000 to $140,000, then in increments of $50,000 thereafter), there is a concomitant increase of 0.07 practices. The impact of producer actions is quite predictive of the number of tradi-

Table 1. OLS Regression: Number of Traditional IPM Practices Used.

Variable	Full Model Coefficient	T-value	Truncated Model Coefficient	T-value
Intercept	3.989	2.930***	4.190	6.728***
Washington	-1.007	-2.173**	-1.063	-2.559***
California	-1.82	-2.714***	-1.183	-3.031***
Income	0.073	1.821**	0.074	1.923**
Age	-0.003	-0.204		
Education	0.106	0.664		
Computer Use	0.371	2.032**	0.395	2.354***
Information Sources	0.172	1.515*	0.167	1.545*
Promote IPM	0.555	2.777***	0.573	3.228***
	F= 3.625***	Adjusted R^2 =.297	F= 6.252***	Adjusted R^2 =.337

*=0.10 one-tailed significance level;
**=0.05 one-tailed significance level;
***=0.01 one-tailed significance level.

tional IPM practices utilized. For every increased use of computers, there is an associated increase of 0.4 in IPM practices. Likewise, the more information that is gathered, the greater the number of IPM practices that are used, as every additional information source led to use of an additional 0.17 practices. Finally, and quite as expected, promotional activities in IPM are associated with a significant and powerful increase in IPM tool use in that each activity is tied to an increase of 0.57 in IPM tool use.

GEOGRAPHIC INFORMATION SYSTEMS AND GEOGRAPHIC POSITIONING SYSTEMS

While both geographic information systems (GIS) and geographic positioning systems (GPS) rely on mapping technologies, they are separate technologies with different capabilities. GIS, succinctly defined, is "a computer system capable of holding and using data describing places on the earth's surface," and may be used as part of production management systems to assist decision-making (5). GPS, on the other hand, is a positioning system using 24+ satellites to pinpoint a location on the earth's surface (6). In other words, GPS may be used as a tool within the rubric of GIS.

GIS and GPS provide producers with the ability to accurately map everything in a production area, including pest populations. The software can be linked to agrichemical application equipment in order to precisely apply pesticides and fertilizers. Decreasing the chemical load in the environment and providing better pest control lowers overall costs. In addition, this technology can be used to refine understanding about the distribution of pest populations and therefore target both monitoring and control tactics. GIS and GPS can also be used to analyze the most efficient travel paths for machinery and delivery of products, and offers the benefit of evaluating crop varieties by location.

While both GIS and GPS have been around for some time, they are still not used extensively by producers, even IPM producers who presumably have greater need for the information that both can provide. These technologies could significantly change the way land is managed, leading to more economically efficient and environmentally sound production. However, even researchers are currently struggling to make sense of the staggering amount of information collected and mapped. Even such fundamentals as the size and frequency of the sample units have yet to be determined. Although these technologies have limitless potential, they remain out of the reach of most producers, not only because of financial constraints, but also because many IPM producers do not know of the capabilities and benefits offered by GPS and GIS.

With both GIS and GPS, producers were asked "Do you use GIS/GPS for weed and pest infestation and field conditions?" Four potential responses were provided: yes; no; not available; or, unaware. While most of the producers answered "yes" or "no" (GIS: yes= 4/ no=39; GPS: yes= 2/ no=49), a few stated that these technologies were unavailable (GIS=5; GPS=4) or that they were unaware of their existence (GIS=13; GPS=6). This suggests that barriers, either to technology access, or more likely, to information and awareness, exist even among those who thrive on, even need, information.

To test this, we ran two models to assess determinants of barriers to the use of GIS and GPS. This was done by combining the "yes" and "no" responses, which indicate the ability to adopt and are thus coded as "1"

(GIS=70.5%; GPS=83.6%) and by combining the "not available" and "unaware" responses, which indicate that a barrier exists and are thus coded as "0" (GIS=29.5%; GPS=16.4%). The dichotomous nature of the dependent variables precludes the use of ordinary least squares (OLS) regression. Instead, we use logistic regression, which transforms the dependent variable by taking its natural logarithm to assess the probability that a barrier does or does not exist.

In addition to the determinants we used in testing how many IPM practices are used by producers in the three sites, we used that equation's dependent variable. This decision was based on the premise that the greater the number of traditional IPM practices used, or "bundled," by producers, the more likely new technologies such as GIS and GPS will have been considered by IPM producers. In other words, active innovators will remain active in their search for new and useful technologies and practices.

Ability to Adopt Geographic Information Systems (GIS). We inferred the existence of barriers to GIS use when respondents stated that the technology was unavailable or that they were unaware of it, and we inferred that no barriers existed when respondents stated that they made a yes/no decision regarding GIS use. Initial analysis of the barriers to GIS use, based on 61 cases, suggests that the full model was poorly specified, as the model Chi-square did not reach significance. However, stepwise removal of non-significant variables, in which the least-significant variables were removed first, resulted in a model achieving a high level of statistical significance. In addition, this model correctly predicted 79% of the cases as belonging in either the "no barrier" or "barrier" category.

Analysis of the variables that remained in the truncated model (age, computer use, and traditional IPM practice use) was highly suggestive. The exponent of B was used to show direction and power of relationships. Exponent values greater than one indicate a positive relationship, while values less than one indicate a negative relationship between the ability to adopt and determinants of the ability to adopt. In other words, exponents of B greater than one predicted for the ability to make adoption decisions and those less than one indicated barriers of availability or awareness. Age, as expected, was highly significant and exhibited a slight negative relationship with use of GIS, as older farmers were more likely to have encountered a barrier to GIS use by being either unaware of it or by seeing the technology as unavailable. Computer use, likewise, was negatively related to the use of GIS. As computer use increased, so too did the likelihood of a barrier being in place. In other

22

words, as computer use increased among IPM producers, the barriers to adoption of GIS technology increased.. This counter-intuitive result has to be viewed with caution, because the great majority of responses were negative regarding GIS use. Finally, as hypothesized, the use of traditional IPM practices had a significant effect, as every extra IPM practice used led to a nearly two-fold increase in the perception that no barrier to adoption is in place. For space considerations, only the significant variables are included in Table 2.

Ability to Adopt Geographic Positioning Systems (GPS). The full model for barriers to the use of geographic positioning systems (GPS) was, like the GIS model, poorly specified and did not reach significance. However, stepwise removal of non-significant variables led to a significant model. In addition, in nearly 84% of the cases under consideration, the truncated model correctly predicts the presence or absence of barriers to adoption of GPS.

Interestingly, in this model, farm location resulted in a large difference in the existence of barriers to the adoption of GPS, a factor not apparent with GIS. Specifically, Washington farmers were nearly three times more likely (2.91) and California farmers were nearly 13 times more likely (12.72) not to have experienced a barrier than were Massachusetts farmers. As expected, farmers promoting IPM were less likely to have

Table 2. Geographic Information System (GIS) Barriers: Logistic Regression Model.

Variable	Truncated Model Exponent of B	Wald Chi-square
Intercept		2.0904*
Age	0.918	7.229***
Computer Use	0.475	3.948*
Traditional IPM Practices	1.805	5.283**
	% Correctly Predicted=78.69	**Model Chi-square = 13.604***

*=0.10 one-tailed significance level;
**=0.05 one-tailed significance level;
***=0.01 one-tailed significance level.

23

Table 3. Geographic Positioning System (GPS) Barriers: Logistic Regression Model.

Variable	Truncated Model Exponent of B	Wald Chi-square
Washington	2.914	1.465
California	12.721	3.887**
Promote IPM	2.296	1.934*
Traditional IPM Practices	1.615	2.166*
	% Correctly Predicted=83.61	Model Chi square=10.102**

*=0.10 one-tailed significance level;
**=0.05 one-tailed significance level;
***=0.01 one-tailed significance level.

experienced barriers to use of GPS. For each promotional activity engaged in, there was a greater than two-fold increase in the likelihood that barriers to GPS use were not experienced. Finally, use of traditional IPM practices was statistically significant, as there was a little more than a 1.5 times greater probability of not experiencing a barrier to the adoption of GPS for each additional IPM practice applied in the fields. Again, only significant variables are included in Table 3.

New IPM Technologies and Practices

As with any endeavor, new technologies and practices loom on the horizon. There is no guarantee that any one of these will be the "next big thing" or that other technologies and practices will not come to the forefront in the near future. However, analysis of trends in agriculture suggests that the new biotechnology promises a wave of new technologies that will lead to changes in practices. The following adaptations of the new biotechnology stand out as particularly consequential for producers using IPM methodologies.

DNA DIAGNOSTICS

Since every living creature is unique, each has a unique DNA recipe. Individuals within any given species, breed, or hybrid can usually be identified by minor differences in their DNA sequences; as few as one difference in a million polypeptide bonds can be detected. Using the techniques of DNA fingerprinting and PCR (polymerase chain reaction, which rapidly duplicates specific DNA molecules in response to temperature changes), scientists can diagnose viral, bacterial, or fungal infections in plants; distinguish between closely related insects; or map the locations of specific genes along the length of DNA molecules. Because information is the basis for IPM, added clues about pests and weeds provided by these techniques might suggest control strategies.

Although these techniques have great potential for pest, weed, and disease control, complexity and cost may well put them out of the reach of the typical IPM producer. DNA diagnostics have to be performed in a laboratory, are relatively time consuming, and are expensive.

BIOPESTICIDES

Biopesticides are a relatively new class of pesticides that are generally derived from naturally occurring substances. They include not only Bt products, but also baculoviruses and fungi. Biopesticides provide narrow-spectrum pest control that does not affect beneficial insects, wildlife, or humans. They may also be effective when used in conjunction with resistance management strategies. However, the costs and benefits to farmers from biopesticides have not been adequately evaluated. Just because biopesticides are being used is not proof of their necessity. The high cost of producing biopesticides, coupled with regulatory constraints and problems with formulation and marketing, have led to serious disappointments. However, much of this can be traced back to unrealistic expectations concerning their role in pest management. Too often, we expect slow-acting biological control agents to compete directly with powerful, fast-acting synthetic chemicals.

RESISTANCE, THE NEW BIOTECHNOLOGY, AND ADOPTION

Pest resistance to pesticides has become a major problem in U.S. and world agriculture. Crop rotation still presents a viable alternative to chemicals when dealing with pests and weeds. However, there are instances where rotations do not prevent resistance, as demonstrated by the corn rootworm, which has developed behavioral strategies to circumvent soybean/corn rotations (5). Genetically modified crop varieties are new products that may be integrated into IPM programs as a new technology. Economic analyses performed on the effects of European corn borer control using genetically transformed varieties indicate that good results can be achieved. In addition, with these transgenic crops, the cost of scouting or pesticide application on a crop may be eliminated, so the only additional cost will be the premium placed by the company on a bag of genetically altered seed. However, a major concern with transgenic crops is that resistance may still develop. Most insect populations, if they are excessively exposed to a toxin, can develop resistance, making pest control products less effective. Potato growers are fully aware of this problem, especially concerning the Colorado potato beetle. Resistance is also common in many other crop pests, including key pests of cotton and apples. Also, resistance in the field to Bt products has been documented in the diamondback moth. Fortunately, the European corn borer has not shown resistance to chemical control to date. It is in everyone's best interest – biotechnology companies, growers, and agriculture in general – to maintain the usefulness of this valuable technology. The U.S. Environmental Protection Agency (EPA) supports the position that 1) resistance may occur and already has in a few cases, and 2) resistance management is the key to preserving the technology of transgenics expressing the Bt toxin as a vital tool for producers. Refuges are critical in achieving this goal (11). The placement and size of a refuge is dependent on target pest biology, as well as other factors. Although a refuge size of 20% sprayed and 40% unsprayed was generally acceptable and recommended by the EPA, the debate regarding these numbers goes on (9).

These recommendations may all be valid; however, the theory of innovation diffusion suggests that refugia strategies may be tough to implement. Additionally, producers need to understand their own pest and weed management problems in order to determine whether they even need Bt seeds. It should be insect and weed management, not genetic engineering, that drives management decisions..

Yet another challenge to genetic engineering has just been documented. Crops that have been genetically engineered to resist herbicides can pass genes conferring herbicide tolerance on to their weedier cousins, producing hybrid strains of "super weeds" (10). Bergelson (1), professor of ecology and evolution at the University of

Chicago, asserts that genetic engineering can substantially increase the chances of "transgene escape," or the spread of certain traits from one plant to another. For example, corn, which is a grass, can cross with timothy grass, an abundant weed. If the corn contains a gene that confers resistance to a pesticide, the resultant "weedy" hybrid may become a pesticide-resistant nuisance that can compete with crops for water and nutrients. Bergelson warns that the widespread use of transgenic crops may directly cause the creation of weeds with traits intended to increase the fitness of crops, spurring a need for new pesticides.

The abundance of crop varieties is also creating challenges. There are more and more seed varieties marketed on a continual basis. In fact, two to three new varieties are entering the market every year, and producers have less and less experience with these new varieties. Producers are generally operating with a certain comfort level when they plant their seed. However, the comfort level with these new seeds is diminishing as producers become more and more overwhelmed. More important, producers now have less margin for error, since the profit margins of agricultural producers almost across the board are shrinking. Thus, there is more pressure for producers to "get it right the first time." But by pushing new varieties, we are diminishing whatever safety net producers still have. Thus, whether it is new seed varieties or other technologies, producers may feel compelled to use these "silver bullet" technologies in the hopes of pest and weed control panaceas.

Information – The Importance of Dealerships and Crop Consultants

Over the years, many producers develop a relationship of some sort with their chemical supplier. Historically, chemical dealers have provided the necessary pesticides and application guidance, since this is where profits reside. Recent developments, namely increased competition and less-expensive chemicals, have decreased this profit margin. As a result, it is estimated that 20% to 30% of the 12,000 agrichemical dealers in the United States will go out of business during the next decade. Dealers must now diversify their products, in both selling chemicals and providing information. To survive, dealers must become more service oriented and provide site-specific information required by IPM producers.

Chemical dealers face several challenges when they start providing information-based services to producers. For example, they must compete with independent and unaffiliated crop consultants. The dealers have to contend with the potential for conflict of interest, i.e., providing agrichemical consulting while selling the same products they sometimes have to advise against using. Additionally, existing infrastructure has to be integrated with the new task of adding a service-oriented side of the business, namely the provision of information (12).

There is a need, however, to strengthen the relationship between chemical dealers and producers. In fact, dealer associations are advocating an active role for dealers in providing information and increasing the focus on service. Chemical dealers need to take a proactive approach regarding public perception by relaying a positive image of themselves, particularly with respect to environmental performance of IPM. However, as we saw from anecdotal evidence from a targeted survey sent to IPM producers, none of the respondents rely on agrichemical representatives for their information. It can be tentatively concluded that there is still a lack of trust between chemical representatives and IPM producers. On the other hand, producers do rely on their crop consultants for information.

Regardless of whether the source of information regarding IPM technologies is agrichemical companies, crop consultants, or academia, the technologies that are actually adopted by producers do not always reflect the information that is provided. There is strong evidence that IPM producers are being bombarded with technical information about technologies that are being "sold" to producers as the solution to their problems, promising the ultimate cost-saving technique, or the newest technology that can increase yields while at the same time control pest damage. In essence, the new technologies sweep away everything we know and have practiced for decades. Some of them are more environmentally friendly and economical, such as the traditional IPM tactics that have been practiced by producers for a long time. However, "new" technologies in IPM, such as GIS, GPS, biopesticides, and DNA diagnostics, are all tools that may or may not be helpful or even necessary for producers in their current IPM program. Thus, the hypothesis stands that producers are adapting their IPM systems to fit these new IPM technologies rather than matching the new technologies to their IPM systems and needs.

Evidently, producers do not trust what their information sources say. Most IPM producers consult trade journals, extension personnel, and, not surprisingly, their neighbors.

Policy Implications and Conclusion

For all intents and purposes, economic factors dictate the decisions that producers make concerning adoption of new IPM technologies or practices. As we have seen, there are risks other than economic that producers must take if they are to adopt a new technology. Most producers are relatively risk averse. In fact, "insurance spraying" of pesticides is not uncommon for many producers in case some external factors threaten to damage the crops and reduce the yield. Additionally, the producers are entirely responsible should a crop consultant make the wrong recommendation. Consequently, producers often ignore the reduced input recommendation by crop consultants, because they know that they are the bearers of the financial loss should "something" happen.

New technologies in IPM are not fail-safe. There is always a probability, however small, that they will fail, the producers will not need them, or that the IPM producers who adopt them will incur financial losses because the technologies provide no benefits. A hypothetical example would be a producer who purchases satellite images to detect increased weed populations when he can just as well inspect his fields without such a technology and come up with the same result. The resulting financial loss could be devastating to many producers.

There are ways to reduce the risks and barriers to adopting new IPM technologies. One incentive over and above the crop insurance that the federal government offers is IPM insurance. Producers would enter an IPM program, adopting the new technologies that IPM offers. If external factors occur, such as excessive rainfall, extreme temperatures, even the failure of IPM test plots on their fields, a producer's damages and losses in yield would be covered by private IPM insurance. The same would apply should a producer use a different, biologically engineered variety and resistance develops. IPM insurance would work essentially in the same way as ordinary insurance policies. The producer pays a premium, can collect on damages if they occur, and will save money because he/she will save on pesticide inputs. Large farms may not perceive a need for IPM insurance to manage risk associated with adopting new IPM technologies, since they are in a much better position to manage risk by themselves. However, smaller operations may profit from this kind of insurance. Additionally, this sort of IPM insurance could be promoted by crop consultants, manufacturers of chemicals, extension personnel, and even GIS companies who offer remote sensing of fields to producers.

29

Access to the right information for every producer is essential for decision-making as well as for assessing the risks and benefits of new IPM technologies. Over the years, agricultural production has become more technologically sophisticated, and researchers and policy-makers have the task of identifying ways to convey more and more complex technologies to producers. What we have to continue to guard against is allowing new IPM technologies to drive producers' pest management systems instead of having the systems determine the technologies. Given the increasing complexity of technologies, the overwhelming need for information, the downsizing of our key information providers, and the diminishing of profit margins, IPM has a tough road ahead.

Literature Cited

1. Bergelson, P. 1998. Engineered Crops Spawn Super Weeds. Environmental News Network, Archive. **http://www.enn.com/enn-news-archive/1998/09/090398/superweed.asp**
2. Brown, L. A. 1981. Innovation Diffusion. Page 21. Methuen & Co. New York, NY.
3. Coble, H. 1998. A New Tool for Measuring the Resilience of IPM Systems -- The PAMS Diversity Index. Pages 42-45 in: Proc. IPM Measurement Systems Workshop. Esther Day, ed. Chicago, IL.
4. ESRI (Environmental Systems Research Institute, Inc.). 1997. Understanding GIS: The ARC/INFO Method. New York: John Wiley and Sons.
5. Gray, M. Presentation on the Corn Rootworm Problem in the Midwest. Pages 19-26 in: Proc. IPM Measurement Systems Workshop. Esther Day, ed. Chicago, IL.
6. Hurn, Jeff. 1989. GPS: A Guide to the Next Utility. Sunnyvale, CA: Trimble Navigation.
7. Nowak, P. 1987. The adoption of agricultural conservation technologies: economic and diffusion explanation. Rural Sociology, 52(2): 208-220.
8. Nowak, P., Padgett, S., and Hoban, T. J. 1996. Practical Considerations in Assessing Barriers to IPM Adoption. Pages 93-114 in: Proc. Third National IPM Symposium/Workshop. Broadening Support for 21st Century IPM. S. Lynch et al., eds. United States Department of Agriculture, Economic Research Service, Miscellaneous Publication Number 1542. Washington, DC.

9. Panel Member. 1999. Verbal Comments on Refuge Design and Deployment. EPA/USDA Workshop on Bt Crop Resistance Management. June 18, 1999. Chicago, IL (Proceedings forthcoming).

10. Spangler, S. and Calvin, D. 1996. Genetically Engineered Bt Corn. Field Crop Newsletter. Department of Agronomy, Pennsylvania State University. December, 1996.

11. U.S. EPA and USDA. 1999. EPA and USDA Position Paper on Insect Resistance Management in Bt Crops. Office of Pesticide Programs. pp. May 5, 1999 (minor revisions June 16, 1999) **http://www.epa.gov/opbppd1/biopesticides/otherdocs/bt_positio n_paper_816.htm**

12. Wolf, S., and P. Nowak. 1994. The Status of Information-Based Agrichemical Management Services in Wisconsin's Agrichemical Supply Industry. Pages 1-14 in: Proc. 21st International Conference of Site-Specific Management of Agricultural Systems. P. C. Robert, R. H. Rust, and W. E. Larson, eds. Bloomington, MN.

Shifting Agricultural and Ecological Context for IPM

Paul S. Zorner
Mycogen Corporation, San Diego, CA 92121 USA

I would like to begin by relating a story to you that I recently heard about a fellow riding a motorcycle one cold winter day. The wind coming in through the buttons of his jacket was making him very cold, so he finally stopped, turned his jacket around, put it on backward, and took off again. It looked a little silly, but it was functional in that it solved the wind problem. Then he hit a patch of black ice, and since his arms were somewhat restricted because of his jacket, he skidded into a tree and was knocked unconscious. After a period of time, the paramedics arrived and had to elbow their way through a crowd that had gathered to see what they could do to help. The paramedics asked what had happened. A big fellow in the crowd said, "I don't know, sir. When I got here, he seemed to be all right, but by the time I had his head turned around straight, he was dead!"

Those good Samaritans weren't too bright, or more likely, they provided a quite disastrous solution to a non-existent problem because the innovative solution by the biker to his problem could not be recognized by the crowd as something that made sense. It did not fit their perception of reality, which is not an uncommon problem in today's complex world. Introducing new ideas is not always easy. People are comfortable with what they know.

Ecologically, those of us pushing adoption of integrated pest management beyond the current plateau may be in a similar situation. I know quite a few practitioners of IPM who feel as though they've had their necks in at least one vise over the years, as both chemical and biological camps have tried to orient people's activities in a desired direction. And frankly, after 10 years of developing biological pesticides at Mycogen, I would have to say that the chemical vise is still a lot stronger and larger. Let's face it, our basic agronomic pest control practices are still predominantly based on chemicals. They are faster, better, cheaper, and easier to use than their biological counterparts. Prevailing agricultural mentality is that fast, broad-spectrum pest control is better.

I grew up on a farm in Oregon, where solutions to controlling an infestation of armyworms, aphids, or any other pest meant the difference between paying the rent or not paying the rent. My parents were not interested in complex solutions to a simple problem, which to them was "kill the damn bugs as fast as possible before they eat next month's tax payment." Fast, short-term pest control solutions are popular and easy to invest in because you can see an immediate benefit. However, what has various people concerned is the size of the ecological "balance due" from implementing, on a massive scale, short-term solutions that focus solely on the presence or absence of the pest species alone.

The objectives for this conference address this in a statement that paraphrases a recent report from the National Research Council, which expressed concern that a focus on pesticides and individual pests continues to dominate IPM. The NRC calls for a shift in the IPM paradigm away from managing components of individual organisms to an approach that examines process flows and relationships among organisms (1).

The report itself is entitled *Ecologically Based Pest Management: New Solutions for a New Century* and promotes consideration of pest management strategies in the context of whole-farming systems. The report is excellent. If you have not read it, you can find a good summary on their Web site (2).

The basic message from this report is that we need to broaden our perspectives, and I gather from the organizers of this session that this is also my job today. They've asked me to touch on shifting agronomic and ecological factors that may accelerate the adoption of IPM. This was a difficult task for me to come to grips with, because in reading to prepare this talk, I came across opinions from a variety of people that suggest the concept of IPM may actually be part of the ecological problem due to the emphasis it puts on <u>pest control</u>. We likely have the emph<u>asis</u> on the wrong syll<u>able</u>. An array of remarkably diverse people appears to believe that we have to expand our thinking to something more akin to integrated <u>habitat</u> management,

33

because in the ending analysis, they think that putting too much emphasis on the pest itself prevents us from thinking about parameters of our farming systems that allow various pests to get out of hand in the first place.

Bob Zimdahl at Colorado State University wrote an opinion on this topic that was published in the most recent issue of *Weed Science* (5). He discusses weeds in his article, but I think the point is valid across all pest management disciplines, and his point is that we are all asking the wrong question.

He states an opinion that we manipulate the natural world to produce food, and weeds are an inevitable part of food production. However, he feels that the emphasis on control has obscured the right question, which is, Why is the weed where it is in the first place? And a reasonable second question is, What is it about the way we practice agriculture that allows a specific weed to be so successful? He continues in his dialogue to conclude that until we understand the answers to those questions, we will continue to recommend employment of short-term solutions to pest problems that resolve immediate productivity issues. But when those same solutions are viewed from a larger ecological perspective, they may have taken us several steps in the wrong direction.

I think the primary question that Dr. Zimdahl raises is quite valid, the question being, What is it about the way we practice agriculture that allows specific pests to be so successful? In our zeal to control what is often just one or two recalcitrant species, I think we're learning that we're getting whole systems out of balance, and that the resulting imbalance, in turn, allows a variety of other organisms to become pests as well. The point here is that rather than focusing on pest control, we ought to be turning our attention to diversity preservation through habitat management; this new focus, in turn, may allow us to stabilize pest populations with fewer inputs, which likely is a higher-order solution to our problems.

To illustrate this concept, I offer an example that comes from our experiences at Mycogen in developing biological pesticides. We happen to make Bt products and insecticidal soaps as foliar sprays. However, historically and in today's market, our experience indicates that foliar Bts are not seen by a majority of people as effective primary insecticides. They are too slow, have too narrow a spectrum of control, and are too difficult to use effectively. But in reality, Bt products are not slow acting. They simply don't knock things down like an organophosphate insecticide. Bt toxemia results in insect paralysis within minutes of ingestion and stops foliar feeding quickly. But from a grower's visual perspective, the now-comatose insect is still evident in the field for a few days.

To people whose perspective is shaped by the immediate lethality of synthetic insecticides, this is not good, because they may not recognize that the insect is causing no further damage. However, to beneficial insects, the slow death associated with Bt products is like opening a smorgasbord. There are no toxic residues to influence their population levels, and suitable prey are available longer and are much more lethargic and easier to attack because of the Bt toxemia. The problem is that not everyone sees these subtle benefits as positive. IPM, at least with Bt products and insecticidal soaps, is a lot of management work, and is perceived as being low tech and high risk because there is so much perceived management involved. Growers need good scouting, diverse application tools, and threshold calculations. A great deal of knowledge is required to make things fit together. IPM today is a lot like modern warfare: no massive attacks, but lots of surgical strikes with smart bombs based on intensive reconnaissance. However, intensive reconnaissance performed by enlightened crop consultants may not always be available or can be perceived by the grower as an extra cost that is hard to justify.

Mycogen scientists, such as Dr. George Soares, attempted to address these perceptions in various trials around the world. One interesting example of these trials was with tomatoes grown for fresh harvest in Mexico. George set up comparative trials and demonstrated that an IPM system was an economic investment equivalent to a strict chemically based system, it supported high population levels of beneficial insects, and crop yields were actually much greater (Mycogen, unpublished).

Yet, despite what seemed like solid data to us, George had a difficult time getting our management to support the concept of marketing an integrated system because it was seen as too management intensive and too knowledge intensive to be widely adopted. As a company, we were focused on selling specific insect control products and not a larger system, because we did not know how to capture value from marketing that larger system. It was simply easier to focus on selling our specific products. In this specific case, we may know the answer to Dr. Zimdahl's question, "What is it about the agricultural system we practice that allows specific pests to get so out of control?" In this instance, it is an over-reliance on synthetic chemicals, too much impact on beneficial insects, perhaps an operating system too complex for biological control, and insufficient attention by our industry as a whole to market systems for pest control instead of specific control products. This situation is not good for the grower, it is not good for the biological control industry, and it is not good for the long-term use and efficacy of synthetic chemicals. This is not a sustainable system ecologically or economically. However, for people to understand this, they have to have the time and the information to appreciate the larger ecological picture.

One such person looking at the larger picture is Dr. Edward O. Wilson. Dr. Wilson is a renowned entomologist-turned-population-ecologist from Harvard University. He has written several books and been honored by many organizations for his contributions to our ecological knowledge base. I read an interview with him recently (4) in which he summarized his perception on mankind's activities in our environment as being represented by an ecological footprint, and that the size of this footprint for any specific activity was proportional to resources consumed and impact on other species. Bigger in this case is not better. One highlight of this interview was his contention that the resources needed to support the standard of living in the United States is 12 times the average for people in the rest of the world. We appear to regard energy and natural resources as having no limit. At this point, with the way these resources are priced, the economics favor such a philosophy, but it is just not sustainable. We're not only heading for problems; we already have problems, and the early signs of this are upon us, in that "arable land, usable water and productive waste management techniques are getting harder and harder to find" (4). These are critical points from an agricultural perspective; arable land, usable water, and productive waste management are critical elements in food production. E.O. Wilson states that either we change or some future generation will starve. He also introduced the concept that species diversity is key to maintaining balanced ecosystems that can sustainably provide a foundation for the biological resources we depend on to support our standard of living.

In that regard, I think the issues that face those of us at this conference go far beyond simple pest control. I have been talking with Dr. Werner Nader, who works with the Instituto Nacional de Biodiversidad in Costa Rica (abbreviated as INBio). INBio is an organization established by the Costa Rican government to save, know, and rationally use the country's species biodiversity, which according to Dr. Nader is among the richest in the world (W. Nader, personal communication). Dr. Nader states that Costa Rica is estimated to have more than 500,000 individual species, of which only 85,000 have been described. However, there is a problem. Costa Rica has lost more than 75% of its rain forest since the 1940s. Those areas being preserved at this point exist as isolated islands, which is an ecological concern because it limits the amount of genetic variability in breeding populations of an individual species (W. Nader, personal communication). Surprisingly, at least to me, the major source of this deforestation was not urban growth, timber harvest, or even agronomic uses, but clearing of land for cattle grazing. The short-term economics of clearing land and grazing cattle for shipment of hamburger back to the United States were such that people, such as ourselves, could buy our hamburgers for a few cents less if the cattle were raised in Central America rather than on ranches in the USA. However, as you might suspect, the long-term economic impact on the environment is not good. These fragile rain forest soils erode, which destroys their productivity and leads to further deforestation as new grazing sites are needed. The loss of habitat also leads to irreparable loss of species diversity, estimated to be up to 75 species a day (W. Nader, personal communication), and thereby takes away potential sources of genetic information that could benefit humankind to a much greater degree than a marginally cheaper hamburger. The ecological footprint, as measured by all ecosystems impacted by this process, is huge.

Fortunately, sensitivity to the magnitude of this problem has stopped the mass clearing of the rain forest. In this case, people did listen. The government has stepped in and preserved roughly 22% of the land mass of Costa Rica as a species preserve (W. Nader, personal communication), but only after realizing that the long-term economics of what was happening did not make sense despite the attractive short-term benefit. The export value of this meat industry to Costa Rica is $30 million, while the value of tourism drawn by their diversity base is nearly $1 billion (W. Nader, personal communication). I think that this is an excellent example of people taking a bad situation and making it much better by factoring in a broad array of ecological and economic parameters to change a practice that, on a simpler scale, seemed quite good for the local economy but in reality was destroying the basic productivity of an entire nation. They measured the complete footprint, which is something we involved in IPM should be doing as well. People bought into a more ecologically sound practice because the long-term benefits were tangible and they could share in the proceeds. Gene-mining within the rain forest by major corporations, under the auspices of INBio, and tourism are putting money back into many levels of society in Costa Rica (W. Nader, personal communication). This is a success story driven by thinking on a higher, more complex plane that needs and is receiving more adherents.

Robert Shapiro, the Chief Executive Officer of Monsanto, provided some additional perspective to this line of thought in an articulate interview that appeared in the *Harvard Business Review* in 1997 (3). The focus of the interview was "sustainability." When asked why sustainability was becoming an important component of Monsanto's planning process, he responded with principles and ideas that are voiced by many people, but Mr. Shapiro's comments are worth paraphrasing here.

Said Mr. Shapiro, "1.5 billion people live a subsistence life, spending their days trying to get food and firewood to make it to the next day...without radical change we will live in a world of mass migrations and environmental damage on an unimaginable scale...at best...a few islands of privilege and prosperity will survive in a sea of misery and violence." He continued later in the article with the thought that "our nation's economic system evolved in an era of cheap energy and careless waste disposal, when limits seemed irrelevant. None of us today...running a house or managing a business...is living in a sustainable way...the whole system has to change..." Obviously, Mr. Shapiro presented very strong imagery, very keen awareness, and a very strong corporate purpose in terms of looking at their business from a macro-system's perspective. He also provided in his article a more specific and interesting example of the ecological impact made by traditional insect control in potatoes that helps provide an understanding of what impact pest control might have if one looks at it simply in terms of resource consumption (3).

Perhaps you have seen this example used before, but I found the magnitude of the numbers absolutely amazing. In general, the numbers estimate the resource requirements of treating the U.S. potato crop for control of Colorado potato beetle. Approximately 2 million kg of raw materials, 1 million barrels of oil, and 1.7 million kg of inert ingredients are consumed to make 2 million kg of formulated product and 900,000 kg of waste. Growers use another 550,000 liters of fuel to apply the insecticide and generate 180,000 containers that need to be disposed of when empty. These are tangible numbers that represent a really remarkably large footprint for a relatively small market. It would be interesting to see a similar calculation for all pesticides (synthetic and natural). I imagine the numbers would be staggering. I think calculations like this show that modern pest control really is a residue of cheap energy and abundant natural resources. It is not a situation that is going to continue. The economics will change, because the abundant natural resources supporting cheap energy are becoming limited, and practitioners of integrated pest management need to be prepared with systems and tools to adapt to a new reality.

Now, speaking of tools, Mr. Shapiro makes a very simple statement via his balance sheet about the potential impact of biotechnology on that ecological footprint (3). The net drain on natural resources by engineering pest control into the plant according to the numbers provided in the interview is zero. Granted, this number is likely not to be exactly correct and it does not account for concerns people might have about resistance management and other ecological issues, but it does make a significant point, which is that the tools of biotechnology offer a powerful means to influence the ecological footprint of modern agriculture. Debating whether this influence is a benefit or a liability depends on specific circumstances and people's personal views on biotechnology. My personal opinion is that the tools of biotechnology have finally evolved to the point that there is tremendous potential to place characteristics into crops that will allow us to adopt ecologically based pest management and more sustainable farming practices. I recognize that this opinion is not universally held, but that is not the purpose of my presentation. My purpose is to introduce those parameters that I feel are going to influence the adoption of integrated pest management systems, and one would have to have been living in a cave for many years not to recognize that agricultural biotechnology has a significant place in our discussions. Many other speakers are addressing this issue today, and I encourage you to listen carefully and reach your own conclusions, but please think broadly. This is new technology with distinct issues to resolve, but also with tremendous potential to decrease inputs, increase unit productivity, and decrease the complexity of farm management systems. Dr. Shapiro's potato plants represent an example of reduction in agricultural inputs. A second example would be the elimination of the complexity and cost of using foliar-applied, biologically derived pest control. Ten years of experience developing biological pesticides tells me that retooling plants to afford them better genetic means of repelling pests will allow people to shift to sophisticated IPM systems, because it will eliminate the complexity of IPM systems and lower the cost. To me, the relevant question associated with the application of biotechnology to agriculture has changed from "What can be done with the tools of biotechnology?" to "What should be done?" and "How quickly can it be implemented?" Furthermore, regardless of how you personally measure the word "should" in economic, ecological, or ethical frames, my point is that the tools of biotechnology have tremendous potential to reduce the ecological footprint that agriculture makes on the face of the earth. We will need to use these tools if we are to feed a growing world without depleting a number of critical natural resources. So, let us engage in educated debate about integrated pest management, chemical pesticides, and agricultural biotechnology, but in that debate, let us ponder more complex issues that take us beyond questions of pest control alone and consider the impact as measured by principles such as that established by Dr. Wilson with his concept of "ecological footprints."

I'd like to close with a quote that I obtained from Ms. Elaine Brooks (Elaine Brooks, personal communication). The quote is from a student paper in a class at San Diego City College designed to study human impact on the environment by mapping species diversity in a variety of environments around San Diego. The results of this mapping were interesting, but what struck me about these students and their papers was their sense of awe on seeing for the first time the complexities of the world around them and their influence on that web of life. This quote is credited as being originally from Baba Dium, a noted African environmentalist:

"In the end, we will conserve only what we love,
We will love only what we understand,
and we will understand only what we are taught."

My experience during the past 20 years of working in agriculture has taught me that there are people out there trying to grow food and fiber for the rest of us who can do a great deal with new technology only if they can understand what we are talking about. They can only understand if we are willing to teach them. Let us strive to make our work relevant to people other than those sitting in this room. Let us strive to make our messages simple, and let us strive to take our research beyond a simple focus on pest control to developing information on how our farming practices may lead to certain organisms becoming pests.

Literature Cited

1. Kennedy, G. G. 1999. Introduction to the Conference. Pages 2-11 in: Emerging Technologies for Integrated Pest Management: Concepts, Research, and Implementation. G. G. Kennedy and T. B. Sutton, eds. North Carolina State University. Raleigh, NC.
2. National Research Council. 1996 Ecologically Based Pest Management. Web site summary of report. **www.nap.edu/bookstore/isbn/0309053307.html**
3. Shapiro, Robert As interviewed by Joan Magretta. Growth through Global Sustainability. Pages 79-88 in: Harvard Business Review, 1997 .January-February
4. Wilson, Edward O. As interviewed by Scott Lafee in the San Diego Union Tribune. May 6, 1998. Section E. Page 1.
5. Zimdahl, R. L. 1999. My View. Weed Science 47:1.

Section 2

New Diagnostic Techniques for IPM

Diagnostic techniques have long been used in seed testing and certification programs and in IPM, primarily to diagnose pest and pathogen problems. The importance of diagnostic techniques in crop protection will certainly increase as the risk of pest introductions increases due to the continuing internationalization of trade, as concern over food safety and environmental pollution increases, and as the need to monitor the deployment of transgenes and the development of resistant pests and pathogens increases.

The field of diagnostic testing has advanced tremendously during the last decade, primarily in association with advances in immunology, molecular biology, and the biomedical sciences. Although advances in diagnostic testing have had their greatest applications in the area of human health, many of these advances are being applied to diagnostic needs in agriculture, including: detection of plant pathogens and diseases, monitoring pesticide residues on foods and in the environment, monitoring levels of naturally occurring and genetically engineered toxins in plants, identification of pest species, and the detection of pesticide resistance in pest organisms and pathogens.

The field of diagnostics is especially challenging because of the rapidity with which new applications and detection formats are being developed. Many of the newest diagnostic technologies have tremendous potential to enhance IPM. In the first chapter of this section, Chester Sutula provides an overview of recent advances in agricultural diagnostic technologies. In the next chapter, Robert Henkens et al. describe an array of low-cost, high-throughput sensors and diagnostic equipment that can be used to assay gene expression. Although developed for biomedical applications, the technologies they describe provide a perspective on the technologies that are available and the potential that they have for use in IPM-related diagnostics. In the third chapter, Michael Roe et al. discuss the use of biochemical, immunochemical, molecular, and bioassay approaches for detecting resistance to insecticides in pest insects. The final chapter of this section, by Sandy Stewart, provides a discussion of the primary factors that must be considered when selecting and using diagnostic tests for the detection of plant pathogens and decision-making in IPM. The chapter emphasizes the importance of selecting a diagnostic test that is appropriate to the task at hand, and of sampling and field validation of diagnostic tests to ensure that test results can be reliably interpreted.

General Overview of Diagnostic Technologies

C. L. Sutula
Agdia Inc. Elkhart, IN 46514 USA

Many different technologies are available to measure and detect thousands of analytes of interest in agriculture, but only a few of these have been particularly useful to a broad range of workers interested in plant pathogens and insects. These technologies may be simply described as methods that employ (1) antibodies, (2) nucleic acids, or (3) bioassays.

During the last 20 to 25 years, all three of these technologies have been used to detect and measure attributes of plants, pests, and the environment. In this article, I will discuss mainly those methods that use antibodies and nucleic acids. (See chapter by Roe et al. in this volume for a discussion of bioassays.)

Methods based on antibodies

Antibodies are proteins produced in animals in response to the recognition of something "foreign," and have been known for nearly 100 years. Antibodies to anything capable of acting as a foreign material can be deliberately produced, and they can be harvested, purified, and then used as a wonderfully specific material able to bind the substance (antigen) that was used to produce them. Because they are too small to be visible in most experiments, antibodies are labeled so that their binding to the target antigen can be followed. The most common methods for labeling antibodies are:

- With enzymes - as in ELISA
- With colored particles - as in lateral flow immuno-strip tests

- With materials such as fluorophores, radioisotopes, biotin, luminol, and metal chelates

The general approach in which antibodies are used can be illustrated with an example. To observe the binding of an antibody to a target antigen, it helps to present the antigen bound to a solid surface. This can be an antigen present in tissue on a microscope slide or an antigen that has been attached to a membrane, such as nylon or nitrocellulose, or adsorbed to a solid plastic, such as polystyrene. After blocking any remaining active sites, exposing the bound antigen to a solution of antibody produces binding of its specific antibody. If the antibody is labeled with an enzyme, evidence of antibody binding is obtained by washing away the excess antibody and contacting the object with a solution of a substrate for the enzyme. As the enzyme processes the substrate, a color change is produced. The experiment can be arranged to produce color only when the enzyme is present. In turn, this means antibody was bound and the antigen is present. No color can mean antigen was not present at a certain level of detection.

The trick employed to make the antibody visible gives the method its name, ELISA: Enzyme Linked to antibody (Immuno) solid (Sorbent) Assay. A very popular method to produce binding of many antigens is to first attach a specific antibody to a solid, where it acts as a monolayer affinity device capturing the antigen. After washing away the sample, the capture of antigen is made visible by the process already described.

The color response can be remarkably clear in such immunoassays. Positive samples often can produce colored solutions with optical densities greater than 2.0, and negative samples may exhibit optical densities of 0.05 or lower. The colors can be read visually or recorded with a spectrophotometer. Tests for large protein antigens, such as coat proteins of plant viruses, can often achieve a sensitivity of several nanograms of virus/ml of plant extract. For smaller antigens and with different assay arrangements, 10 pg of analyte/ml can often be detected and quantified. Most of these tests are done in a microwell format, using arrays of 96 microwells arranged in eight rows of twelve wells each in polystyrene or polyvinylchloride plates.

This antibody methodology has been applied very successfully in ever-increasing scale and variety since it was first developed during the 1970s. It has had a major impact on the production of plants. The use of this technology in conjunction with tissue culture, with the excision of meristems, with heat and chemical therapy, and with flush-through production methods has resulted in dramatic decreases in the amount of pathogens in crop seeds and propagules. The data from many seed crops produced by these methods emphasize an important IPM principle: early disease incidence is often related to the quality of the seed that is planted. Before these methods became widely used, it was customary to consider weeds and other hosts to be the sources of inoculum for troublesome infections. Now, it is clear that planting infected material can be a major source of disease problems.

The role of the diagnostic method in these applications is to define the negative plant. Since the young plant or the plant in tissue culture often does not express symptoms of disease, laboratory methods using antibodies, and more recently nucleic acids, are necessary to determine if a pathogen is present. Similarly, for crops started from biologically true seed, the seed itself can be tested, often without germinating it. Another obvious use of the diagnostic method is to detect the presence of a pathogen or to confirm a diagnosis of disease in a growing crop.

During the past 20 years, this technology has become routinely used in more than 50 crops. The use of these diagnostic tools has created a worldwide diagnostic business that is small but very important for the subset of agriculture that it serves. Three companies pioneered this activity in the early 1980s: Inotech, Agdia, and Boehringer Mannheim. Since then, tests for more than 200 pathogens have become available. The market is served by seven established, worldwide suppliers aided by many smaller laboratories and "do-it-yourselfers."

While this business is very complex, the market is quite small. It is about 100 to 1000 times smaller than human diagnostics. The market is very fragmented, by crop, state, country, and cultural practice. The items tested have a small unit value with a narrow window for testing per crop/season. Thus, this business challenges us to serve rather than simply make money. Of course, the small scale of the business has a major impact on the rate of technology, product, and market development. This prompts the question, How do you make such a business work? The answer lies mainly in the unselfish cooperation and encouragement from members of the worldwide research community and their supporting governments. Many individuals have provided biomaterials, time, and expertise to make this business succeed.

The antibody technology is easily extended to detect and measure many plant materials (quality factors). The Bt endotoxins provide a recent example in which specific assays are appearing for a series of Cry proteins expressed in transgenic plants. The example is also interesting because of the impetus it provides to extend the availability of this technology to the field site in addition to the laboratory or field station.

Agdia's test to detect Bt-Cry1Ac in transgenic cotton is present-ed as a pouch for extraction and a flow immuno-strip test. To begin the test, the pouch is cut open and a piece of plant tissue approximately 3 cm^2 is placed into the extraction liquid inside the mesh saddle of the pouch. The pouch is then placed on a hard, flat surface, and the tissue fragment is rubbed with a pen to extract Bt-Cry1Ac. Next, the test is inserted at the side of the pouch in a way that just contacts the extract. About 10 minutes later, a colored test line appears in the Bt-Cry1Ac positive sample. No line appears when the plant is Bt-Cry1Ac negative. The test also includes a control line to demonstrate that the test is working properly.

This Bt test is an example of an important trend initiated by the wide acceptance of pregnancy tests that first appeared in this format. It works in the following way: Sample enters a conditioning pad and contacts a porous material that contains antibody labeled with colored particles. The antibody reacts with Bt-Cry1Ac in the sample as it moves in the flow in the porous components of the strip. Upon reaching the test line, it encounters more antibody, which binds Bt-Cry1Ac. In this way, particle-labeled antibodies are captured producing a colored line as the process continues. After passing the test line, some of the labeled particles react with the control line before the excess fluids are absorbed. Tests employing this type of immuno-strip format are appearing for many applications outside of human or veterinary diagnostics. They will play a key role in IPM as well.

The antibody technology has also been applied to insects. Shortly after the demonstration that ELISA could be used to detect plant viruses, research reports began to appear on the use of antibodies to identify insects and insect parts. It was only natural that these immuno-methods would be applied to the vectors of the pathogens. Even as early as 1980-82, ELISA was used to test leafhoppers for beet curly top virus, as part of an effort to spray insecticides only on those leafhopper flights that contained the virus (2). More recently, ELISA has been used to test thrips for tomato spotted wilt virus and impatiens necrotic spot virus as a way to determine if control measures have been effective (1). And several years ago, Trowell and co-workers at CSIRO developed a technology that produced the Lepton kit--a way to identify the cotton bollworm in the egg stage (3,5). Last year, Agdia introduced Hel-ID, a kit using monoclonal antibodies developed at Mississippi State University that can identify both the budworm (*Heliothis virescens*) and the bollworm (*Helicoverpa zea*) from the eggs of these insects (4,6). The idea developed by Trowell and associates was that if you know the identity of a field sample of the eggs, you might be able to predict the field population of the insects that will follow (5). In this application, it is not practical or really possible to make this judgment from visual inspection of the eggs.

Agdia's kit labels antibodies for eggs of the two insect species with separate enzymes: peroxidase for the budworm, alkaline phosphatase for the bollworm. After breaking and introducing the contents of individual eggs into separate wells, a distinct color is produced only if the budworm is present. After recording the results and washing out the microwells, the bollworm test is concluded. Since the test identifies both insects separately, a negative result has a definite meaning. Either a microwell was missed, or some other insect egg was sampled. Hel-ID tests conducted with laboratory-reared insects showed very good agreement with known populations, i.e., about 99%. Tests conducted in the field gave very similar results. Both studies were employed with multiple operators and sources of eggs.

Methods based on nucleic acids

The useful methodology, so far, has been hybridization using nucleic acid probes labeled with substances such as biotin, fluorescein, digoxigenin, metal chelates, and enzymes. However, the introduction of PCR (polymerase chain reaction--a technology that greatly increases the concentration of a selected, target nucleic acid sequence) revolutionizes our ability to detect nucleic acid sequences at a very high level of sensitivity. The same technology is also very useful for many other purposes, such as preparation and labeling of nucleic acid probes and even labeling nucleic acid targets of samples.

The methodology introduced by PCR-assisted processes allows us to use even less sensitive methods to detect the products, such as electrophoresis. At the same time, this methodology has produced very novel approaches, such as the PE BioSystems' Taqman® or Igen's Origen®.

A simple example will help illustrate the utility of these nucleic acid tools. In 1984, Agdia introduced a hybridization test for the potato spindle tuber viroid (PSTVd) using P^{32} radiolabeled probes. Now the test is performed using a digoxigenin-labeled probe. After the hybridization, the bound probe is detected using an antibody to digoxigenin labeled with alkaline phosphatase. The presence of the enzyme is made visible at very high sensitivity by using a substrate that undergoes an enzyme-catalyzed chemiluminescence, which is detected by very sensitive x-ray film. The results are very easy to interpret, since positive samples form dense black spots, and negative samples are clear with no background.

Agdia's Testing Service conducts this test in a shared way with its customers. This is done by sending a blank, pretreated membrane to the customer, who collects the samples and spots a small amount of sample extract on the membrane. The customer then returns the membrane to Agdia for processing. There are very few analytical methods that can be performed using this type of cooperative group approach and provide results that are reliable at a level of several picograms per applied spot. This method has been a major factor in eliminating PSTVd from seed and commercial potato production in the USA.

The PSTVd example also illustrates two other characteristics that seem to be required of diagnostic methods in agriculture, simplicity of format and low unit cost. The membrane result gives an elegant answer with simple operations. The cost is about $1.50 to $2.00 per applied sample. A sample can represent 10 to 20 separate plants extracted together.

Finally, I would like to indicate some of the interesting directions this nucleic acid technology is taking. Lambdatech, a Belgian diagnostics company, has introduced a method they call ELOSA. This methodology combines nucleic acid probes with antibody recognition to make the results of a PCR process visible. While Lambdatech is only one of many companies developing such tests, they are one of the first to produce diagnostic tests that use this format and are practical in agriculture. In this case, the successful amplification of a target nucleic acid is signaled by the production of a distinct color in an ELISA-like microwell plate.

Hybridization of nucleic acid probes to sample targets is also being used to develop elegant sensing systems by constructing arrays of thousands of microelements containing probes on small silicon wafers and other substrates (see chapter by Henkens et al. in this volume). This technology promises rapid, simultaneous detection of many hundreds of target nucleic acid sequences in one analytical operation.

With these last examples, it appears that arrays should have been included as a separate, especially useful technology that promises to make many contributions to IPM. In fact, my last example is an array idea that results in a very useful diagnostic device to identify bacteria. BioLog, a company in California, produces this diagnostic. Their devices use arrays of distinct metabolic reactions contained in special 8x12 microwell plates followed by pattern recognition to identify the organism. The products received an IR100 award in 1991.

Literature Cited

1. Bandla, M. D., Westcot, D. M., Chenault, K. D., Ullman, D. E., German, T. L., and Sherwood, J. L. 1994. Use of monoclonal antibody to the nonstructural protein encoded by small RNA of tomato spotted wilt tospovirus to identify viruliferous thrips. Phytopathology 84:1427-1431.
2. Mumford, D. L. 1982. Using enzyme-linked immunosorbent assay to identify beet leafhopper populations carrying beet curly top virus. Plant Disease 66:940-941.
3. Ng, S. S., and Cibulsky, R. J. 1997. Development of *Heliothis* antibodies for the Lepton HTK insect diagnostic test kit. Pages 883-885 in: Proc. Beltwide Cotton Conf. Vol. 2.

4. Sutula, C. L., Boyan, W. W., and Xia, J. 1999. Hel-ID: a tool for bollworm and budworm management. Pages 77-78 in: Proc. Beltwide Cotton Conf. Vol. 1.

5. Trowell, S. C., Lang, J. C., and Garsia, K. A. 1993. Heliothis identification kit. Pages 176-179 in: Pest Control and Sustainable Agriculture. S. A. Carey, D. J. Dall, and W. A. Milne, eds. CSIRO Publications, Melbourne.

6. Zeng, F., Ramaswamy, S. B., and Pruett, S. 1998. Monoclonal antibodies specific to tobacco budworm and bollworm eggs. Ann. Entomol. Soc. Amer. 91:677-684.

Use of DNA Technologies in Diagnostics

Robert Henkens, Celia Bonaventura, Valentina Kazantseva, Mario Moreno, John O'Daly, Rebecca Sundseth, Steven Wegner, and Marek Wojciechowski

AndCare, Inc., Durham, NC 27713 USA

Pathogenic organisms cause problems in both humans and plants. Molecular controls of pathogens are becoming possible as a result of technological advances in the area of DNA diagnostics. When scientists can identify genes that are unique to a stage of growth of a particular parasite, bacterium, or virus, drugs can be targeted to these genes. Similarly, knowledge of the products of these genes can be incorporated into the design of improved treatments against human or plant pathogens. AndCare, Inc., is developing new tools for nucleic acid analysis to make this process of drug targeting and treatment development more cost effective.

We are developing a new line of products for rapid assays to correlate disease state with load, genotype, and gene expression of pathogens. These are based on the use of low-cost, self-contained, disposable nucleic acid sensors and desktop-sized analytical instruments.

The overall goals of our research are 1) to develop advanced DNA sensors and sensor arrays for direct, genetic analysis of pathogens; and 2) to demonstrate the utility of electrochemical sensors and sensor arrays for gene-expression studies of disease. Because of the parallels between the need for better pest controls for agriculture and for better solutions to human health problems, we envision that the AndCare Electrochemical Gene Assay System will be widely used in both arenas for research on disease prevention.

Significance

Advances in molecular biology have led to a rapidly increasing store of knowledge on the genomes of disease-causing organisms. This knowledge can potentially be used to identify nucleic acid sequences that are specific to pathogens, to locate therapeutic targets, and to develop new and improved drugs to treat or prevent disease. To achieve this potential, AndCare is engaged in innovative research to apply advanced DNA technology to the problem of quantitative and qualitative molecular analysis of pathogenic organisms and their gene expression during disease progression.

Human health problems and problems requiring better pest control in agriculture can benefit from use of DNA technologies in diagnostics. The increasing worldwide problems of disease caused by parasites, bacteria, and viruses have been reviewed (**www. niaid. nih.gov**). The emergence of drug-resistant strains, a breakdown of public services, changes in climate, greatly increased population mobility, a lack of understanding of the threat, and a focus on other diseases, such as cancer, have all led to the resurgence of disease-causing organisms throughout the world at a time when infectious diseases were thought by many of us to be controlled. Moreover, newly emerging diseases, drug-resistant pathogens, insecticide-resistant carriers of pathogens, and the growth of disease-susceptible human and plant populations will continue to pose an enormous challenge for improvement of diagnosis, treatment, and prevention of diseases associated with pathogenic organisms.

For example, one of the most deadly infectious diseases--malaria--has made a comeback with the appearance of insecticide-resistant mosquitoes and drug-resistant strains of the malaria parasite. Malaria continues to claim an estimated 2 to 3 million lives annually and to account for unknown morbidity in the approximately 300 to 500 million people infected annually (14). Furthermore, it has been predicted that malaria will increase dramatically in the world's temperate regions—where it is now practically non-existent—and that it will also increase in tropical regions (9). As a result of the spread of drug-resistant parasites and insecticide-resistant mosquitoes, our tools for dealing with this threat are diminishing at the same time that the threat is increasing.

The overall goal of our work at AndCare, Inc., is not an incremental advance in gene analysis of pathogens, but instead a major increase in assay performance in terms of reduced time, complexity, and cost in an assay format that can be automated and multiplexed for high-throughput applications. The tools that we intend to produce may have additional application areas, some of which are listed here: 1) kinetics of gene expression in different cell types under different stimulatory conditions; 2) comparison of pathogenic and non-pathogenic strains or comparison of different organisms that share the same mechanisms of pathogenesis; 3) low-cost, high-throughput genetic tools for studies to relate gene expression of the pathogen with disease progression; and 4) studies of genotype and gene expression in non-infectious diseases, such as cancer.

AndCare's new electrochemical sensor technology is expected to overcome problems of increased sensitivity and throughput without requiring complicated techniques or large, expensive instruments. We anticipate that the AndCare Electrochemical Gene Assay System will find use in a wide range of laboratories and clinics concerned with human and plant disease and disease prevention.

SIGNIFICANCE OF TECHNICAL APPROACH

Our research is focused on development of a novel analytical assay using electrochemical sensor arrays to detect genes expressed by pathogens during disease progression. The purpose of the assay is to facilitate development of drugs targeted at specific genes or gene products and for identification of gene products for new vaccines and drugs. We believe this to be a highly innovative solution to the need for economical determination of pathogen load and genotype and for detection of pathogen gene expression during growth. The assay incorporates electrochemical DNA probe sensors for messenger RNA (mRNA), ribosomal RNA (rRNA), and genomic DNA (gDNA) into individual tests and into arrays for the measurement of a large number of probes using a low-cost, high-throughput instrument.

The rationale for development of the DNA sensor assays and instruments for malaria is that new candidate vaccine antigens will increasingly be discovered as the complete sequence of the genome of *Plasmodium falciparum* is revealed by researchers. These will lead to new drug and vaccine candidates. This effort will be

54

facilitated by advanced technology for economical identification of genes expressed during the life-cycle stages of the parasite, which in turn will permit development of therapies targeted to pre-erythrocytic, blood, and sexual stages of the parasite. Moreover, genetic analysis and directed mutations may lead to altered mosqui-toes that cannot support the growth of the malaria parasite.

As envisioned, the AndCare Electrochemical Gene Assay Sys-tem is aimed at the research market and would provide a key capability not yet available in assays for pathogen gene expression.

Technical and Performance Characteristics of the Electrochemical Gene Assay System Compared with Current State-of-the-Art Methods

SIGNIFICANCE OF ELECTROCHEMICAL SENSOR TECHNOLOGY

As outlined below, the assays used in the Electrochemical Gene Assay System are based on the hybridization of nucleic acid probes with pathogen rRNA, mRNA, or gDNA. Target nucleic acid sequences complementary to the probe sequences are detected electrochemically with an enzyme-linked assay. A proprietary colloidal-gold sensor is used to measure the product of the enzyme-catalyzed reaction as a change in current, proportional to the concentration of probe target hybrids captured on the sensor.

The following compares cost, complexity, and throughput of the AndCare Electrochemical Gene Assay System technology to that of conventional DNA chip assays. In recent years, DNA assays that do not require radioactive labels have been developed for the research and clinical markets to eliminate hazards of handling radioactive materials. The most common non-radioactive labels in use today are fluorescent, chemiluminescent, or enzyme types. Several companies have developed automated instrument/assay systems based on non-radioactive labels, but realizing the necessary sensitivity and high throughput has required complicated techniques and expensive equipment, such as large laser-driven scanners used in the analysis of assays using gels, blots, and current-generation DNA chips.

AndCare's new electrochemical sensor technology is expected to meet the needs for increased sensitivity and throughput without requiring complicated techniques or large, expensive instruments.

Furthermore, unlike current DNA chip assays, researchers using this "open" system can synthesize and use their own probes for the assays, rather than relying only on probe systems provided by the manufacturer. The AndCare system could find advanced applications in a wide range of biotechnology/biomedical laboratories and hospital clinics and in agricultural research institutions. We envision that the system will become widely used in evaluation of gene expression, clinical diagnostics, DNA/RNA probe technology, DNA sequencing, and advanced DNA/RNA chips.

The electrode sizes (diameter) of commercially available microelectrode arrays are on the order of 100 to 1000 nm for each electrode in the array, which could conceivably represent the size of a single sensor for detection of DNA/RNA targets using electroactive labels. The detection of submicroliter biological samples on a sensor array could reduce the size of each sample significantly, thereby reducing sample-processing times and reagent costs. In addition, for the specific assays for electrochemical detection of gene expression, the hybridization times required are less than 20 min. This is dramatically reduced over the times required for current DNA chip analysis, thereby representing a major improvement that can increase sample throughput.

TECHNICAL ASPECTS

As described below, AndCare electrochemical sensing technologies for DNA diagnostics are currently available for research applications (**www.andcare.com**).. The new tools include disposable electrochemical sensors, portable electrochemical monitors, and detection kits for native and PCR-amplified nucleic acids. We are working on new technologies for DNA diagnostics in which sensors will be arranged in arrays, with each element in the array capable of quantifying the level of a specific nucleic acid target. The assay and instrumentation for handling sensor arrays will provide the basis for developing a very high-throughput, automated, economical assay for the gene-expression studies required for drug- and vaccine-development studies. It could be used for blood or tissue samples in human, plant, or animal studies, or directly in cell-culture studies.

For example, using the Electrochemical Gene Assay System, gene expression in cell cultures could be measured directly, without reverse transcriptase polymerase chain reaction (RT-PCR) amplification. To measure gene expression, we work with dual sensors that

56

can be incorporated into arrays in a micro-titer plate format. This format is widely used both in small biotechnology/biomedical research laboratories and in large, highly automated facilities for high-throughput assays in drug discovery and development. We will initially use a 96-well plate format, but may later extend the work to larger plates.

The current sensitivity of AndCare's electrochemical detection system is approximately 5×10^5 targets. In cell cultures in 96-well plates, assuming about 50,000 cells per well, we can currently detect 10 copies per cell with our electrochemical detection technology. We expect to increase our detection sensitivity to 1 to 5 copies per cell, with 1 hr or less hybridization time.

The general electrochemical detection method is described by the following equations, together with Fig. 1:

probe + nucleic acid target → specific hybrid
specific hybrid + enzyme conjugate → enzyme-labeled species.

The enzyme-labeled species is captured at the sensor surface, as shown schematically for the Direct-Detect format in Fig. 1A, and a current is generated upon addition of enzyme substrate, as shown schematically in Fig. 1B.

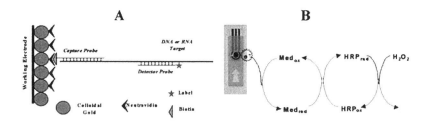

Figure 1. (A) Schematic of the configuration of probes and labels for the DIRECT-DETECT method of capture and detection of DNA or RNA on the surface of AndCare's electrochemical sensors. (B) Series of oxidation/reduction reactions initiated by horseradish peroxidase (HRP) and resulting in generation of mediated catalytic current on the surface of the sensor (center of the figure) for a three-electrode-sensor format.

AndCare's Current Commercial Products

AndCare has developed a number of electrochemical sensing technologies and has commercialized several new tools for diagnostics, electrochemical research, and molecular biology. The commercial tools are described below briefly and more fully elsewhere (**www.andcare.com**).

AndCare's sensor-based quantitative electrochemical assays are based on broad enabling technology (1,2,3,4,5,6,7,8,10,11,12,13,15, 16,17; Henkens et al., U.S. Patent Numbers: 5,217,594; 5,225,064; 5,334,296; 5,368,707; O'Daly et al., U.S. Patent Number: 5,391,272). By employing patented colloidal-gold electrodes and a novel electrochemical measurement method, AndCare's assay systems can deliver accurate, quantitative test results on complex and optically opaque samples in a fraction of the time required to conduct ordinary assays.

The sensors (Fig. 2) are produced in sheets of up to 40 individual sensors per sheet by screen-printing methods. The electrode circuits of the sensors are formed by screen-printing carbon inks and silver onto plastic, which is die-cut into individual disposable sensor strips. Modified forms can be used for different applications.

The coating of colloidal gold onto the working electrode(s) of the sensor strips (Fig. 3) results in an electrode studded with a microarray of nanometer-sized gold particles that can be further modified to accomplish specific assays. The incremental cost increase associated with use of colloidal-gold-modified electrodes is

Figure 2. Photograph of two types of AndCare sensors.

Figure 3. Controlled deposition of nanometer-sized colloidal-gold particles onto a sheet of printed sensors.

negligible. The performance obtained is far better than that from bulk gold electrodes (5; Wegner et al., U.S. Patent Number 5,468,366).

AndCare also produces portable potentiostats for research, as illustrated in Fig. 4. These are used in a commercially available three-step test for quantitative measurement of blood-lead levels, as shown in Fig. 5.

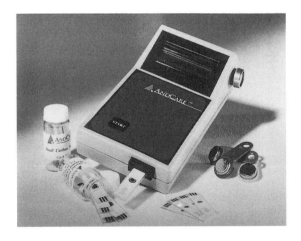

Figure 4. Portable electrochemical monitors.

These portable electrochemical monitors for research have the following characteristics:

- Inexpensive, hand-held potentiostats;
- Powered with a 9V battery or AC power supply;
- Modes of operation: Staircase and Square Wave Voltammetry (including anodic and cathodic stripping) or DC, Differential and Intermittent Pulse Amperometry;
- Broad operational flexibility via RS232 interface and computer software;
- Stand-alone operation without computer; ideal for teaching and research.

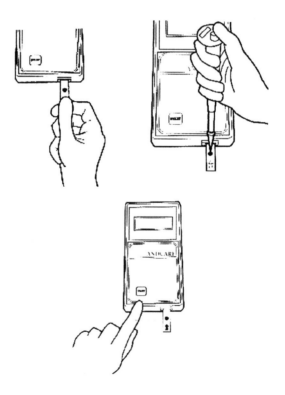

Figure 5. AndCare's 3-Step Blood-Lead Test System.

NEW ELECTROCHEMICAL DETECTION SYSTEMS FOR NATIVE AND PCR-AMPLIFIED NUCLEIC ACIDS

AndCare has developed a family of fast and sensitive DNA and RNA detection assays using hand-held electrochemical monitors, disposable screen-printed sensors, and inexpensive reagents (Fig. 6). With the monitor operating in the Intermittent Pulse Amperometric mode, nucleic acids can be detected with attomol sensitivity in 10 seconds or less. Our nucleic acid detection systems eliminate the need for gels, radioisotopes, and blotting. They are simple to use, inexpensive, and faster than conventional approaches.

The RAPID PCR-DETECT System is intended for detection and quantification of double-stranded PCR products directly, i.e., without a hybridization step.

The HYBRID PCR-DETECT System incorporates added specificity through selective hybridization of detector and/or capture probe.

The DIRECT-DETECT System offers fast and sensitive detection and quantification of non-amplified nucleic acids. Direct hybridization of capture and detector oligonucleotides with the target allows sensitive and specific results in a timely fashion.

The versatility, as well as sensitivity and specificity, of our DNA detection systems were studied in several research labs. The confirmatory assays involved detection of bacteria (e.g., *E. coli* and *Salmonella*), viruses (e.g., polio and hepatitis A), and single base mutations of medical importance (e.g., Factor V Leiden, BRCA1, Sickle Cell Anemia).

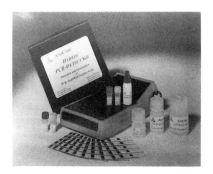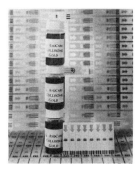

Figure 6. Nucleic acid analysis systems.

(A)   (B)

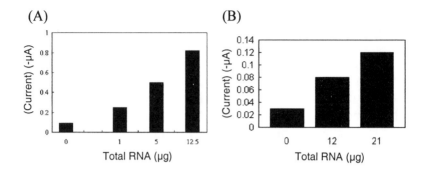

Total RNA (µg)   Total RNA (µg)

Figure 7. Sensitive and specific direct detection of mRNA within total human RNA. (A) Detection of the moderately abundant ApoA1 mRNA in human liver. (B) Detection of the low abundance p53 mRNA from cultured human cells. Specific mRNAs were detected electrochemically following hybridization to sequence-specific probes and captured on AndCare sensors.

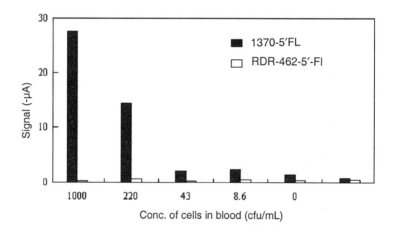

Conc. of cells in blood (cfu/mL)

Figure 8. Specificity in detection of the RT-PCR Products of *E. coli* rRNA. The RT-PCR-amplified products were detected electrochemically following hybridization in a universal bacterial rRNA probe (1370-5'FL), but were undetectable following hybridization to a heterologous, *Streptococcus*-specific probe (RDR-462-5'-FL). This experiment demonstrates the potential for specificity in discrimination among very closely related target sequences.

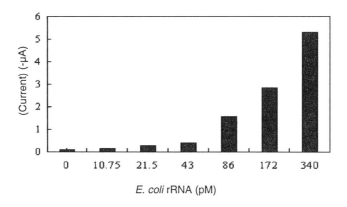

Figure 9. Direct detection of *E. coli* rRNA. *E. coli* rRNA was hybridized to sequence-specific capture and detector probes, captured on AndCare's disposable electrodes, and quantified electrochemically.

Results Obtained Using AndCare
Electrochemical Gene Assay System

Results on measurement of gene expression and qualitative and quantitative analysis of pathogens using AndCare sensor technology have been obtained, as described above (Fig. 7, 8, 9).

Conclusions

The agricultural problems associated with pest management are paralleled by the increasing worldwide problems of human disease caused by pathogenic organisms. The emergence of drug-resistant strains, newly emerging diseases, insecticide-resistant carriers of pathogens, and the growth of disease-susceptible populations all pose enormous challenges to improve diagnosis, treatment, and prevention of human and plant diseases associated with pathogenic organisms. DNA diagnostics offer the potential for improved pest management in agriculture, as well as improved methods of combating human diseases associated with pathogenic organisms.

Advances in molecular biology have led to a rapidly increasing store of knowledge on the genomes of disease-causing organisms. This knowledge can potentially be used to identify nucleic acid sequences specific to pathogens, to locate therapeutic targets, and to develop new drugs against pathogenic organisms. To accomplish this potential, AndCare is engaged in innovative research to apply advanced DNA technology to the problem of quantitative and qualitative molecular analysis of pathogenic organisms and their gene expression during disease progression.

AndCare's new electrochemical sensor technology is expected to meet the needs for increased sensitivity and throughput without requiring complicated techniques or large, expensive instruments. Furthermore, unlike current DNA chip assays, researchers using this "open" system can synthesize and use their own probes for the assays, rather than relying only on probe systems provided by the manufacturer. These considerations suggest that the AndCare Electrochemical Gene Assay System will find use in a wide range of laboratories and clinics concerned with human and plant disease and disease prevention.

Literature Cited

1. Crumbliss, A. L., O'Daly, J. P., Perine, S. C., Stonehuerner, J., Tubergen, K. R., Zhao, J., and Henkens, R. W. 1992. Colloidal gold as a biocompatible immobilization matrix suitable for the fabrication of enzyme electrodes by electrodeposition. Biotechnology and Bioengineering 40:483.

2. Crumbliss, A. L., Stonehuerner, J., Henkens, R. W., Zhao, J., and O'Daly, J. P. 1993. A carrageenan hydrogel stabilized colloidal gold multi-enzyme biosensor electrode utilizing immobilized horseradish peroxidase and cholesterol oxidase/cholesterol esterase to detect cholesterol in serum and whole blood. Biosensors and Bioelectron. 8:331-337.

3. Crumbliss, A. L., Stonehuerner, J., O'Daly, J., Zhao, J., and Henkens, R. W. 1994. The use of inorganic materials to control or maintain immobilized enzyme activity. New J. Chem. 18:327-339.

4. Henkens, R. W., Kitchell, B. S., O'Daly, J. P., Perine, S. C., and Crumbliss, A. L. 1987. "Bioactive electrodes using metallo-

proteins attached to colloidal gold. Recl. Trav. Chim. Pays Bas. 106:298.

5. Henkens, R. W., O'Daly, J. P., Perine, S. C., Stonehuerner, J., Tubergen, K. R., Zhao, J., and Crumbliss, A. L. 1991. Electrochemistry of colloidal gold supported oxidase enzymes. J. Inorg. Biochem. 43:120.

6. Henkens, R. W., Johnston, T. E., Nevoret, M. L., O'Daly, J. P., Stonehuerner, J. G., and Zhao, J. 1992. Determination of lead by its inhibition of isocitrate dehydrogenase and diagnosis of lead poisoning. Pages 275-280 in: Metal Ions in Biology and Medicine (Vol. 2). J. Anastassopoulou, Ph. Collerly, J. C. Etienne, and Th. Theophanides, eds. John Libbey Eurotext, Paris.

7. Henkens, R. W., Zhao, J., and O'Daly, J. P. 1992. Convenient enzymatic determination of trace mercury in water. Pages 317-318 in: Metal Ions in Biology and Medicine (Vol. 2). J. Anastassopoulou, Ph. Collery, J. C. Etienne, and Th. Theophanides, eds. John Libbey Eurotext, Paris.

8. Henkens, R. W. 1996. Detection of microbial pathogens using gene-probe sensors. Scientific conference on chemical and biological defense research. Edgewood Area Conference Center, Aberdeen Proving Ground, MD.

9. Martens, W. J. M., et al. 1995. Potential risk of global climate change on malaria risk. Environmental Health Perspectives, May 1995.

10. O'Daly, J. P., Zhao, J., Brown, P. A., and Henkens, R. W. 1992. Electrochemical enzyme immunoassay for detection of toxic substances. Enzyme Microb. Technol. 14:299-302.

11. O'Daly, J. P., and Henkens, R. W. 1995. Applications of electrochemistry to enzyme-linked immunoassay. Reviews in Pesticide Toxicology 3:45-60.

12. Stonehuerner, J. G., Zhao, J., O'Daly, J. P., Crumbliss, A. L., and Henkens, R. W. 1992. Comparison of colloidal gold electrode fabrication methods: The preparation of a horseradish peroxidase enzyme electrode. Biosensors & Bioelectronics 7:421.

13. Sundseth, R., Thompson, D. M., Taylor, G., Kasensteva, V., Kasensteva, A., Zhang, H., Wojciechowski, M., Naser, N., Wojciechowski, M., Herfst, C., Castillo, S., O'Daly, J., Wegner, S., and Henkens, R. 1998. Electrochemical detection and quantitation of DNA and RNA. Oral presentation at Cambridge

Healthtech Institute DNA and RNA Diagnostics meeting. Washington, D.C.

14. WHO: 1996. Weekly Epidemiological Record of the World Health Organization 71:17-24.

15. Zhang, H., Thompson, D., Sundseth, R., Naser, N., O'Daly, J. P., Wegner, S., Henkens, R. W., Wojciechowski, M. 1997. Disposable sensor-based pulse amperometric detection of pathogens and DNA mutations. Gordon Research Conference on Electro-chemistry. Ventura, CA.

16. Zhao, J., Henkens, R. W., Stonehuerner, J. G., O'Daly, J. P., and Crumbliss, A. L. 1992. Direct electron transfer at horseradish peroxidase/colloidal gold modified electrodes. J. Electroanal. Chem. 327:109-119.

17. Zhao, J., Henkens, R. W., Crumbliss, A. L. 1996. Mediator-free amperometric determination of toxic substances based on their inhibition of immobilized horseradish peroxidase. Biotechnology Progress 12:703-708.

Detection of Resistant Insects and IPM

R. M. Roe, W. D. Bailey, F. Gould, C. E. Sorenson,
G. G. Kennedy, J. S. Bacheler,
Department of Entomology, North Carolina State University,
Raleigh, NC 27695-7647 USA

R. L. Rose, E. Hodgson
Department of Toxicology, North Carolina State University, Raleigh,
NC 27695-7633 USA

C. L. Sutula
Agdia Inc., Elkhart, Indiana 46514 USA

Damage resulting from insects feeding on plants continues to be a major factor limiting agricultural production throughout the world. To a limited extent, insects are managed by cultural and biological control practices. However, in most cases, these approaches have not been adequate to provide the desired level of crop protection, particularly in annual crops, and insecticides are required. Without insecticides, the economic production of food and fiber on the scale required by the world's population would not be possible. For some major insect pests, multiple applications of pesticides are frequently required each year.

Herbivorous insects have been engaged in an evolutionary "arms race" with plants for eons. Through natural selection, insects have evolved numerous, highly adaptive mechanisms that enable them to cope with the defensive chemistry of their host plants. The genetic diversity that has enabled insects to adapt to their hosts' defenses provides the raw material upon which selection for insecticide resistance operates. Multiple insecticide applications over consecutive seasons rapidly eliminate susceptible individuals and leave behind a population of insecticide-resistant insects. Insecticide resistance has been documented in more than 450 species of arthropods, and is a major concern of entomologists. Unfortunately, resistance is usually first detected by growers when they experience control failures, at which time the genes responsible for resistance are widespread in the population, and resistance cannot be easily reversed. Once resistance to a particular class of insecticides is established in the field, new compounds with novel modes of action must be developed. It is critical to the continued success of agriculture and the maintenance of a clean and healthy environment that we prevent the evolution of resistance to our newest and safest insect control systems. An important aspect to achieving this goal is the development of better techniques for resistance diagnosis.

Resistance to Our Safest and Newest
Biological Insecticides is Possible

INSECT RESISTANCE TO BT

The microbial insecticide *Bacillus thuringiensis* (Bt) has revolutionized our thinking about insect control. Bt produces an array of protein endotoxins that attack the gut tissue of a narrow range of insect species, including some of the most important economic, lepidopteran pests of agriculture. These endotoxins have no adverse effects on other organisms. Since Bt endotoxins are proteins, they can be genetically cloned into agricultural crops to provide the plant with an endogenous protection from selected pest insects. During the summers of 1996, 1997 and 1998, respectively, 0.72, 0.93 and 1.13 million hectares of genetically engineered Bt cotton were grown in the Southern United States. The future expansion of this technology in cotton and in other crops is expected to have important environmental benefits due to a reduction in the use of the older and more toxic broad-spectrum insecticides (16). A primary concern of many scientists (e.g.,17,25,38) and environmental groups regarding the use of Bt genes in transgenic crops is the enhanced risk that insect populations will evolve resistance to Bt endotoxins more rapidly than would occur with the traditional use of the Bt toxin in foliar sprays. Although there may have been some skepticism about the potential of insects to evolve resistance to Bt endotoxins (11,21), there is now a significant body of laboratory (9,18,38) and field data (39) indicating that insect pests are capable of evolving high levels of resistance to this insecticide. Genetic adaptation of pests to transgenic Bt toxin-expressing crops is also expected to result in resistance to commercial formulations of fermented strains of Bt, such as Dipel®, which is an important management tool for organic and conventional farmers.

INSECT RESISTANCE TO SPINOSAD

A new biological insecticide, spinosad, was recently developed by Dow AgroSciences (40). Spinosad is derived by fermentation from the bacteria *Saccharopolyspora spinosa*. Unlike Bt, spinosad (spinosyns A and D) is an apolar, complex, polycyclic insecticide that is applied by traditional spray methods to plants. The spinosyns are especially significant because of their high toxicity to the economically important Lepidoptera and their safety to non-target organisms and the environment. These compounds are also important because they represent a new class of insecticide chemistry with a novel mode of action. Spinosad alters nicotinic acetylcholine receptor function and, perhaps, the GABA-gated chloride channels at a novel site of action.

Spinosad has only recently (in 1996) come into widespread use as an insecticide, and it was deemed reasonable by some to assume that insects would not demonstrate natural resistance to this new compound (23). Resistance, however, is at least theoretically possible. A field strain of the tobacco budworm was selected for 11 successive generations by topical application of technical spinosad (8). The first indication of resistance was noted in the sixth generation and by the ninth generation, a maximum topical dose of 0.8 micrograms of spinosad per third instar produced no mortality at 12 days after application. After 11 generations of selection, the selected budworms demonstrated a resistance ratio greater than 355-fold, compared to the LD_{50} of the unselected parental strain. In the first generation of selection, a dose of 0.05 micrograms per larva produced 78% mortality, while in the eleventh generation, a dose of 60 micrograms per larva produced only 39% mortality (a 1,200-fold increase in the selection dose). The selected insects are also resistant when fed spinosad formulated as Tracer®.

THE RESISTANCE CHALLENGE

Largely in response to the development of transgenic Bt crops, renewed effort is being made by the EPA (24,41), the USDA, and private companies (15,41) to maximize the long-term economic and environmental benefits that can be gained from the newest insect control systems. One critical step in decreasing the risk of resistance is the development of systems for detecting low frequencies of resistance alleles in regional pest populations (41) or high frequencies of resistance alleles in local populations. Such a monitoring system could be used to signal a potential problem and the need for a remedial control strategy. Some of the newest biological insecticides such as transgenic Bt crops and spinosad, present challenges that are different from those associated with traditional insecticide chemistries (organophosphates, carbamates, pyrethroids), because the new toxins either penetrate across the insect cuticle very slowly or do not penetrate at all. Consequently, for example, the adult vial test, commonly used to detect insecticide resistance in tobacco budworm (28), is not applicable to Bt, which is a gut poison and does not penetrate the cuticle.

This presentation concentrates on the approaches to resistance detection, which must be simple, inexpensive and reliable. For more information on resistance management, some recent papers are pertinent (4,19,22,25,26,27). For additional information on resistance diagnosis, the following references are available (2,10,12,13, 14,29,34,35,36,37).

Approaches to Resistance Detection

There are a number of possible mechanisms of insecticide resistance in insects. Insecticide metabolism, sequestration, and altered target site are usually the most important mechanisms of insect resistance that eventually result in insecticide failure. Sequestration usually involves metabolic enzymes such as esterases, which can be assayed by measuring their esteratic activity. For the most part, excretion and reduced penetration are less important in insect resistance, as is behavioral resistance. Insecticide transport and the role of p-glycoproteins in insect systems are not yet fully appreciated, but may be critical for some of the newer insect control agents like spinosad.

In most examples of resistance in the field, more than one of these mechanisms has been selected simultaneously, even in cases where one mechanism may predominate. For this reason, a particular mechanism, such as increased metabolism, can be used as a marker for resistance, even though the major mechanism might be target-site insensitivity. The preferred approach in resistance monitoring and detection, however, would be to assay simultaneously for all possible mechanisms of resistance. In some cases, as will be discussed later, this approach is not practical. In cases in which the predominant mechanism for resistance to a particular insect control system is known, assays can be designed more specifically.

There are several factors that require consideration in designing a resistance monitoring strategy for insects. These are listed in Table 1. Especially for new insect control methods, the mechanisms of resistance are usually unknown. In these cases, assessing insect susceptibility is the only method that can be used to detect resistance, and the route of insecticide exposure to the insect in resistance detection assays should be as relevant as possible to field conditions. On the surface, this may seem straightforward but actually can be quite complex. For example, some insecticides enter the insect at different rates through contact and ingestion, and the rates of penetration are affected by an array of factors including the insecticide formulation, temperature, humidity, and light, as well as the nature of the insect's interaction with its host plant, and other factors.

Table 1. Factors to Consider in Resistance Detection

- Single or multiple resistance mechanisms possible
- What are the resistance mechanisms?
- Contact or oral insecticide
- Most appropriate developmental stage to assay
- Assay individual insects or a population
- Sample size needed for assay
- Field versus laboratory assay
- Cost per assay/specialized equipment
- Assay time
- Training

Duplicating these conditions in the laboratory or under field-test conditions may not be practical or even possible for routine resistance monitoring. It is usually most appropriate to measure resistance in the developmental stage of the insect that typically is the target for insect for insect control. However, this may not always be practical. For example, in Lepidoptera, although insecticides typically are used to control the larval stage, most resistance monitoring programs target the adult stage, which is more easily collected by using light or pheromone traps.

Other factors to consider in resistance monitoring are: sample size, field versus laboratory assays, cost, assay time, level of training needed to conduct the assay, and whether to evaluate individual insects or populations for resistance (Table 1). One important factor associated with agricultural diagnostics, and especially with resistance detection, is expense. Unlike medical diagnostics where cost may not be a limiting factor (especially in life and death situations), cost is the primary factor in agricultural diagnostics. This greatly limits the type and amount of information that can be collected. The most innovative new technologies are being developed in medical diagnostics, and the challenge for us is to adapt these new approaches to agriculture. The cost difference between medical and agricultural diagnostics can be 1000-fold or greater.

Limited resources also affect sample size and the type of sampling that is possible. Although it would be best to sample large numbers of insects frequently and in many geographical locations, this is usually not practical. Important goals in designing assays for resistance are to keep the cost per assay as low as possible and to simplify the sampling and assay techniques so that highly trained technicians and/or expensive monitoring equipment are not required to conduct the assay.

Other important factors in resistance detection are assay time and location. Ideally, assays should be simple to conduct on insect life stages that are easy to collect from the field, and they should produce results rapidly, either in the field or at the home of the grower or consultant. The time required to obtain results can be important when the results are used to decide on the most appropriate insect control method to use against pests that are anticipated to exceed the economic threshold for the crop.

The issue of whether to develop a field or a laboratory (or home) assay is not always clear and often depends on personal preference, amount of time available before an answer is needed, the purpose of the assay, and cost. If the assay is being conducted as part of a regional monitoring program for resistance detection and not for insecticide treatment decisions, then it may be preferable to send samples to a central testing facility. With overnight mailing and e-mail communication, test results can be made available to the grower in 2-3 days or less. The cost of such a system would likely be higher than for assays conducted in the field by growers and consultants.

The decision of whether to use a field or a laboratory assay is also affected greatly by the assay type. Four approaches to resistance detection will be discussed (Table 2). Biochemical, immunochemical, and molecular approaches will be considered together because the advantages and disadvantages of these techniques are similar. Insecticide bioassays have been used for many years to detect resistance. Although we will discuss bioassay technology in general, this presentation will focus on a novel bioassay technique being developed for Bt resistance, the feeding disruption assay.

Table 2. Approaches to Resistance Detection

- Biochemical
- Immunochemical
- Molecular
- Bioassay

Biochemical, Immunochemical and Molecular Approaches to Resistance Detection

Increased metabolism (increased P450, esterase, phospho-triesterase, and glutathione transferase activity) and increased sequestration by high molar concentrations of esterases are common mechanisms of insect resistance to insecticides. These have been documented by numerous laboratories. In our own research experience, increased metabolism was associated with or responsible for organophosphate, carbamate, and/or pyrethroid resistance in the

73

tobacco aphid (1), tufted apple budmoth (20), tobacco budworm (42), Colorado potato beetle (6), and German cockroach (5). The optimum method for measuring increased insecticide metabolism would be to assay for the rate of degradation of the insecticide itself. This is not usually practical, especially for routine resistance monitoring, because the assay would most likely require radiolabeled pesticide and radiochromatography. One exception is the measurement of increased phosphotriesterase activity. Metabolism of the insecticide, paraoxon, which is colorless, results in a colored product, paranitrophenol. Recently, Devorshak and Roe (unpublished) developed a microtiterplate microassay to monitor paraoxon activity associated with insecticide resistance in the tufted apple bud moth and other insects.

Some common spectrophotometric, non-insecticidal substrates used for resistance detection are listed in Fig. 1. The advantage of these substrates is that the color development is continuous and non-destructive to the enzyme being studied; this can provide not only a measure of the enzyme activity but also information about the kinetics of the enzyme-substrate interaction. Abdel-Aal et al. (3) synthesized naphthalene thioesters as a successful alternative substrate for measuring 1-naphthyl acetate (1-NA) esterase activity. 1-NA esterase activity is usually measured with an end-point assay using Fast Blue B salt, and has been correlated with resistance in a number of insect species. Again, the advantage of using naphthalene thioesters is that the metabolite can be monitored spectrophotometrically in a continuous assay using Ellman's reagent. Rose et al. (33) and Zhao et al. (42) developed a microassay for P450 activity, which was used successfully to diagnose resistance in the tobacco budworm in cotton. Other non-insecticidal substrates for P450 activity, i.e., methoxyresorufin, benzo(a)pyrene, and benzphetamine, have been successfully used for resistance diagnosis but are mostly applicable to laboratory research.

Biochemical, immunochemical and molecular approaches to resistance detection are similar in that the assay is usually measuring the over-expression of a specific gene or gene family that is associated with or directly responsible for insecticide resistance. The only difference between these three approaches is the method of detection, i.e., substrate metabolism, an antibody-antigen interaction, or detection of a specific sequence of nucleic acid (either RNA or DNA). Biochemical assays can be read simply by observing a color change in a solution or of immobilized insect enzymes on a solid

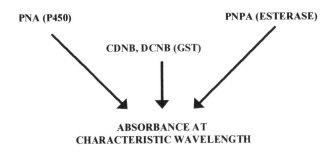

Figure 1. Common spectrophotometric substrates for resistance diagnosis. CDNB, chlorodinitrobenzene; DCNB, dichloronitrobenzene; PNA, paranitroanisole; PNPA, paranitrophenyl acetate.

support (e.g., filter paper (1). Alternatively, precise and accurate measurements are possible at a single wavelength and under controlled temperatures in a microtiterplate. With discriminating doses of insecticides, it is possible, at least theoretically, to examine an altered target site like that found in acetylcholinesterase using the biochemical approach.

The principles of enzyme-linked immunoassay are reviewed by Roe (30). With this technique, the concentration of the analyte is detected directly by antibody-antigen binding and the technique is not limited to measuring only enzyme activity. Assays to measure nucleic acids include methods such as DNA-DNA and DNA-RNA hybridization, and PCR (polymerase chain reaction) DNA amplification (also applicable to RNA). The most important advantages of these molecular approaches are the ability to see even single base pair substitutions and to detect mutations that might lead to resistance, even before the genes are expressed in an individual insect. Single base pair substitutions can be detected using PNA (protein nucleic acid) probes or by PCR using primers with 3′ mismatches.

Immunodiagnostics have traditionally been very adaptable to field uses, with a variety of formats too numerous to cover in this presentation. To date, molecular techniques have been mostly confined to the laboratory. However, the latest applications in immuno- and DNA-DNA electrochemistry may make both immunodiagnostics and DNA analyses trivial to use in the field. Biochemical, immunochemical, and molecular arrays will also likely increase the sophistication of resistance detection in the future.

The advantages and disadvantages of the biochemical, immuno-chemical, and molecular approaches to resistance detection are outlined in Table 3. The most important advantages, especially for biochemical and immunochemical assays, are their low cost, rapid assay time (in seconds or minutes), and portability to the field. Biochemical and immuno-assay technologies in general are highly advanced. However, it appears that their use in resistance diagnosis is limited. The significant disadvantage of these approaches (Table 3) is that the exact mechanism(s) of resistance must be known *a priori*. This is often not practical or even possible, especially for novel insecticide chemistries. Once reliable markers for resistance are established for a given pesticide at a defined locality, these assay approaches may have greater utility.

Bioassays for Resistance Detection

TRADITIONAL BIOASSAY APPROACHES

Probably the most common method for resistance diagnosis is the adult vial test. In this assay, the inside of a vial is coated with a diagnostic dose of an insecticide, the adult insect is placed into the vial and intoxication is assessed after some period of exposure. Intoxication is usually judged as knockdown or death. This method is especially useful for Lepidoptera, which can be easily collected from the field using light or pheromone traps. The other advantages of this technique are that the assays are conducted on individuals, the results are available in a few days, and the assay is inexpensive. The tests are most often conducted in the laboratory under controlled environmental conditions. One possible disadvantage of the adult vial test is that the assay is conducted on the adult stage, while the pest must be controlled as a larva. The insecticide only enters into the insect in this test by topical exposure, which may not accurately mimic field conditions. In most cases so far, there has been some degree of positive correlation for resistance between both life stages. However, there is also evidence that resistance levels vary between insect life stages (20). The major disadvantage of the vial test is that

Table 3.   Advantages and Disadvantages of Biochemical, Immunochemical, and Molecular Approaches to Resistance Detection

Advantages

- Assay for a specific resistance mechanism
- Assay individual insects
- Assay is highly adaptable to field use
- Assay is fast and inexpensive

Disadvantages

- All mechanisms of resistance are not assayed simultaneously
- Assays use non-insecticidal substrates

it does not work for insecticides such as Bt that are only toxic by ingestion.

Another approach similar to the adult vial test, which produces basically the same results, involves incorporation of insecticides directly into the sticky surface of sticky traps and assessment of mortality of the moths in the trap after some period following capture.   Insects can also be collected from the field, a colony maintained in the laboratory for multiple generations on either artificial diet or on plants, and a log-probit analysis of insecticide dose versus mortality determined.   This assay approach is costly, requires insect-rearing facilities, takes months to conduct, and is not practical for more than a few sample populations.   However, it produces an accurate description of population susceptibility to a particular insecticide chemistry, as long as the number of insect generations reared in the laboratory is kept to a minimum.

The most significant advantages of using bioassays to detect resistance are that the assay is conducted with the insecticide of interest and that all possible resistance mechanisms, except perhaps behavioral, are detected simultaneously. The goal of our research in this area has been to combine the best qualities of bioassay technology with some of the advantages of the biochemical assays, i.e., the short assay time, the assay of individual insects, the ease of reading an end-point, and field applicability. At the same time, any

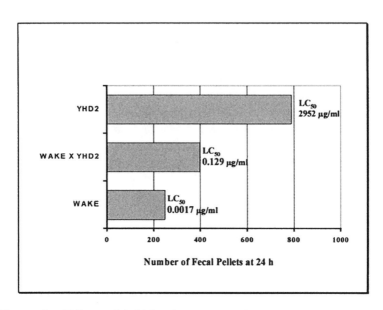

Figure 2. Effect of 0.004 micograms of MVP per ml artificial diet on the production of colored feces by susceptible (Wake), F1 hybrid, and highly resistant (YHD2) third instar tobacco budworms.

new resistance detection system should also be compatible with insecticides, such as Bt, which do not penetrate the cuticle.

FEEDING DISRUPTION ASSAY FOR BT

The feeding disruption assay (7,31,32) uses a patented, artificially colored indictor diet that includes a discriminating dose of any insecticide. The assay was originally designed for diagnosis of Bt resistance in tobacco budworm and cotton bollworm larvae collected as eggs from cotton, but the same approach has been used to assay tobacco budworm for resistance to Larvin® and Tracer®. Feeding disruption (a measure of intoxication) in larvae fed a diet containing a diagnostic dose of the insecticide is determined by the absence of colored feces after a prescribed time interval, which can be as brief as 4 hours, but more typically is 24 hours. As shown in Fig. 2, the level of fecal production can also be used to determine the level of resistance to Bt by this method. Due to differential toxicity of Bt to the tobacco budworm and cotton bollworm, feeding

78

disruption assays can also be used for species diagnosis (7,31,32). Because high population densities of the cotton bollworm may not be effectively controlled by transgenic Bt cotton, species diagnosis in young larvae is important. Because of the short time required for the feeding disruption assay, the larvae remain viable and the same insects can be used for both species diagnosis and diagnosis of resistance to other insecticides. A co-technology is being developed, hydrateable meal pads, that will greatly extend the versatility of this assay approach (patent pending).

The advantages of the feeding disruption assay are listed in Table 4. The most significant of these is its application in diagnosis of Bt resistance, given that traditional assay approaches, such as the adult vial test, do not work for Bt. The feeding disruption assay provides results relatively quickly and has an easy-to-read end-point, as compared to a Bt mortality assay. Other significant features include the ability to measure the level of resistance (Fig. 2) and the ability to use the assay in the field or in the home or office of the grower or consultant without any specialized or costly equipment.

Table 4. Advantages of the Feeding Disruption Assay

- More rapid than traditional mortality assays, especially for Bt
- Applicable to contact and oral insecticides
- Monitors the relative importance of the most relevant mechanisms of resistance simultaneously
- Targets the larval stage where the insecticide is used
- Better simulation of natural field exposure
- Assay is non-destructive, allowing for other diagnostic procedures on the same insect, e.g., species diagnosis or multiple resistance diagnosis
- Measures the level of resistance
- Measures resistance in individual insects

Summary

The sustainability of agricultural production will depend greatly on our ability to preserve our safest and most efficient insecticides. Although industry has had continued success in the discovery of new insecticide modes-of-action, there are no assurances that this will be

79

possible in the future. The one thing that is absolutely clear is that insects will continue to develop resistance to even our newest biological insecticides. The early detection of resistance is an important part of resistance management. Resistance diagnostics will be greatly affected in the future by our increased understanding of the molecular biology of resistance and new technologies that are being developed in medical diagnostics.

Acknowledgments

The feeding disruption assay research was supported by grants to R.M.R. from the North Carolina State University/NSF Research Center for Integrated Pest Management (IPM94-007), the North Carolina Agricultural Research Service, Cotton Incorporated (98-632), the Southern Regional IPM Program (98-34103-6269), and Agdia, Inc. (and the USDA SBIR Program Phase I and Phase II).

Literature Cited

1. Abdel-Aal, Y. A. I., Wolff, M. A., Roe, R. M., and Lampert, E. P. 1990. Aphid carboxylesterases: biochemical aspects and importance in the diagnosis of insecticide resistance. Pestic. Biochem. Physiol. 38:255-266.
2. Abdel-Aal, Y. A. I., Ibrahim, S. A., Lampert, E. P., and Rock, G. C. 1993. Detection methodology of esterase-mediated insecticide resistance: from bioassay to biotechnology. Pages 13-33 in Reviews in Pesticide Toxicology Volume 2. R. M. Roe and R. J. Kuhr, eds. Toxicology Communications, Raleigh, NC.
3. Abdel-Aal, Y. A. I., Lampert, E. P., Wolff, M. A., and Roe, R. M. 1993. Novel substrates for the kinetic assay of esterases associated with insecticide resistance. Experientia 49:571-575.
4. Alstad, D. N., and Andow, D. A. 1995. Managing the evolution of insect resistance to transgenic plants. Science 268:1894-1896.
5. Anspaugh, D. D., Rose, R. L., Koehler, P. G., Hodgson, E., and Roe, R. M. 1994. Multiple mechanisms of pyrethroid resistance in the German cockroach, *Blattella germanica* (L.). Pestic. Biochem. Physiol. 50:138-148.
6. Anspaugh, D. D., Kennedy, G. G., and Roe, R. M. 1995. Purification and characterization of a resistance-associated

esterase from the Colorado potato beetle, *Leptinotarsa decemlineata* (Say). Pestic. Biochem. Physiol. 53:84-96.

7. Bailey, W. D., Zhao, G., Carter, L. M., Gould, F., Kennedy, G. G., and Roe, R. M. 1998. Feeding disruption bioassay for species and *Bacillus thuringiensis* resistance diagnosis for *Heliothis virescens* and *Helicoverpa zea* in cotton (Lepidoptera: Noctuidae). Crop Prot. 17:591-598.

8. Bailey, W. D., Young, H. P., and Roe, R. M. 1999. Laboratory selection of a Tracer®-resistant strain of the tobacco budworm and comparisons with field strains from the southeastern US. Pages 1221-1224: in Proc. Beltwide Cotton Conferences. National Cotton Council, Memphis, TN.

9. Bauer, L. S. 1995. Resistance: a threat to the insecticidal crystal proteins of *Bacillus thuringiensis*. Fla. Entomologist 78:414-443.

10. Bolin, P. C., Hutchison, W. D., Andow, D. A., and Ostlie, K. R. 1998. Monitoring for European corn borer (Lepidoptera: Crambidae) resistance to *Bacillus thuringiensis*: logistical considerations when sampling larvae. J. Agric. Entomol. 15:231-238.

11. Briese, G. J. 1981. Resistance of insect species to microbial pathogens. Pages 511-545 in: Pathogenesis of Invertebrate Microbial Diseases. E. W. Davidson, ed. Allanheld Osmun, Totowa, New Jersey.

12. Coustau, C., and ffrench-Constant, R. 1995. Detection of cyclodiene insecticide resistance-associated mutations by single-stranded conformational polymorphism analysis. Pestic. Sci. 43:267-271.

13. ffrench-Constant, R. H., and Roush, R. T. 1990. Resistance detection and documentation: the relative roles of pesticidal and biochemical assays. Pages 4-38 in: Pesticide Resistance in Arthropods. R. T. Roush and B. E. Tabashnik, eds. Chapman and Hall, New York.

14. Field, L. M., Anderson, A. P., Denholm, I., Foster, S. P., Harling, Z. K., Javed, N., Martinez-Torres, D., Moores, G. D., Williamson, M. S., and Devonshire, A. L. 1997. Use of biochemical and DNA diagnostics for characterizing multiple mechanisms of insecticide resistance in the peach-potato aphid, *Myzus persicae* (Sulzer). Pestic. Sci. 51:283-289.

15. Fischhoff, D. A. 1996. Insect-resistant crop plants. Pages 214-227 in: Biotechnology and Integrated Pest Management. G. J. Persley, ed. CAB International, Wallingford, UK.
16. Gasser, C. S., and Fraley, R. T. 1989. Genetically engineering plants for crop improvement. Science 244:1293-1299.
17. Gould, F. 1988. Evolutionary biology and genetically engineered crops. BioScience 38:26-33.
18. Gould, F., Anderson, A., Reynolds, A., Bumgarner, L., and Moar, W. 1995. Selection and genetic analysis of a *Heliothis virescens* (Lepidoptera: Noctuidae) strain with high levels of resistance to *Bacillus thuringiensis* toxins. J. Econ. Entomol. 88:1545-1559.
19. Jutsum, A. R., Heaney, S. P., Perrin, B. M., and Wege, P. J. 1998. Pesticide resistance: assessment of risk and the development and implementation of effective management strategies. Pestic. Sci. 54:435-446.
20. Karoly, E. D., Rose, R. L., Thompson, D. M., Hodgson, E., Rock, G. C., and Roe, R. M. 1996. Monooxygenase, esterase, and glutathione transferase activity associated with azinphos-methyl resistance in the tufted apple budmoth, *Platynota idaeusalis*. Pestic. Biochem. Physiol. 55:109-121.
21. Krieg, A., and Langenbruch, G. A. 1981. Susceptibility of arthropod species to *Bacillus thuringiensis*. Pages 837-896 in: Microbial Control of Pests and Plant Diseases 1970-1980. H. D. Burges, ed. Academic Press, New York.
22. Lemon, R. W. 1994. Insecticide resistance. J. Agric. Sci., Cambridge 122:329-333.
23. Leonard, B. R., Graves, J. B., Burris, E., Micinski, S., and Mascarenhas, V. 1996. Evaluation of selected commercial and experimental insecticides against lepidopteran cotton pests in Louisiana. Pages 825-830 in: Proc. Beltwide Cotton Conferences. National Cotton Council, Memphis, TN.
24. Matten, S. R. 1998. EPA regulation of resistance management for Bt plant-pesticides and conventional pesticides. Res. Pest Manag. 10:3-9.
25. McGaughey, W. H., and Whalon, M. E. 1992. Managing insect resistance to *Bacillus thuringiensis* toxins. Science 258:1451-1455.
26. McKenzie, J. A., and Batterham, P. 1994. The genetic, molecular and phenotypic consequences of selection for insecticide resistance. Trends Ecol. Evol. 9:166-169.

27. Oberlander, H. 1996. Insect science in the twenty-first century. Amer. Entomol. 42:140-147.

28. Plapp, F. W. Jr., McWhorter, G. M., and Vance, W. H. 1987. Monitoring for resistance in the tobacco budworm in Texas in 1986. Pages 324-326 in: Proc. Beltwide Cotton Conferences. National Cotton Council, Memphis, TN.

29. Price, N. R. 1991. Insect resistance to insecticides: mechanisms and diagnosis. Comp. Biochem. Physiol. C. 100:319-326.

30. Roe, R. M. 1991. Enzyme-linked immunosorbent assay of small molecular weight toxicants. Pages 273-287 in: Pesticides and the Future: Toxicological Studies of Risks and Benefits. E. Hodgson, R. M. Roe and N. Motoyama, eds. North Carolina State University, Raleigh, NC.

31. Roe, R. M., Bailey, W. D., Gould, F., and Kennedy, G. G. 1999. Insecticide resistance assay (patent pending). US patent application 09/112,274.

32. Roe, R. M., Bailey, W. D., Zhao, G., Young, H. P., Carter, L. M., Gould, F., Sorenson, C. E., Kennedy, G.G., and Bacheler, J.S. 1999. Assay kit for species and insecticide resistance diagnosis for tobacco budworm and bollworm in cotton. Pages 926-930 in: Proc. Beltwide Cotton Conferences. National Cotton Council, Memphis, TN.

33. Rose, R. L., Barbhaiya, L., Roe, R. M., Rock, G. C., and Hodgson, E. 1995. Cytochrome P450-associated insecticide resistance and the development of biochemical diagnostic assays in *Heliothis virescens*. Pestic. Biochem. Physiol. 51:178-191.

34. Roush, R. T., and Miller, G. L. 1986. Considerations for design of insecticide resistance monitoring programs. J. Econ. Entomol. 79:293-298.

35. Scharf, M. E., Bennett, G. W., Reid, B. L., and Qui, C. 1995. Comparisons of three insecticide resistance detection methods for the German cockroach (Dictyoptera: Blatellidae). J. Econ. Entomol. 88:536-542.

36. Shearer, P. W., and Riedl, H. 1994. Comparison of pheromone trap bioassays for monitoring insecticide resistance of *Phyllonorycter elmaella* (Lepidoptera: Gracillariidae). J. Econ. Entomol. 87:1450-1454.

37. Sims, S. R., Greenplate, J. T., Stone, T. B., Caprio, M. A., and Gould, F. L. 1996. Monitoring strategies for early detection of Lepidoptera resistance to *Bacillus thuringiensis* insecticidal proteins. Pages 229-242 in: Molecular Genetics and Evolution

of Pesticide Resistance. T. M. Brown, ed. American Chemical Society, Washington D. C.

38. Tabashnik, B. E. 1994. Evolution of resistance to *Bacillus thuringiensis*. Ann. Rev. Entomol. 39:47-79.

39. Tabashnik, B. E., Cushing, N. L., Finson, N., and Johnson, M. W. 1990. Field development of resistance to *Bacillus thuringiensis* in diamondback moth (Lepidoptera: Plutellidae). J. Econ. Entomol. 83:1671-1676.

40. Thompson, G. D., Michel, K. H., Yao, R. C., Mynderse, J. S., Mosburg, C. T., Worden, T. V., Chio, E. H., Sparks, T. C., and Hutchins, S. H. 1997. The discovery of *Saccharopolyspora spinosa* and a new class of insect control products. Down to Earth 52:1-5.

41. Troyer, P. 1995. National Biological Impact Assessment Program News Report. pp 1-3.

42. Zhao, G., Rose, R. L., Hodgson, E., and Roe, R. M. 1996. Biochemical mechanisms and diagnostic microassays for pyrethroid, carbamate, and organophosphate resistance/cross-resistance in the tobacco budworm, *Heliothis virescens*. Pestic. Biochem. Physiol. 56:183-195.

Pathogen Detection and Pesticide Use

Sandy J. Stewart
Immunovation, Inc., Research Triangle Park, NC 27709 USA

The utilization of diagnostics in agriculture has existed in some form for centuries and probably as long as man has cultivated crops. However, recent advancements in diagnostic technology coupled with other tools of IPM can provide significant enhancements to world food production.

Diagnostics can be very valuable tools when utilized carefully and knowingly. Unfortunately, many diagnostics are used to answer questions that they were not meant to answer, and many diagnostics are sold with no claims concerning their utility except to be able to detect the pathogen. It is therefore up to the users to educate themselves on the various diagnostics' abilities and the answers they are capable of providing. Of course, the basic reason for utilizing diagnostics, as well as IPM, is economics. Can diagnostics result in greater efficiencies and a satisfactory financial return? The answer is a qualified "yes," as long as the diagnostics are used properly to answer the appropriate questions. Additionally, the pathogen and the crop must be considered. Not all pathogens result in a significant crop yield loss. In these cases, the utilization of a specific diagnostic might not be warranted. However, there are certainly many major crops in which specific diseases can be devastating if not controlled. Once again, it is up to the users to educate themselves.

In recent years there has been an explosion of agricultural diagnostics utilizing many of the technologies previously available only in the human diagnostic market. There are a plethora of immunologically based tests for viral and bacterial pathogens, as well as, more recently, various tests for fungal pathogens. Most tests are performed in laboratories, but there are an increasing number of field tests being developed that will enable the grower to make quick decisions when necessary. There are even PCR (polymerase chain reaction) assays that detect a specific pathogen's DNA, but these must be performed in a laboratory. There are tests that can detect a pathogen pre-symptomatically, quantify its presence, distinguish dead from living pathogens or spores from hyphae, and even detect multiple pathogens within the same test. However, there are advantages and disadvantages to all of them, and more important, one must know their limitations to obtain valid results.

Questions

There are a myriad of parameters that should be considered when choosing a diagnostic tool and deciding how it should be used, if at all. For example, are you attempting to check the effectiveness of a pesticide? Are you attempting to time pesticide applications or determine whether you even need to apply pesticide? Are you trying to identify a specific pathogen and track its progression? Are you simply looking to verify your decision? Once these questions are answered and you know why you want to use the diagnostic, questions arise concerning how the diagnostic should be used. What constitutes a sample? Is a piece of one leaf in a field representative of what is occurring in the field in general? Obviously not, but consideration must be given to the spatial and temporal aspects of the sampling plan, as well as the appropriate sample unit and sample size. Other questions include: Is the observation of a specific fungus on a given leaf relevant information? If you can see symptoms, is it already too late to stop the disease? What level of infection constitutes a diseased state, and what is the sensitivity of the diagnostic to detect pre-symptomatic infection? How has this been verified?

The only way to answer such encompassing questions for use in IPM is through field trials. This obviously is very costly and time consuming and therefore is rarely accomplished by agricultural diagnostic companies today. Unlike human diagnostics, agricultural diagnostics do not require any form of registration or certification. One is simply taking the manufacturers word that it works. In most cases, the diagnostic does work, at least under the parameters of the manufacturer's testing program, which very likely does not include actual field trials similar to those required for registration of a crop protection product. Simply running the test in a field does not provide any information to answer the above questions. However, it is also important to consider the old saying, "you get what you pay for." In most cases, agricultural diagnostics are considerably less expensive than human diagnostics simply because the market will not bear a higher price. With these constraints, it is difficult for most manufacturers to justify the costs of full-blown field trials when the return on the product does not even cover the cost of the trials. Hopefully, with time and acceptance, this will change. An excellent source of information is through university research programs, where well controlled experiments are designed to provide detailed results pertaining to many of the questions above (3,6). Fortunately, links between industry and universities are providing both with what they need most -- quality products and funding.

Sampling

HOW AND WHERE

Probably one of the most important issues concerning agricultural diagnostics is the sampling method to use. In agriculture, the sampling issue is much more severe than in human or veterinary diagnostics. In human diagnostics, samples are typically composed of blood or urine, whereby the target of interest is already in a soluble aqueous and pH-maintained fluid representing the entire test area. Agricultural diagnostics are literally attempting to get "blood out of a turnip."

In some cases, a parameter can be fairly easy to determine, such as whether soil testing versus plant testing should be done. Simply from the diagnostic point of view, it is much easier to test plant tissue than soil. Also, in many cases, pathogens are present in the soil at various levels throughout the year and only become a problem when environmental conditions permit. Processing plant tissue for testing requires careful consideration as well. As discussed above, many of these considerations can be determined by the question that one wants to answer. The easiest answer to obtain is whether a specific single lesion on a plant leaf or stem is a specific anticipated pathogen or not. This assumes that the diagnostic is indeed specific and does not cross-react with other plant pathogen species, and also that the diagnostic is sensitive enough to detect that level of infection. Most commercial diagnostics on the market today are capable of the sensitivity required to detect a pathogen in a visible lesion, but some also cross-react with other pathogen species and genera. The last assumption requires that the investigator be skilled enough to reasonably determine possible causes of the lesion. Unlike human diagnostics, there is not a large panel of tests that this single sample will be tested against. Therefore, a negative answer simply eliminates that pathogen but gives no additional information.

Unfortunately, many investigators utilizing diagnostics as a tool for IPM want to have a much more global view of a field or geographic area. It then becomes increasingly important to identify the correct questions and the correct diagnostic to answer those questions. Investigators often want to know if a "disease state" exists in their field prior to visible symptoms. This is somewhat analogous to many human pre-symptomatic diagnostics, such as pregnancy, HIV, and now even rheumatoid arthritis and insulin-dependent diabetes. However, as discussed above, the sampling method is considerably different, and the definition of a "disease state" is equally different. A "disease state" in agriculture can exist when a foliar pathogen, which is generally present as mycelia or spores on debris in a field, has moved into the upper leaf levels of a crop. In this case, detection of the pathogen in soil, or even on lower leaves, is not indicative of subsequent yield loss or future movement into the upper leaf levels. Only when the pathogen is present on the upper leaves is a disease state considered. However, pathogen presence in quantifiably high levels on Feeke's growth stages 2 to 4 can also constitute a disease state. All of this is obviously contingent upon the crop, pathogen, growth stage, and environmental conditions.

From a sampling point of view, one must first define the physical sampling area of interest. This is dependent upon a number of factors, such as the crop, the pathogen of interest, the current environmental conditions, the growth stage, and the desired level of testing. The next step is to determine a random sampling method. However, considering the myriad of variables in the process, most individuals simply resort to walking a "W" pattern or using some other conventional sampling method. The next several variables interact closely with one another and should have been statistically determined and field-validated by the diagnostic manufacturer. Specifically, these variables are: how many samples, how much tissue, and which tissue (which leaf level or what part of the stem). For instance, for *Septoria* testing, a sample consisting of 25 wheat leaves from the same leaf level and cut to the same length longitudinally from the center of the leaf constitutes a representative and reproducible sample for testing (5). However, it was also determined that two leaf levels must be tested, typically the flag leaf and Feeke's growth stage 2 (F-2). This was determined in field trials in which yield loss correlated most significantly with disease incidence in the uppermost leaves, and even pre-symptomatic detection of pathogen on F-2 was a good indicator that a serious infection could result should the environmental conditions become favorable (4,5).

Once the sample has been taken, it should be tested as soon as possible. This eliminates additional growth of the pathogen throughout the sample and minimizes drying of the sample, which makes solubilization of the pathogen more difficult and less reproducible. Some diagnostic kits include a device for maceration but others do not. In any case, it is most important to always use the same amount of extraction buffer per unit sample and to always perform the extraction in the same way, with the same maceration device, and to the same extent. Without these considerations, there is no valid expectation of comparing results from one sampling to the next, and therefore tracking disease progression is impossible.

WHEN AND WHAT

The question of "when" corresponds to the intended use of the diagnostic, whereas "what" should be defined by the diagnostic itself. Intended uses might include: timing pesticide application, measuring pathogen presence, or measuring disease progression. All of these will determine the most appropriate time to sample and how often samples should be taken.

For instance, a grower intending to monitor disease progression may choose to take weekly samples throughout the growing season or possibly only until the last pesticide application has been made. However, results may also be obtained by initially monitoring throughout an entire growing season and then applying those results in subsequent seasons. In this case, one would sample on the same day each week and at the same time of the day to avoid the variables of dew-soaked morning tissue versus dry afternoon tissue. One would also monitor at least two leaf levels at each sampling and record the weekly weather conditions, as well as any pesticide applications. Assuming the diagnostic is quantifiable, one can easily graph disease progression throughout the season and determine the effects of weather and pesticide application on pathogen level and movement (2,4). This information can assist in improving the efficiency of future pesticide applications.

Another use might be to determine if a pesticide application is warranted or not. This type of "spot" check can be very effective when combined with other tools of IPM, such as weather prediction. In this case, a reasonable time to test is just following a significant rainfall, when a fungal or bacterial pathogen has a good environment for growth coupled with its potential for spread due to splashing. Another reasonable testing time is just prior to the last legal date for pesticide application. A grower can determine if there is a significant level of pathogen already present that could easily bloom with favorable weather conditions, at which point, it would be too late for application. One may also want to test just prior to a conventional application date or growth stage to determine if application is warranted at this time. Many growers are also interested in testing the efficiency of a specific product. Therefore, testing just following pesticide application can help answer this question. Of course, this is dependent upon the mode of action of the pesticide itself. These are just a few of the questions that can easily be answered by the proper diagnostic and should be considered by every diagnostic user.

Technologies

The two most widely used technologies in diagnostics are immunologically based tests and PCR-based tests. They both have advantages and disadvantages, although both are quite capable formats for diagnostics. Immunologically based tests have been in existence much longer than PCR tests and therefore still dominate both the human and agricultural markets. Immunologically based tests are typically of the ELISA (enzyme linked immunosorbent assay) format and are based upon the natural reaction of antibody production and recognition against foreign substances. The specificity of antibody recognition and the binding capacities are some of the greatest within nature and some of which man cannot reproduce synthetically. Therefore, the strength of an immunoassay can be its specificity and ease of quantifiable reactions. The disadvantage is that development of antibodies is both costly and time consuming. However, an immunoassay is very inexpensive to perform, can be performed by minimally trained technicians, and can even be adapted to a field test for use by growers. We developed an "on-site" field test for *Septoria tritici* and *Septoria nodorum*, which can be performed within 5 minutes, excluding harvest. We also utilized our field-tested, quantitative ELISA for validation and calibration of the on-site test (7).

Molecular-based testing, such as PCR, uses pieces of DNA that recognize the exact complementary piece of DNA in a sample. This reaction is generally then amplified by making thousands of copies of the target DNA such that detection is easier. This can also be a very specific recognition. The major advantage to PCR assays is that they can be relatively quick and inexpensive to develop. The major disadvantage is that they are costly and must be performed by a well-trained technician in a laboratory. They are also only semi-quantitative compared to the true quantitative reaction of antibody for its antigen. Unfortunately, PCR is also not a well-accepted practice in agriculture today simply because of its recent intro-duction. To circumvent this problem, we introduced our first PCR-based assay in conjunction with an immunologically based assay detecting the exact same pathogen (1). This type of validation is certainly proof-positive for specificity and can be used to cross-calibrate each test.

Diagnostic Validation

The validation process that each agricultural diagnostic manu-facturer utilizes can vary considerably and may not address many of the concerns of educated users. One of these concerns is sensitivity. Many manufacturers report sensitivity based upon units of antigen, or O.D. (optical density of the sample on the test), per volume of sample. What does this mean? How does this relate to an infection level of the pathogen in the field? The sensitivity of a test is very important, since a negative response simply means that, for a specific sample, there was no detectable pathogen within the detection limits of the test, sampling procedure, and extraction procedure. The level of pathogen present can easily be just below the detection limits of the test. Many manufacturers do not specify what constitutes a sample, which makes any measurement even more obscure. At the very least, as mentioned above, the sampling process and the sample itself have to be strictly consistent. For example: Harvest 25g of flag leaves and add 25ml of extraction buffer, macerate for 2 minutes, and run an assay on the sample. The volume of extraction buffer to unit sample must always remain the same in order to expect consistent results. Different proteins solubilize at different concentrations, so this must be accurate. Unfortunately, the only way to accurately relate a specific signal in a diagnostic to what is actually occurring in the field is through field trials. In order to assess the sensitivity of a diagnostic for detecting a pathogen both prior to and after symptoms have been expressed, results must be compared to measures of disease incidence and severity in the field. A weekly regime of testing using the strict sampling methods described above will visually identify previous pre-symptomatic states that should have been identified by the diagnostic. The following is an outline of one method of validation:

- Determine type of diagnostic required (typically immunological or PCR)
- Determine sample and sampling method
- Develop quantitative laboratory diagnostic
- Test with greenhouse and field-grown infected tissue
- If applicable, adapt to a field diagnostic
- Use laboratory assay to validate/calibrate field test
- Perform field trials with artificial inoculum at various sites
- Perform field trials with natural infection and unknown conditions
- Put into user's hands and make adjustments as necessary

The above procedure generally requires two growing seasons to accomplish because of the various conditions addressed. The end result is a diagnostic that has truly been tested and is capable of effectively addressing a large number of issues. Most important is that this diagnostic can be supplied with instructions detailing the sampling procedure, interpretation of results and possible uses.

Summary

The concept of IPM is broad and encompassing and cuts across many conventional fields of study. Diagnostics constitute one tool in this constantly evolving management strategy. The proper diagnostic held in educated hands can be a very potent tool. The many variables and scenarios described above demonstrate the complexity of analysis. A diagnostic that correctly identifies the cause of a visible lesion on a plant leaf gives only a brief snapshot of what is actually occurring in the field. As users become more knowledge-able about the use and power of diagnostics, they will force manufacturers to provide even better products, and they will be able to justify the higher costs for better products.

Literature Cited

1. Beck, J. J., Beebe, J. R., Stewart, S. J., Bassin, C., and Etienne, L. 1996. Colorimetric PCR and ELISA diagnostics for the detection of *Pseudocercosporella herpotrichoides* in field samples. Brighton Crop Protection Conference--Pests and Diseases 1:221-226.
2. Etienne, L., Steden, C., and Suter, J. 1996. A new diagnostic tool for *Mycosphaerella spp.* in banana leaves. Brighton Crop Protection Conference Symposium--Diagnostics in Crop Production 65:281-286.
3. Gudmestad, N. C., Baer, D., and Kurowski, C. J. 1991. Validating immunoassay test performance in the detection of *Corynebacterium sepedonicum* during the growing season. Phytopathology 81:475-480.
4. Meyer, B., and Etienne, L. 1997. Septoria Watch, an immuno-based diagnostics system for efficient disease management. 15th Long Ashton International Symposium - A focus on Septoria.
5. Mittermeier, L., Dercks, W., West, S. J. E., and Miller, S. A. 1990. Field results with a diagnostic system for the identification of *Septoria nodorum* and *Septoria tritici*. Brighton Crop Protection Conference--Pests and Diseases 2:757-762.
6. Ricker, R. W., Marois, J. J., Dlott, J. W., Bostock, R. M., and Morrison, J. C. 1991. Immunodetection and quantification of *Botrytis cinerea* on harvested wine grapes. Phytopathology 81:404-411.
7. Stewart, S. J., and Jehan-Byers, R. 1996. Development of a rapid on-site test for the presymptomatic detection of *Septoria tritici* and *Septoria nodorum*. Brighton Crop Protection Conference--Pests and Diseases 1:215-220.

Section 3

Genetic Engineering for IPM

Genetic engineering, defined as the insertion of novel DNA into the genome of a target organism, represents a powerful component of modern biotechnology that will almost certainly have a major impact on agriculture. The inserted DNA may be obtained from other organisms or artificially synthesized, and may code for novel proteins in the target organism, interfere with or enhance the expression of normal genes, or merely serve as an inert marker on a chromosome. Recombinant DNA technology is being used to incorporate value-added traits into crops, including insect and pathogen resistance and herbicide tolerance. It is also being used to genetically modify insect pathogens to enhance their effectiveness as insect control agents, and to enhance the effectiveness of biological control agents for plant pathogens. The potential also exists to use this technology to increase the efficiency of autocidal control strategies to manage insect pests.

The first five chapters in this section focus on the present status and potential role in pest management of crops genetically engineered to express insect or pathogen resistance or herbicide tolerance. In the first chapter, Stewart Sherrick and Graham Head provide a general overview of transgenic crops in IPM. In the next chapter, Roger Beachy and Mohammed Bendahmane provide a brief overview of the use of pathogen-derived resistance to develop transgenic crops that are resistant to virus infection. They also discuss regulatory concerns regarding virus-coat-protein-mediated resistance. In the third chapter, Gary Fitt and L. J. Wilson review briefly the current status of insect-resistant transgenic crops. They provide an in-depth discussion of the experience in Australia with using insect-resistant transgenic varieties expressing *Bacillus thuringiensis* endotoxin to manage the bollworm complex in cotton. Dan Hess and Stephen Duke next discuss the current status of and rationale for using transgenic herbicide-tolerant crops, and provide an analysis of the problems and issues associated with their use in IPM. David Bridges then provides an insightful discussion of the implications of pest-resistant and herbicide-tolerant crops for IPM. In the final chapter devoted to transgenic plants, Janet

Andersen and Elizabeth Milewski describe the background for and current status of the regulation of pesticidal transgenic plants in the United States.

The next two chapters focus on the use of genetic engineering to enhance the effectiveness of natural enemies and competitors of insects and plant pathogens in pest management. Brian Federici discusses how recombinant DNA technology is being used to overcome the limitations of wild-type insect pathogens that are currently registered for use as insect control agents. This is followed by David Ferro's cautionary description of factors that led to the rise and fall of foliar *B. thuringiensis* sprays as important tools in the management of Colorado potato beetle. The history of the foliar-applied *B. thuringiensis* products on potato contains many lessons relative to the introduction of new technologies for IPM.

In the final chapter of this section, Fred Gould and Paul Schliekelman discuss the concepts of autocidal control of insect pests, and describe how advances in molecular genetics and genetic engineering may enhance the feasibility of using autocidal control strategies in insect management.

General Concepts, Status, and Potential of Transgenic Plants in IPM

[1]Stewart Sherrick and [2]Graham Head
Monsanto Company, [1]Youngsville, NC 27596 and
[2]St. Louis, MO 63198 USA

The advent of biotechnology has resulted in a new phase of agricultural production that will continue to gain in significance over the years to come. The specific properties of commercial transgenic plants are what account for the significant value of this new technology. For example, the ability to have high levels of expression of a plant-produced toxin can provide excellent and specific efficacy against a particular pest. An example is the high level of *Bacillus thuringiensis* (Bt) toxin produced in Bollgard® cotton and its efficacy on *Heliothis virescens*. The new traits can be expressed in all relevant tissues to provide complete protection from an insect pest in the case of insect-protected crops or from a specific herbicide with herbicide-tolerant crops. This provides a significant advantage over conventionally applied pesticides, which may be present only on tissues where the spray is deposited rather than in all tissues that can be attacked by a pest. Expression of a trait can occur throughout the season, providing season-long protection from a particular pest rather than protection for a short window of time, as with a conventional pesticide. Biotechnology has provided the ability to insert genes with highly specific and targeted effects. Traits can also be combined to produce a crop plant with more than one desired characteristic.

Biotechnology has already had global ramifications. There has been a six-fold increase worldwide in the number of hectares planted to commercial transgenic crops from 1996 to 1997. While the majority of hectares have been in the United States, global expansion is expected to continue. Expectations for 1998 are for more than 20 million ha in the USA and more than 30 million globally. Monsanto has been a major player in agricultural biotechnology and is rapidly expanding its global product portfolio. Herbicide-tolerant Roundup Ready® crops include soybeans, corn, cotton, sugarbeets, and canola. Insect-protected crops include YieldGard® corn, Bollgard cotton, and NewLeaf® potatoes. Additional traits, such as disease resistance, are being developed that will provide solutions to several of today's unsolved disease problems.

While agronomic trait development and commercialization are a reality, the second and third waves of biotechnology are still in the early stages of development. The second wave, which is already starting to emerge in the marketplace, contains quality traits, such as canola and corn with enhanced nutritional qualities. Examples include canola oil rich in beta-carotene that may help with night blindness and corn that produces a substance that may be useful in cancer therapy. A third wave of plant biotechnology products is in development in which plants may replace factories as production facilities. Examples include production of pharmaceuticals and plastics from plants.

Transgenic plants and IPM

INSECT-PROTECTED PLANTS

Transgenic plants have already resulted in new approaches to integrated pest management. Bollgard cotton has altered pesticide use for insect control in cotton. For example, in 1997, Bollgard cotton eliminated the use of an estimated 1.2 million liters of synthetic insecticide. Benefits of such reductions include reduced worker exposure, decreased environmental contamination, and fewer container disposal problems. Growers' time and fuel also are saved by reducing the number of trips across a field. In addition, beneficial insects are preserved in the system, and, in fact, they appear to be a significant component in the success of Bollgard cotton. In areas where resistance to conventional insecticides has developed, Bollgard has become an integral part of cotton cropping systems and has allowed continued production of cotton that otherwise would not have been economically feasible.

As a result of expression patterns and the value of current insect-protected transgenic technology, there has been an increased emphasis placed on resistance management for Bt toxins. Sophisticated plans specific to Bollgard cotton, NewLeaf potatoes and Bt corn varieties have been developed and implemented to manage resistance development. Components include grower stewardship, implementation of a spatial refuge for susceptible insects, monitoring programs, and the eventual combination of multiple genes with differing modes of action. Differences in performance of genetic events in Bt. corn varieties have resulted in different resistance management plans associated with different commercial products. Cotton and corn production in the same geographic area adds another layer of complexity to the design and implementation of effective management programs. At a minimum, insect-protected transgenic plants have created substantial scientific discussion and extensive research into resistance management strategies that are very different from those used with conventional crop protection chemicals.

HERBICIDE-TOLERANT TRANSGENIC PLANTS

The development of herbicide-tolerant crops has enabled rapid adoption of conservation tillage practices. These new technologies have minimized weed control problems, which were once considered a limitation to adoption and success of conservation tillage practices. Perennial weeds that have typically been problems in many conservation tillage systems are now controllable with new transgenic technologies. There is significant flexibility in how these new tools can be used, which allows growers to utilize the system that works best with their farming practices. The benefits of conservation tillage include both environmental and economic savings to growers. As these tillage systems evolve and adoption continues, additional research on integrated pest management practices will be needed to mesh these two important aspects of crop production.

Herbicide-tolerant plants have also provided new tools for managing existing weed resistance. Many of the herbicide-resistant weeds that exist today are now controllable using transgenic systems in most major crops (corn, soybeans, and cotton). Transgenic plants have provided a second mode of action for managing resistant weed problems, such as triazine-resistant *Chenapodium album* and *Amaranthus* spp., and imidazolinone-resistant *Xanthium strumatium*.

As with insect-protected plants, herbicide-tolerant plants (such as plants incorporating Roundup Ready technology) may require specific management programs to ensure the sustainability of this technology. Although a glyphosate-resistant population of ryegrass has been identified in Australia, it is generally felt that resistance to glyphosate will not be a significant issue. This is based on several factors, including more than 20 years of commercial use, an exhaustive search for glyphosate-resistant genes, and the unique mode of action of glyphosate in plants. Thus, the risk of glyphosate resistance in other weed species is expected to be low and manageable. Monsanto will continue to recommend weed control programs based upon agronomic needs. Treatments may include Roundup Ultra® alone, as well as the promotion of tank- or pre-mixtures for specific agronomic situations (e.g., graminicide mixtures for volunteer Roundup Ready corn). Nonetheless, resistance is not to be taken lightly and will be rigorously monitored, and management programs will be instituted as necessary to ensure the long-term viability of this technology.

While the development of transgenic plants has increased the emphasis placed on resistance management, it has also provided an opportunity to employ new management options. The institution of grower agreements provides an avenue for education on resistance management issues and stewardship by growers. This has allowed the implementation of programs that will be required both now and with future technologies to ensure sustainability.

Research Needs

Additional research needs have been created with the introduction of transgenic crops. The effectiveness of the technologies is such that changes in weed and insect pest spectra will occur. Crop monitoring and scouting techniques have already been modified for Bollgard cotton, and thresholds for supplemental treatments have been developed and are still undergoing adjustment as more is learned about these systems. Research to understand these changes will be important in developing future integrated pest management plans. As the land area occupied by these crops continues to expand, there also will be an increased need to understand relationships across multiple cropping systems on a regional scale.

The Future

The future of agriculture will certainly include transgenic crops, just as integrated pest management will continue to play an important role in crop production. Precision agriculture, as it evolves, will provide another layer of sophistication complementing transgenic plants and the pest management opportunities they afford. As these technologies advance and systems continue to develop, we will see an increased emphasis on integrated crop management as a practice that will encompass both current and future integrated pest management techniques. The potential of transgenic plants to provide precise, targeted responses enhances the possibilities for the future of agricultural crop production.

®   Registered trademark of Monsanto Company

Genetic Engineering in IPM:
A Case History for Virus Disease Resistance

Roger N. Beachy and Mohammed Bendahmane
Donald Danforth Plant Science Center
St. Louis, MO 63105 USA

It has long been known that prior infection of a plant with a mild, non-lethal strain of virus can provide protection against other, more severe strains of the pathogen if the viruses are closely related to one another. In 1986, that concept was extended to the use of transgenic plants when scientists at Washington University, in collaboration with researchers at Monsanto Company, developed transgenic plants that carried the gene encoding the coat protein of tobacco mosaic virus (TMV) (15). In these studies, plants that contained high levels of coat protein (CP) were resistant to TMV infection, and those with low levels of CP exhibited low levels of resistance.

This work was followed by similar studies in which genes encoding the CP of TMV or of tomato mosaic virus (ToMV; a tobamovirus) were introduced into tomato plants. Like the other studies, the tomato plants showed resistance against TMV and tomato mosaic virus (14,17). Interestingly, both the tomato plants and the tobacco plants that produced TMV CP showed a substantial degree of resistance against ToMV and closely related tobamoviruses. However, these plants showed only moderate to low level resistance against less closely related tobamoviruses (13). The first field trials of transgenic tomato plants were held in 1987 following approval by the U.S. Department of Agriculture. The study, which was conducted on a small field in Jerseyville, Illinois, was highly successful, and the results, published in 1988 (14), are considered the first example of genetically modified plants that exhibit disease resistance. Such resistance is referred to as coat protein-mediated resistance.

101

These studies were subsequently confirmed by using coat protein genes of other viruses, including Alfalfa Mosaic Virus (AMV), Cucumber Mosaic Virus (CMV), Potato Virus X, Potato Virus Y, Potato Leaf Roll Virus (PLRV), and others. Subsequently, the degree of resistance to infection was shown to be quite variable with different viruses and different plants. Furthermore, with more detailed analysis, it became apparent that the mechanisms of disease resistance were different in different host:gene systems (3,5,8,10,12).

In the late 1980s, it was determined that the Environmental Protection Agency (EPA) would have jurisdiction over the regulation of these plants, because they showed resistance to a pest, i.e., a disease. The EPA took the position that disease resistance by genetic transformation, but not by more standard plant breeding approaches, was subject to their review and approval. In subsequent years, plants were developed that exhibited commercial levels of resistance. The Asgrow Seed Company (now Seminis Corp.) developed transgenic lines of squash plants and other cucurbits that exhibited resistance to multiple viruses, and commercialized a variety of virus-resistant squash plants in the mid-1990s. Subsequently, the U.S. Department of Agriculture deregulated coat protein genes in general, and the EPA later exempted coat protein genes from a list of genes that required regulation. The U.S. Food and Drug Administration decided not to regulate the foods derived from such transgenic plants, because they saw no unique risks associated with the plants. It has long been known that viruses frequently infect vegetables and fruits, and it was agreed that the addition of a coat protein gene did not provide a unique risk, based upon historical practice.

In addition to coat protein genes, a variety of other virus gene sequences have been employed in transgenic plants in an attempt to block virus infections. These include genes encoding replicase proteins and genes that do not encode a protein *per se*, but produce viral RNA sequences. The expression of certain genes activate a hitherto unknown type of biochemical activity in which the plant destroys certain types of RNA sequences. In these cases, the transgenic plants destroy the viral RNAs that are produced during virus replication (1,2).

At present, several different kinds of transgenic plants with virus resistance are being developed for commercial release. Under the brand name NatureMark®, potatoes that contain a *Bacillus thuringiensis* gene (to confer resistance to the Colorado potato beetle) and a replicase gene to protect the plants against the potato

102

leaf roll virus (PLRV) are currently being marketed. PLRV causes significant damage to potato crops around the world. The NatureMark Company is currently developing a plant variety that contains the PLRV gene plus genes for coat proteins of other potato viruses. It was recently reported that potato plants possessing resistance to PLRV or to potato virus Y (based on coat protein-mediated resistance), and resistance to Colorado potato beetle (based on the *B. thuringiensis* Cry 3A gene for protein toxin) require much less insecticide for control of aphids (which transmit virus infection) and other insects than do non-transgenic plants. Keniewski, et al. (11) reported that transgenic potato crops might save about 2.3 million kg of formulated insecticide per year in the USA alone.

Scientists at the University of Hawaii, in collaboration with scientists at Cornell University, recently developed a variety of papaya plants that contain the coat protein gene of papaya ring spot virus (9). This virus has caused significant damage to papaya production in certain locations in Hawaii, and the new virus-resistant varieties are expected to enable production to be regained in those areas. The disease-resistant varieties have undergone significant field trials and were released for production in 1999.

Several companies are experimenting with various approaches to develop virus resistance in small fruits, including raspberries and strawberries, and in tree fruits, including plums, apples, and grapevines. Other types of virus-resistant crops, including alfalfa, corn, and tomatoes, have undergone extensive experimental applications and are being evaluated for commercial development. Economic factors will determine whether or not these will be brought to market in the short term as well as in the long term.

Regulatory Concerns Regarding Coat Protein-Mediated Resistance

During the course of the development of transgenic plants that expressed coat protein-mediated resistance, a number of questions were raised by members of the scientific and regulatory communities, including those who questioned the use of the new technologies in general, as well as members of the Environmental Protection Agency. One of the first issues raised was the possibility that coat protein produced in the transgenic plant would encapsidate or trans-encapsidate other messenger RNAs or virus RNAs. It was argued that trans-encapsidation may expand the host range of the virus. Experimental evidence has shown that during some virus infections, certain virus coat proteins can, indeed, trans-encapsidate other RNAs, including other viral RNAs. Other studies have provided evidence that trans-encapsidated RNAs could lead to increased insect transmission of normally non-vectored viruses. However, biological impact of encapsidation is considered to be negligible for several reasons. First, there are many examples of mixed virus infections in which trans-encapsidation occurs in natural settings. There have been few or no examples in which this has led to increases in virus diseases caused by trans-encapsidation. Second, the capacity of a virus to infect and replicate depends not on the capsid protein *per se*, but on the nature of the viral RNA and whether or not the viral RNA can self-replicate in the new host. This is a property of the viral nucleic acid rather than the capsid protein (4,5,16).

The concerns about insect transmission and the potential for recombination between transgenic RNA and viral RNA has led scientists to conduct several types of experiments to eliminate the potential safety concerns that have been raised. In the case of coat protein-mediated resistance against tobacco mosaic virus, recent experiments have shown that it is possible to eliminate the capacity for virus coat protein to self-assemble while retaining coat protein-mediated resistance. As a result, the transgene can provide strong levels of coat protein-mediated resistance but is unable to form a virus particle. A second approach involves eliminating sequences from the coat protein that are required for acquisition of the virus by insect vectors. Removing such sequences reduces the potential risk that a trans-encapsidated RNA could be acquired and transmitted to a new host. A third approach that has been used involves reducing the possibility of RNA recombination by identifying and removing the sequences in viral RNA that are important for recombination. Removing those sequences has shown that it is possible to reduce the likelihood of intermolecular recombination, an issue that has been raised as a point of potential concern (4,5, and references cited within).

These and other studies have also made it possible to develop transgenes that confer very high levels of resistance. For example, in our laboratory, we modified the structure of the amino acid sequence of the coat protein of TMV and developed molecules that dramatically increased the efficacy of coat protein-mediated resistance while simultaneously reducing the possibility of virus assembly (6,7). These modifications, which were made based upon the known three-dimensional structure of tobacco mosaic virus coat protein, provide a striking example of how structural and molecular knowledge of a genetic mechanism can lead to increased performance of a gene while simultaneously addressing issues of bio-safety concern.

Summary and Conclusions

Pathogen-derived resistance can be efficiently used to develop transgenic crops that are resistant to virus infection. Applications of genetic engineering to develop transgenic crops significantly reduced the damages caused by viruses, and can substantially reduce the use of pesticides. During the development and application of pathogen-derived resistance, a variety of concerns have been raised relative to bio-safety risks of the use of transgenic plants that express virus-derived gene sequences. It is, therefore, important to gain a more complete understanding of the molecular, cellular, and structural mechanisms of virus transmission, assembly and disassembly, replication, and movement of the infection in the host. The knowledge generated by such studies will help in designing new genes that increase the efficacy of pathogen-derived resistance and reduce or eliminate the bio-safety concerns related to the use of pathogen-derived resistance in agriculture.

Literature Cited

1. Baulcombe, D. C. 1996. Mechanisms of pathogen-derived resistance to viruses in transgenic plants. Plant Cell 8:1833-1844.
2. Baulcombe, D. C. 1999. Viruses and gene silencing in plants. Arch. Virol. Suppl. 15:189-201.
3. Beachy, R. N. 1997. Mechanisms and applications of pathogen-derived resistance in transgenic plants. Curr. Opin. Biotech. 8:215-220.
4. Beachy, R. N. and Bendahmane, M. 1999. Pathogen-derived resistance and reducing the potential to select viruses with increased virulence. Pages 87-94 in: National Agricultural

Biotechnology Council, Vol. 10. R. W. F. Hardy and J. B. Segelken, eds.

5. Bendahmane, M., and Beachy, R. N. 1999.   Control of tobamovirus infection via pathogen-derived resistance. Advances in Virus Research 53:369-386.

6. Bendahmane, M., Fitchen, J. H., Zang, G. H., and Beachy, R. N. 1997.   Studies of coat protein-mediated resistance to TMV: Correlation between assembly of mutant coat proteins and resistance. J. Virol. 71:7942-7950.

7. Clark, W. G., Fitchen, J. H., and Beachy, R. N. 1995. Studies of coat protein-mediated resistance to TMV: i. The PM2 assembly defective mutant confers resistance to TMV.   Virology 208:485-491.

8. Fitchen, J. H., and Beachy, R. N. 1993. Genetically engineered protection against viruses in transgenic plants.   Annu. Rev. Microbiol. 47:739-763.

9. Gonsalves, D. 1998. Control of papaya ringspot virus in papaya: A Case Study. Annu. Rev. Phytopathol. 36:415-437.

10. Hackland, A. F., Rybicki, E. P., and Thomson, J. A. 1994. Coat protein-mediated resistance in transgenic plants. Arch. Virol. 139:1-22.

11. Keniewski, W., Zalewski, J., and Rogan, G. 1999. From concept to commercialization of virus and insect resistant potatoes. XIth International Congress of Virology, [Abstract V57RS.1].

12. Lomonossoff, G. P. 1995. Pathogen-derived resistance to plant viruses. Annu. Rev. Phytopathol. 33:323-343.

13. Nejidat, A., and Beachy, R. N. 1990. Transgenic tobacco plants expressing a coat protein gene of tobacco mosaic virus are resistant to some other tobamoviruses.   Mol. Plant-Microbe Interact. 3:247-251.

14. Nelson, R. S., McCormick, S. M., Delannay, X., Dubé, P., Layton, J., Anderson, E. J., Kaniewski, W., Proksch, R. K., Horsch, R. B., Rogers, S. G., Fraley, R. T., and Beachy, R. N. 1988. Virus tolerance, plant growth and field performance of transgenic tomato plants expressing coat protein from tobacco mosaic virus. Bio/Technology 6:403-409.

15. Powell-Abel, P., Nelson, R. S., De, B., Hoffmann, N., Rogers, S. G., Fraley, R. T., and Beachy, R. N. 1986. Delay of disease development in transgenic plants that express the tobacco mosaic virus coat protein gene. Science 232:738-743.

16. Robinson, D. J. 1996. Environmental risk assessment of releases of transgenic plants containing virus-derived inserts. Transgenic Research 5:359-362.
17. Sanders, P. R., Sammons, B., Kaniewski, W., Haley, L., Layton, J., Lavallee, B. J., Delannay, X., and Tumer, N. E. 1992. Field resistance of transgenic tomatoes expressing the tobacco mosaic virus or tomato mosaic virus coat protein genes. Phytopathology 82:683-690.

Genetic Engineering in IPM: Bt Cotton

Gary P. Fitt[1] and L. J. Wilson[2]
CSIRO Entomology[1] and CSIRO Plant Industry[2],
CRC for Sustainable Cotton Production, Narrabri, NSW, Australia

Biotechnology is rapidly producing a suite of new crops with enhanced resistance to insect pests and diseases. With many other potential transgenic traits being developed, there are numerous opportunities for future improvements in crop production. Insect-resistant transgenic crops have the potential to revolutionize pest management through reductions in the requirement for disruptive pesticides and for the novel integrated approaches that can then be implemented. At the forefront of these developments are crops that express specific insecticidal proteins derived from *Bacillus thuringiensis* (Bt). Insect-tolerant Bt crops will not only require less insecticide for targeted pests, but will allow better survival of predators and parasites that are normally suppressed by pesticide sprays, and hence, there will be less propensity for outbreaks of secondary pests (46). As such, Bt plants should provide a sound foundation for future IPM systems in which a range of control tactics can be integrated. Prospects for reducing environmental impact and losses to pests are outstanding for several crops in many parts of the world, but there remain a number of regulatory and logistical issues for successful long-term implementation of this technology.

While the concept of integrated pest management (IPM) has a long history, it is difficult to honestly describe the pest management practices used in many intensive production systems as IPM. Many components of IPM are being used, such as sampling systems, thresholds, and cultural practices in some cases, but for the most part, management of key pests relies strongly on intervention with pesticides. This is certainly true of cotton production in Australia, the

USA, China, India, Pakistan, and parts of South America, where pesticides represent a significant component of the cost of production. Reliance on pesticides brings significant environmental liabilities with increasing concern over off-target drift, pesticide resistance, and chemical residues in waterways, soils, and livestock. The imperative to move away from reliance on pesticides is a strong one. Bt crops offer one component to assist with this.

Adoption of Bt crops worldwide

Most successful of the insecticidal transgenic crops to date, and the only ones yet commercialized, are those expressing the delta-endotoxin genes from *Bacillus thuringiensis* (Bt). A huge number of Bt crystal proteins have been identified (35), with many more likely to be discovered. Individually, the Bt proteins deployed in transgenic crops show specific activity against narrow groups of pest species and usually have little or no direct effect on non-target species, including beneficial insects. This specificity offers real advantages over most conventional pesticides as tools for IPM.

Worldwide, Bt proteins have been commercialized in three crops to date: cotton (expressing the CryIAc crystal protein), maize (CryIAb), and potatoes (Cry3A). In 1998, approximately 9.6 million ha of Bt crops were grown (corn - 8.3 million, cotton - 1.1 million, potatoes - 0.2 million) out of a total of 28 million ha of transgenic crops (the remainder are mostly herbicide-tolerant crops ISAAA Web page **http://www.isaaa.cornell.edu/**). The CryIAC protein in Bt cotton provides activity against many lepidopteran species, with the main targets being Heliothine moths (e.g., *Heliothis virescens, Helicoverpa armigera*), which plague production in many parts of the world. Bt maize meets a previously unmet need in targeting the European corn borer (*Ostrinia* sp.) and similar species, for which there were previously no effective controls. Potatoes have been engineered for control of Colorado potato beetle (*Leptinotarsa decemlineata*), a significant pest in many production areas. Many Heliothine moths and *Leptinotarsa* have a history of evolving resistance to pesticides, an issue for deployment of this technology that is discussed below. Many more Bt crops are under development, including rice, sorghum, lupins, peas and other legumes, and several tree crops (8,37). The first non-Bt transgenic crop to be commercialized may target the pea weevil (*Bruchus pisorum*), using an alpha-amylase inhibitor (36).

Wherever it is grown, cotton production relies heavily on pesticides. Bt cotton offers perhaps the most significant step forward in cotton pest management, and, by reducing the need for pesticide intervention to control primary pests, may provide a foundation for development of IPM systems. For this reason, Bt cotton will be the main focus of the remainder of this chapter, with emphasis on its introduction to Australian cotton systems.

Development and Implementation of Transgenic Bt Cottons

Bt cotton varieties have been developed and commercialized as BOLLGARD® in the USA, China, South Africa, and Argentina, and as INGARD® in Australia. In the USA, approximately 900,000 ha of Bt cotton were grown in 1998, with a total of 220,000 ha in other countries (R. Deaton, Monsanto, personal communication). While the introduction of Bt cotton in the USA has been unrestricted, and some areas now grow 90%+ Bt cotton, Australia has adopted a more conservative approach, with the area of use restricted through cooperation between the cotton industry and regulatory authorities. This is justified in part because the *Helicoverpa* species targeted in Australia (*H. armigera* and *H. punctigera*) are at least 10 times less sensitive to the IAc protein than is *Heliothis virescens*, which is the primary target of Bt cotton in the USA. Importantly though, *Helicoverpa zea*, which is occasionally a pest of cotton in the USA, shows a response to Bt similar to the Australian species (18; N. Forrester, personal communication).

In the first commercial season (1996-97), 10% (30,000 ha of a total of 300,000 ha) of the total cotton crop in Australia was Bt cotton, distributed in this proportion across all production regions. In the following year, this increased to 15% (60,000 ha of a total of 400,000 ha), with plans to increase production in 5% increments to a cap of 30% until more efficacious transgenic varieties with a second insecticidal gene become available, possibly early in the next decade.

The introduction of Bt cotton has raised several issues for ecological, environmental, and management research. These include

110

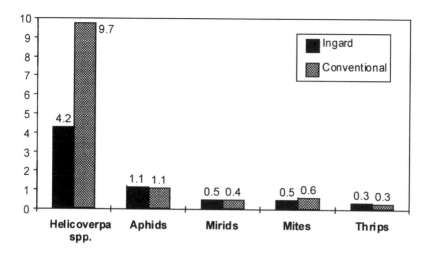

Figure 1. Average number of pesticide sprays used against various pests on INGARD and conventional cotton crops in Australia (1996-97 season). Data from Pyke and Fitt (27).

field efficacy, impacts on non-target species and beneficials, potential gene escape, development of pest management systems, and, most important, the development of a pre-emptive resistance management strategy. Some of these issues are discussed below.

EFFICACY OF BT COTTONS

All Bt cotton varieties currently in commercial use utilize a constitutive promoter (CAMV35S), which gives expression of IAc protein in all tissues throughout the life of the plant. Experience in Australia has shown, however, that the level of expression is not consistent throughout the growing season (9,11,13). Although the efficacy of pre-flowering plants against *Helicoverpa* sp. is generally high, it can be highly variable but typically declines through the boll maturation period. By the end of the season, survival in Bt cotton may be little different from that in non-transgenic cotton (9), although growth rates of survivors on the INGARD crops are still dramatically reduced (Fitt, unpublished). This decline in efficacy begins during flowering, and supplementary pesticide sprays for *Helicoverpa* control have been necessary on INGARD crops, particularly in the last third of the growing season. Similar patterns of declining field efficacy have been seen in parts of the U.S. cotton belt, where *H. zea* has been abundant due to declining levels of Bt in maturing plants and the high tolerance of this species to IAc (18). There is no obvious decline in efficacy of BOLLGARD cottons for the more susceptible *Heliothis virescens*. Greenplate et al. (18) showed declines in the concentration of IAc from 20ppm fresh wt at 40 days after planting to 5ppm fresh wt at 120 days. Holt (22) described similar changes in Bt protein concentration in Australian INGARD varieties.

Despite the seasonal changes in the efficacy of Bt cotton crops to *Helicoverpa* species, there have been significant reductions in pesticide applications in the USA and Australia. Estimates across the U.S. cotton belt, where *H. virescens* is the main target, show reductions of 70% to 80% in total applications (3), while in Australia, reductions of 45% to 50% were recorded in the first two commercial years (Fig. 1). Elsewhere, there is no publicly available data from which to judge the field performance and reduction in pesticide use gained through the use of Bt cottons. Nevertheless, the data from Australia and the USA suggest that, while not removing the need for some pesticide, these reductions represent a significant environmental benefit and a spectacular achievement for any IPM technology.

Substantial variations in efficacy between crops of INGARD, even those in adjacent fields (9), have also complicated management requirements for growers. The continued requirement for some pesticide applications on INGARD, combined with the widespread use of "sprayed conventional cotton" as a refuge option (see below), which causes pesticide drift onto INGARD crops, potentially reduces and to some extent masks the potential gains in conservation of beneficial insect populations that Bt cotton offers. In the future, greater emphasis on using unsprayed refuges (from which there will be no drift) and the development of more effective transgenic crops may increase the realization of the IPM benefits of this technology.

NON-TARGET IMPACTS AND SECONDARY PESTS

Despite the potential benefits, there have been a number of concerns related to the potential environmental impact of Bt crops. The CryIA proteins expressed in Bt cotton and Bt corn have been extensively tested in toxicological studies in the laboratory and field.

These studies strongly support the specific activity spectrums of Bt proteins, which are largely mediated by the gut conditions required for activation of the proteins, and by the need for specific binding to receptors in the mid-gut before toxicity is demonstrated. Field studies in both Australia and the USA have shown no significant impact on most beneficial insects relative to unsprayed conventional controls (11; Wilson and Fitt, unpublished). These results suggest that *Helicoverpa* larvae are incidental prey items for many of the generalist predators found in cotton, which is not surprising given the relatively low abundance of *Helicoverpa* compared with other invertebrates in the crop (Wilson and Fitt, unpublished). Only specialist parasitoids of *Helicoverpa* larvae showed a significant reduction in abundance in Bt cotton in Australia (L. J. Wilson, unpublished) and the USA (25). Significantly though, in all studies, the general abundance of predators and parasitoids in Bt cotton was substantially greater than in sprayed conventional cotton.

Despite widespread acceptance of the specificity of Bt proteins, recent laboratory studies by Hilbeck et al. (20) suggest a negative effect of CryIAb on survival and growth of Chrysopid larvae fed on prey items (*Ostrinia nubialis*) that had consumed tissues of Bt corn. Subsequent studies showed that the lacewing was susceptible to the pure Bt protein as well (21). Whether this effect is significant in the field has not yet been demonstrated. While these results support the need for rigorous evaluation of Bt crops prior to commercialization, they must be judged in the context of overall positive and negative impacts of Bt crops compared with the alternatives.

Changes in the status of what are normally minor pests following the dramatic reduction in use of pesticides against Lepidoptera have been another concern. Many of these are sucking pests (mirids, aphids, pentatomids) that have previously been suppressed by sprays targeting *Helicoverpa* spp. To date, there has been little evidence of such changes in Australia, although aphids occur commonly in unsprayed INGARD fields and often reach high levels earlier than normal. However, some of the aphid outbreaks are probably related to drift onto INGARD crops from nearby commercial cotton, which reduces beneficial populations. In other instances, where drift has not been a problem, aphids have usually been controlled by beneficial insects (Fitt, Wilson, unpublished observation). The green mirid, *Creontiades dilutus*, an early-season pest, appears no more abundant in Bt cotton than in conventional cotton. By contrast, in Mississippi, *Lygus* bugs have required more control on BOLLGARD crops than on conventional crops (23). In some eastern U.S. regions, late-season sucking pests have also emerged as a problem in BOLLGARD. For instance, significant increases in abundance and damage from stinkbugs (e.g., *Nezara viridula*) have resulted in up to 38% yield loss in unprotected Bt cotton (43). The same pattern is seen in Australia in unsprayed INGARD crops, although in commercial INGARD crops, the reduced efficacy of Bt cotton after flowering and consequent pesticide use at that time means that stinkbugs have not been a problem.

Other secondary pests that may show resurgence in the U.S. on Bt cotton crops include *Spodoptera exigua* (beet armyworm) and *S. frugiperda* (fall armyworm). Both species are relatively insensitive to CryIA proteins, and in some instances, they have caused significant damage to Bt crops. However, both are regarded as sporadic pests in which outbreaks result from the interaction of seasonal conditions and disruption to predator populations as part of boll weevil eradication programs. Neither has been a problem in Australia.

RISK OF RESISTANCE TO BT CROPS

The major challenge to sustainable use of Bt crops is the risk that target pests may evolve resistance to the transgenic proteins. Resistance to conventional Bt sprays has evolved in open field populations of *Plutella xylostella* in many parts of the world (41,42). Given the intense selection pressure imposed by the continuous expression of Bt proteins, the possibility of resistance is a real concern, particularly since some of the target pests (e.g., *H. armigera, H. virescens*) have a history of evolving resistance to pesticides (6, 14). For this reason, much effort has been devoted to developing and implementing pre-emptive resistance management plans to accompany the commercial release of transgenic varieties. Options for managing resistance to transgenic plants are dealt with exhaustively elsewhere (2,15,16,29,30,40). The strategy adopted in Australia is targeted at *H. armigera* and based on the use of refugia to maintain susceptible individuals in the population (15,30), but includes other components as well. This strategy takes advantage of the polyphagy and mobility

114

of *Helicoverpa* spp. to achieve resistance management by utilizing gene flow to counter selection in the Bt crops.

Implementation of the resistance management plan in Australia builds on previous experience with a successful pyrethroid resistance management strategy (14) in what is a relatively small and cohesive cotton industry. The industry has established the Transgenic and Insecticide Management Strategy (TIMS) committee with representation of growers, consultants, researchers, seed companies, and chemical industries, to devise, endorse, and implement strategies. The strategy has the following features:

1. effective refuges on each farm growing INGARD cotton
2. a defined planting window for INGARD cotton to avoid late-planted crops
3. mandatory cultivation of crop residues to destroy most overwintering pupae
4. defined spray thresholds for *Helicoverpa* to control survivors in the crops
5. monitoring of Bt resistance levels in field populations

Refuge options for each 100 ha of INGARD cotton include 100 ha of sprayed conventional cotton, 10 ha of unsprayed conventional cotton, or 20 ha of unsprayed sorghum or corn. The major criteria defining effective refugia are that they not be treated with Bt sprays and that they generate enough susceptible adult *Helicoverpa* to ensure that matings between two resistant survivors from a transgenic crop are extremely unlikely events. Refuges should be in close proximity to the transgenic crops to maximize the chances of random mating among sub-populations (5). All aspects of the Insect Management Plan for INGARD cotton are embodied in the label and are part of a grower contract with Monsanto, owners of the Bt gene. At present, the conventional sprayed cotton refuge option is the most popular, but smaller unsprayed refuges may be more appropriate for the future as the proportional area of Bt cottons increases. Much research is underway to better define the value of different refuge options and identify a range of options for growers (12).

The refuge strategy assumes that resistance to Bt is likely to be functionally recessive, that resistance genes are at low frequency in natural populations, and that random mating occurs among individuals from refuges and Bt crops (16,30). These assumptions seem reasonable, based on current knowledge of Bt resistance in field populations of *Plutella* and on laboratory selection of resistance in *Heliothis/Helicoverpa* populations. However, there is some evidence that Bt resistance frequencies may be higher than expected in natural populations (17). Furthermore, the lower-than-expected efficacy of Bt cotton in Australia may increase the likelihood of heterozygotes surviving, thereby increasing the rate at which resistance genes accumulate in the population. Resistance management plans should therefore be conservative, and deployment of transgenic cottons should proceed with caution as more information on the interaction of Bt crops with pest populations is gathered. Every effort is needed to quickly introduce a second insecticidal protein. Resistance management plans have also been devised for the USA (although much less comprehensive than in Australia), and they will be needed in most countries where transgenic Bt cottons may be released. Another issue that will impinge on the long-term sustainability of Bt cottons is the use of the same CryIA genes in other crops that are also hosts of *Helicoverpa* (e.g., Bt corn in the USA). These issues are further addressed elsewhere (8).

Transgenic Bt Cotton as a Foundation for IPM

Transgenic Bt crops are often criticized as being a single-strategy approach to pest management with reliance on one or a few insecticidal protein genes (44). However, despite the "hype" that often surrounds them, Bt crops should not be projected or perceived as "silver bullet" solutions for pest management. Rather, they should be viewed as a foundation on which to build IPM systems that incorporate a broad range of biological and cultural tactics. This potential will be further enhanced as more efficacious transgenic varieties are released. In Australia, Bt cottons expressing two independent Bt proteins (CryIAc and CryIIA) show much more consistent field efficacy and should greatly enhance the sustainability of resistance management (31). Other possibilities for insecticidal genes are also being researched (19,24).

One of the greatest impediments to the development of IPM in cotton has been the lack of tools to control target pests without also disrupting populations of beneficial insects. A reduction in use of disruptive pesticides on Bt cotton will allow more emphasis on the management and manipulation of beneficial species using nursery crops and food sprays or other means of conservation and augmentation. Efforts are underway in Australia to manipulate beneficials through a combination of lucerne (alfalfa) strips as a nursery crop to maintain and increase naturally occurring beneficial species, combined with a predator food spray (Envirofeast[TM]) to draw them into the cotton crop (26). The lucerne strips also act as a trap crop for a secondary pest, the green mirid *Creontiades dilutus*. An IPM system involving Bt cotton, food sprays, and virus sprays has shown great potential to reduce conventional pesticide requirements (Auscott Australia, unpublished). At present, it is unclear whether this relatively complex IPM system will be widely adopted, although current problems with resistance to insecticides and associated escalating costs of pest control ($A400 - $A1000 / ha in 1998-99) are providing the incentive for growers to review their approach to pest management.

Populations of generalist beneficials are indeed higher in Bt cotton than in conventional sprayed cotton. These beneficials will help with management for some secondary pests, particularly those that are largely insecticide-induced pests (e.g., mites and aphids) in sprayed cotton. Early in the growing season, when Bt cottons have high efficacy, the only major pests likely to need supplemental control are the mirids (plant bugs). *Helicoverpa* larvae that survive in Bt cotton as the season progresses have markedly reduced growth rates compared with those on conventional crops (Fitt, unpublished data). If beneficials can be maintained through the season, they should be more effective in controlling these surviving larvae because they will be exposed to predators and parasitoids for longer periods when they are smaller and less damaging.

INTEGRATION WITH SELECTIVE CHEMICALS

A new generation of selective pesticides used only when essential will be an important component for IPM systems based on Bt cotton. Chemical groups in widespread use today (organophosphates, carbamates, synthetic pyrethroids) are inexpensive, have broad-spectrum activity, are significantly disruptive to most predators and parasites (46), in some cases have environmental residue problems, and are increasingly challenged by resistance. In contrast, the new-generation pesticides are more selective, less disruptive, more environmentally benign, and introduce new modes of action to overcome established resistance problems (1). They are also considerably more expensive, which should help to ensure that they are used conservatively and so preserve their long-term usefulness. Highly selective biological insecticides will also have an important role in pest management. Formulations of Nuclear Polyhedrosis Virus (e.g., GEMSTAR®) or, in the future, genetically modified viruses with enhanced speed of kill (28) will provide options for Lepidopteran control in cotton, although they may have a better place in management of Heliothines on other crops growing in agroecosystems where cotton is grown (e.g., sorghum and legumes). Alternative tactics, such as Bt cotton and biological pesticides for other crops, provide options for IPM at the agroecosystem level.

HOST PLANT RESISTANCE AND COMPENSATION

Other components of sustainable IPM systems include genotypes conventionally selected for host plant resistance (HPR) to pests and the plants' capacity to compensate for a degree of injury without yield loss. The cotton plant is not simply a substrate for the interaction of pests and chemicals; it is the template on which a broad range of interactions occurs between the pests and their environment. Cotton has a number of morphological and biochemical traits that impart varying degrees of pest tolerance. The integration of insecticidal transgenes with other HPR characters through conventional plant breeding may enhance efficacy against key and secondary pests. In Australia, the INGARD gene has been incorporated in okra leaf varieties to provide enhanced resistance to both *Helicoverpa* and mites (7,45). Insecticidal secondary compounds of cotton, such as the terpenoid aldehydes (gossypol or the related "heliocides"), reduce survival and growth rates of *Helicoverpa* spp. Sachs et al. (32) showed synergism between IAb protein and high gossypol levels, and some efforts are underway to combine these traits in commercial cultivars. On the other hand, there is some evidence that tannins may reduce the efficacy of Bt transgenes (4).

Cotton has a considerable capacity to compensate, even overcompensate, for insect feeding damage (33). Much greater use could be made of this capacity through the application of appropriate plant damage thresholds in combination with pest thresholds, as pest thresholds alone do not always accurately reflect plant damage. The threshold established for *Helicoverpa* on Bt cotton crops in Australia is two larvae per meter row over two consecutive checks. This threshold was designed to allow time for larvae to feed sufficiently in order to ingest a lethal dose, yet still allow intervention while the larvae are small enough to control with insecticides (generally less than 6 mm for pesticide-resistant *H. armigera*). Thresholds for other pests remain largely unchanged. Cotton genotypes vary in their ability to compensate for pest damage (34). Selection for genotypes with higher compensatory ability in combination with Bt genes could allow the use of higher thresholds for all pests with less risk, therefore reducing the need to intervene with disruptive insecticides. Exploitation of this genetic resource by breeders could provide a more resilient plant background for pest management.

CULTURAL CONTROLS AND OTHER TACTICS

Cultural techniques will integrate easily with Bt cottons. Soil cultivation to destroy any potentially Bt-resistant *H. armigera* pupae surviving in the soil through winter (10) is a mandatory requirement in Australia. The use of trap crops to concentrate *Helicoverpa* or other pests into small areas where they can be controlled at key times through the season, such as the first generation of *Helicoverpa armigera* in the spring, is also being evaluated as part of area-wide approaches to pest population management. Coordinated area-wide approaches have been successful in eradication of boll weevil in much of the eastern U.S. cotton belt (38). Area-wide management of *Heliothis/ Helicoverpa* in the USA, through the use of virus sprays to reduce the first generation (39), has been more problematic. Area-wide approaches to *Helicoverpa* management in Australia, based on Bt cotton combined with trap crops and sacrifice crops, are now being researched.

Conclusions

IPM systems for future production of many agronomic and horticultural crops will, of necessity, be more complex than the pesticide-based systems currently in place, and will require greater effort on the part of crop managers. Transgenic Bt crops with activity against one or more key pests offer great scope to dramatically reduce pesticide dependence and to allow the integration of a wide range of IPM-compatible tactics. The current reliance on Bt endotoxin genes in transgenic crops represents only the first wave of insecticidal proteins for pest management. Alternative transgenes with activity against major lepidopteran and hemipteran pests are under development (24). These include vegetative insecticidal proteins from Bt (VIPs), insecticidal proteins from *Xenorhabdus*, plant expression of a small RNA virus (*H. armigera* stunt virus [19]), various protease inhibitors, alpha amylase inhibitor, and several lectins. These offer possibilities for combining with Bt endotoxin genes to provide more sustainable resistance management or control of minor pests. Few of these alternatives are close to market, but some should appear within the next decade.

Most Bt crops will require a well-researched resistance management strategy based on a thorough understanding of pest and crop ecology in order to maximize long-term benefits and avoid development of resistance to the initial Bt proteins before other technologies become available. A challenge for researchers will be to achieve integrated management systems around Bt crops. In contrast to some expectations, it is likely that these systems will require more, not less, management input than conventional crops.

Finally, a significant challenge for researchers and funding agencies alike is to ensure that research on a range of conventional IPM components continues to be funded alongside the increasing expenditure on various forms of biotechnology. Transgenic Bt crops will not be sustainable technologies alone; they must be supported with other approaches, which will require continued research.

Acknowledgments

We are grateful to many colleagues, including Joanne Daly, Rick Roush, Robert Mensah, Neil Forrester, David Murray, Martin Dillon, and Myron Zalucki, for discussions of the issues covered here.

Literature Cited

1. Bradley, J. R. 1999. Integrating New Insecticide Technologies in IPM. in: G. G. Kennedy and T. B. Sutton, eds. Emerging Technologies for Integrated Pest Management: Concepts, Research, and Implementation. APS Press, St. Paul, MN.
2. Caprio, M. A. 1994. *Bacillus thuringiensis* gene deployment and resistance management in single and multi-tactic environments. Biocontrol Science and Technology 4:487-498.
3. Carlson, G. A., Marra, M. C., and Hubbell, B. J. 1998. Yield, insecticide use, and profit changes from adoption of Bt cotton in the southeast. Pages 973-974 in: Proc. Beltwide Cotton Conferences, Vol 2. San Diego.
4. Daly, J. C., and Fitt, G. P. 1998. Efficacy of Bt cotton plants in Australia - What is going on? Proc. Second World Cotton Conference (in press).
5. Dillon, M. L., Fitt, G. P., and Zalucki M. P. 1998. How should refugia be placed upon the landscape? A modelling study considering pest movement and behaviour. Pages 179-189 in: *Pest Management - Future Challenges.* Proc. 6th Australasian Applied Entomology Conference, Vol. 1, Brisbane. M. P. Zalucki, R. A. I. Drew, G. G. White, eds. University of Qld Press, Brisbane.
6. Fitt, G. P. 1989. The ecology of Heliothis species in relation to agroecosystems. Annu. Rev. of Entomol. 34:17-52.
7. Fitt, G. P. 1994. Cotton pest management: Part 3. An Australian perspective. Annu. Rev. of Entomol. 39:543-562.
8. Fitt, G. P. 1997. Risks, deployment and integration of insect resistant crops expressing genes from *Bacillus thuringiensis.* Pages 273-284 in: Commercialisation of Transgenic Crops: Risk, Benefit and Trade Considerations. G. D. McLean, P. M.

Waterhouse, G. Evans, and M. J. Gibbs, eds. Cooperative Research Centre for Plant Science and Bureau of Resource Sciences, Canberra.

9. Fitt, G. P. 1998. Efficacy of INGARD cotton - patterns and consequences. Pages 233-245 in: Proc. 9th Australian Cotton Conference, ACGRA, Wee Waa.

10. Fitt, G. P., and J. C. Daly. 1990. Abundance of overwintering pupae and the spring generation of *Helicoverpa* spp. (Lepidoptera: Noctuidae) in northern New South Wales, Australia: Implications for pest management. J. Econ. Entomol. 83:1827-1836.

11. Fitt, G. P., Mares, C. L., and Llewellyn, D. J. 1994. Field evaluation and potential impact of transgenic cottons (*Gossypium hirsutum*) in Australia. Biocontrol Science and Technology 4:535-548.

12. Fitt, G. P., and Tann, C. 1996. Quantifying the value of refuges for resistance management of transgenic Bt-cotton. Pages 77-83 in: Proc. Eighth Australian Cotton Conference, Broadbeach, August, 1996.

13. Fitt, G. P., Daly, J. C., Mares, C. L., and Olsen, K. 1998. Changing Efficacy of Transgenic Bt Cotton - Patterns and Consequences. Pages 189-196 in: Pest Management - Future Challenges. Proc. 6th Australasian Applied Entomology Conference, Vol. 1, Brisbane. M. P. Zalucki, R. A. I. Drew, G. G. White, eds. University of Qld Press, Brisbane.

14. Forrester, N. W., Cahill, M., Bird, L. J., and Layland, J. K. 1993. Management of pyrethroid and endosulfan resistance in *Helicoverpa armigera* (Lepidoptera: Noctuidae) in Australia. Bulletin of Entomological Research Supplement No.1.

15. Gould, F. 1994. Potential and problems with multi-gene, high dose strategies for managing resistance to Bt toxins. Biocontrol Science and Technology 4:451-462.

16. Gould, F. 1998. Sustainability of transgenic insecticidal cultivars: integrating pest genetics and ecology. Annu. Rev. of Entomol. 43:701-726.

17. Gould, F., Anderson, A., Jones, A., Sumerford, D., Heckel, D. G., Lopez, J., Micinski, S., Leonard, R., and Laster, M. 1997. Initial frequency of alleles for resistance to *Bacillus thuringiensis*

toxins in field populations of *Heliothis virescens*. Proc. National Academy of Sciences USA 94:3519-3523.

18. Greenplate, J. T., Head, G. P., Penn, S. R., Kabuye, V.T. 1998. Factors potentially influencing the survival of *Helicoverpa zea* on Bollgard cotton. Pages 1030-1033 in: Proc. Beltwide Cotton Conferences, Volume 2. San Diego.

19. Hanzlik, T. N., and Gordon, K. 1998. Beyond Bt genes: plants with stunt virus genes. Pages 206-214 in: Pest Management - Future Challenges. Proc. 6th Australasian Applied Entomology Conference, Vol. 1, Brisbane. M. P. Zalucki, R. A. I. Drew, G. G. White, eds. University of Qld Press, Brisbane.

20. Hilbeck, A., Baumgartner, M., Fried, P. M., Bigler, F. 1998a. Effects of transgenic *Bacillus thuringiensis* corn-fed prey on mortality and development time of immature *Chrysoperla carnea* (Neuroptera: Chrysopidae). Environ. Entomol. 27:480-487.

21. Hilbeck, A., Moar, W. J., Pusztai-Carey. Filippini, A., and Bigler, F. 1998b. Toxicity of *Bacillus thuringiensis* IAb toxin to the predator *Chrysoperla carnea* (Neuroptera: Chrysopidae). Environ. Entomol. 27:1255-1263.

22. Holt, H. E. 1998. Season-long quantification of *Bacillus thuringiensis* insecticidal crystal protein in field-grown transgenic cotton. Pages 215-222 in: Pest Management - Future Challenges. Proc. 6th Australasian Applied Entomology Conference, Vol. 1, Brisbane. M. P. Zalucki, R. A. I. Drew, G. G. White, eds. University of Qld Press, Brisbane.

23. Layton, B., Stewart, S. D., Williams, M. R., and Reed, J. T. 1998. Performance of Bt cotton in Mississippi. Pages 970-972 in: Proc. Beltwide Cotton Conferences, Volume 2. San Diego,

24. Llewellyn, D. J., and Higgins, T. J. 1998. Biotechnological approaches to crop protection: novel sources of insect tolerance genes other than Bt-toxins. Pages 223-229 in: Pest Management - Future Challenges. Proc. 6th Australasian Applied Entomology Conference, Vol. 1, Brisbane. M.P. Zalucki, R. A. I Drew, G. G White, eds. University of Qld Press, Brisbane.

25. Luttrell, R. G., Mascarenhas, V. S., Schneider, J. C., Parker, C. D., Bullock, P. D. 1995. Effect of transgenic cotton expressing endotoxin protein on arthropod populations in Mississippi cotton. Pages 760-763 in: Proc. Beltwide Cotton Conferences, Volume 2. San Antonio.

26. Mensah, R. K. 1998. Habitat diversity: Implications for conservation and use of predatory insects in cotton systems in Australia. Int. J. Pest Management (in press).

27. Pyke, B., and Fitt, G. P. 1998. Field performance of INGARD cotton - the first two years. Pages 230-238 in: Pest Management - Future Challenges. Proceedings 6th Australasian Applied Entomology Conference, Vol. 1, Brisbane. M. P. Zalucki, R. A. I. Drew, G. G. White, eds. University of Qld Press, Brisbane.

28. Richards, A. R., and Christian, P. 1998. Genetically engineered viruses: new insecticides, future challenges - an ecological perspective. Pages 238-246 in: Pest Management - Future Challenges. Proc. 6th Australasian Applied Entomology Conference, Vol. 1, Brisbane. M. P. Zalucki, R. A. I. Drew, G. G. White. eds. University of Qld Press, Brisbane.

29. Roush, R. T. 1996. Can we slow adaptation by pests to insect-resistant transgenic crops? Pages 242-263 in: Biotechnology and Integrated Pest Management. G. Persley, ed. CAB International, London.

30. Roush, R. T. 1997. Managing resistance to transgenic crops. Pages 271-294 in: Advances in insect control: The role of transgenic plants. N. Carozzi and M. Koziel, eds. Taylor and Francis, London.

31. Roush, R. T. 1998. Two toxin strategies for management of insecticidal transgenic crops: Can pyramiding succeed where pesticide mixtures have not? Phil. Trans. Royal Soc. Lond. B. 353:1777-1786.

32. Sachs, E. S., Benedict, J. H., Taylor, J. F., Stelly, D. M., Davis, S. K., and Altman, D.W. 1996. Pyramiding CryIA(b) insecticidal protein and terpenoids in cotton to resist tobacco budworm (Lepidoptera: Noctuidae). Environ. Entomol. 25:1257-1266

33. Sadras, V.O. 1995. Compensatory growth in cotton after loss of reproductive organs. Field Crops Research 40:1-18.

34. Sadras, V.O., and Fitt, G. P. 1997. Resistance to insect herbivory of cotton lines: quantification of recovery capacity after damage. Field Crops Research 52:129-136.

35. Schnepf, S., Crickmore, N., van Rie, J., Lereclus, D., Baum, J., Feitelson, J., Zeigler, D. R., and Dean, D.N. 1998. *Bacillus thuringiensis* and its pesticidal crystal proteins. Microbiol. Molec. Biol Reviews 62:775-806.

36. Schroeder, H. E., Gollasch, S., Moore, A., Tabe, L. M., Craig, S., Hardie, D. C., Chrispeels, M. J., Spencer, D., and Higgins, T. J. V. 1995. Bean alpha-amylase inhibitor confers resistance to the pea weevil (*Bruchus pisorum*) in transgenic peas (*Pisum sativum* L.). Plant Physiology 107:1233-1239.

37. Schuler, T. H., Poppy, G. M., Kerry, B. R., Denholm, I. 1998. Insect-resistant transgenic plants. Trends in Biotechnology (TIBTECH) 16:169-175.

38. Smith, J. W. 1998. Boll weevil eradication - area-wide pest management. Ann. Entomol. Soc. Amer. 91:239-247.

39. Streett, D. A., Bell, M. R., and Hardee, D. D. 1998. Update on the area-wide budworm/bollworm management program with virus in the United States. Proc. World Cotton Research Conference 2.

40. Tabashnik, B. E. 1994. Delaying insect adaptation to transgenic crops: seed mixtures and refugia reconsidered. Proc. Roy. Soc. Lond., Series B. 255:7-12.

41. Tabashnik, B. E. 1994b. Evolution of resistance to *Bacillus thuringiensis*. Annu. Rev. of Entomol. 39:47-79.

42. Tabashnik, B. E., Liu, Y-B., Malvar, T., Heckel, D. G., Masson, L., and Ferre, J. 1998. Insect resistance to *Bacillus thuringiensis*: uniform or diverse? Phil. Trans. R. Soc. Lond. B. 353:1751-1756.

43. Turnipseed, S. G., and Greene, J. K. 1996. Strategies for managing stinkbugs in transgenic Bt cotton. Pages 935-936 in: Proc. Beltwide Cotton Conferences, Nashville.

44. Waage, J. 1996. Integrated Pest Management and Biotechnology: An analysis of their potential for integration. Pages 37-60 in: Biotechnology and Integrated Pest Management. G. J. Persley, ed. CAB International, Cambridge.

45. Wilson, L. J. 1994. Resistance of okra-leaf cotton genotypes to two-spotted spider mites (Acari: Tetranychidae). Jour. Econ. Entomol. 87:1726-1735.

46. Wilson, L. J., Bauer, L. R., and Lally, D.A. 1998. Effect of early-season insecticide use on predators and outbreaks of spider mites (Acari: Tetranychidae) in cotton. Bull. Ent. Res. 88:477-488.

Genetic Engineering in IPM: A Case Study:
Herbicide Tolerance

F. Dan Hess, Sr.
Affymax Research Institute, Palo Alto, CA 94304 USA

Stephen O. Duke
USDA, ARS, Natural Products Utilization Research, National Center
for the Development of Natural Products,
University, MS 38667 USA

This overview of herbicide-tolerant (HT) crops is not a technical review, but rather a general discussion of the advantages and disadvantages of HT crops, as well as the current and projected impact of HT crops on the weed-control market. At this point in time, most of the commercial use of HT crops is in U.S. and Canadian agriculture. Thus, North America will be the primary focus of this overview. For a technical overview of HT crops, consult the book edited by Duke (4) or any of several reviews (3,7,8). Other general reviews and a book emphasizing the effects of HT crops on weed science are available (5,12).

Some commercial HT crops have been obtained by selection of tolerant plants, or tolerant cells in tissue culture, from natural variability in a population or from mutagenized plants or cells. Alternatively, genes conferring tolerance identified in bacteria or other sources have been engineered into crop plants. Examples of herbicide tolerance obtained by selection of herbicide-tolerant plants or cell lines are sethoxydim tolerance in corn (Poast Protected®), sulfonylurea tolerance in soybeans (STS®), and imidazolinone tolerance in corn (imidazolinone-tolerant [IT®] or imidazolinone-resistant [IR®]). Some of the examples of HT crops obtained through genetic engineering are glyphosate tolerance in soybeans (Roundup Ready®), glufosinate tolerance in corn (Liberty Link®), and bromoxynil tolerance in cotton (BXN®). Table 1 lists those HT crops that are now available to farmers in the United States and Canada.

Table 1. Herbicide-tolerant crops available in North America.

Herbicide	Crop	Year available
bromoxynil	cotton	1995
sethoxydim[a]	corn	1996
glufosinate	canola	1997
	corn	1997
glyphosate	soybean	1996
	canola	1996
	cotton	1997
	corn	1998
imidazolinones[a]	corn	1993
	canola	1997
sulfonylureas[a]	soybeans	1994
triazines[a]	canola	1984

[a] not transgenic

The mechanisms of achieving herbicide tolerance are primarily through enhancing herbicide inactivation (metabolism) or through an altered site of action, which reduces or eliminates herbicide sensitivity at the molecular target site. While herbicide tolerance could be achieved by reducing herbicide absorption into the plant or its translocation to the active site, these mechanisms have not yet been used for obtaining a commercial HT crop.

HT crops based on herbicide inactivation mechanisms have had bacterial genes encoding degradative enzymes engineered into them. For example, genes encoding phosphinothricin acetyltransferase (PAT) activity isolated from *Streptomyces hygroscopicus* (*bar* gene) and *Streptomyces viridochromogenes* (*pat* gene) have been engineered into several crop species to achieve glufosinate tolerance. When this enzyme is produced in plants, it converts glufosinate to a non-herbicidal acetylated form by transferring the acetyl group from acetyl-coenzyme A onto the free amino group of glufosinate (2). For bromoxynil tolerance, a specific nitrilase gene isolated from the bacterium *Klebsiella ozaenae* (13) was engineered into cotton. Nitrilase enzymes act on the cyano moiety of nitrile-containing compounds and for bromoxynil forms 3,5-dibromo 4-hydroxybenzoic acid, which has no herbicidal activity. For glyphosate, an amine oxidase enzyme (called glyphosate oxidoreductase [GOX] by Monsanto) was obtained from a bacterium (*Achromobacter* sp.) that converts glyphosate to aminomethyl phosphonate + glyoxylate, neither of which has any herbicidal activity.

Herbicide tolerance in crops due to an altered site of action has been obtained by the selection of surviving plants, or cells from tissue cultures, that have been exposed to the herbicide (sethoxydim, imidazolinones, and sulfonylureas), or from isolating genes from tolerant organisms, usually a bacterium (e.g., glyphosate). The most prevalent method of obtaining glyphosate-tolerant crops has been to engineer resistant 5-enolpyruvylshikimate-3-phosphate synthase (EPSPS) genes from bacteria into sensitive crop plants (see Ref. 1 for review). The best tolerance gene identified to date is named CP4 and came from an *Agrobacterium* species (strain CP4).

Why Are HT Crops Being Developed for Agriculture?

Although the Flavr Savr tomato was the first transgenic crop introduced (1994), HT crops have dominated the transgenic crop market (15). Since their introduction in 1995, the market share of transgenic HT crops has increased dramatically each year. This has occurred against a backdrop of existing weed management options that are effective.

There are a plethora of chemicals available for controlling weeds, and more are entering the market every year. There are few unsolved weed problems, and the environmental profiles of new introductions to the agrichemical market are continually improving. Nearly all herbicides submitted for Environmental Protection Agency (EPA) registration in the USA fall under the "safer pesticide" guidelines issued by the EPA. Thus, why are HT crops becoming an important part of weed control? One important factor is that, in the early development of biotechnology, obtaining herbicide-tolerant plants was one of the more straightforward traits to achieve.

Tolerance by selection was simply a matter of adding herbicide to the test system and selecting cells or plants that survived. Later, single genes imparting tolerance were found to be as effective as tolerance transgenes in crops. Using this relatively simple biotechnology to expand the market for a proprietary herbicide was a clear strategy to increase sales of that herbicide.

Traditional discovery of new herbicides involves a process of synthesis of a large number of molecules, followed by screening on weeds and crops in the greenhouse. The number of compounds that must be synthesized in order to identify a compound of interest continues to increase every year, as does the amount of time and money expended from discovery to commercial introduction. Across the industry, an average of 50,000 compounds screened per new product introduced was the estimate in the early to mid-1990s (14). Now, from several thousand to more than 100,000 compounds (depending on the approach used) are synthesized for every new commercial product introduced. The mature herbicide market, increasingly stringent registration requirements, and relatively low cost of synthesis of large numbers of compounds by combinatorial chemistry will dramatically increase the ratio of compounds screened per commercial product. In addition, crop selectivity associated with a specific molecule is difficult to predict and is usually optimized by substantial additional synthesis and screening once an active lead is discovered. The processes of discovery, structure optimization, toxicology testing, efficacy testing, formulation, government approval, and commercialization can take as long as 10 years.

Because of the considerable costs of discovery, development, and commercial introduction (~$100 million), expansion of the market for existing products is desirable. Production of HT crops for existing herbicides offers the greatest potential for market expansion for some herbicides. If a nonselective herbicide available on the market has an excellent environmental profile, controls a desirable spectrum of weeds, and is priced competitively with crop-selective herbicides, it makes sense to evaluate the technical feasibility and cost of introducing tolerance into one or more major crops with which it is currently incompatible. A consideration in selecting a herbicide for development of HT crops is ownership of the proprietary rights to all herbicides to which the HT crop would be tolerant. There is a major sales advantage if the same company holds all rights to the herbicide and the HT crop technology. In North America, this has been the case for the two major HT crop types (glyphosate- and glufosinate-tolerant). However, glyphosate patent rights expire in the USA in late 2000. These two herbicides are unique in that there are no analogues on the market with which these HT crops might be compatible. Most commercial herbicides belong to classes to which there are numerous analogues sold by several companies. A HT crop for one of these herbicides is likely to be tolerant to most or all analogues of a particular class. Such a situation significantly reduces the incentive for development of HT crops for these herbicides. For these reasons, companies have historically been most interested in nonselective, single-analog herbicides (e.g., glufosinate and glyphosate) for co-development of the herbicide and the HT crop.

There are several benefits for the farmer of having HT crops available for weed control. For one, it allows an expanded use of herbicides that, through extensive testing and use, have been shown to have a desirable environmental profile and spectrum of weeds controlled. To date, herbicides that are used in HT crops are applied postemergence, which allows herbicide application based on need. Unless the weed population is extreme, the application can be delayed until an assessment of the weed population and species present can be made to determine the optimum herbicide type and concentration to use. Another benefit is that the HT trait can be combined with other crop protection traits, such as insect resistance and disease resistance, to provide crop plants with a wide range of pest management options for IPM. Currently, seeds with both HT traits (e.g., glyphosate tolerance) and insect tolerance traits (*Bt* genes) are available (e.g., cotton and corn).

Perhaps the greatest environmental damage caused by agriculture is loss of soil by erosion. The two current major HT crop herbicides (glyphosate and glufosinate) lend themselves to reduced- or no-tillage agricultural practices that dramatically reduce soil erosion. Most other HT crops in development are also for postemergence herbicides. Other benefits of reduced- or no-tillage agriculture include less soil compaction and reduced movement of pesticides from agricultural fields to surface water. Reduced- or no-tillage agriculture can greatly influence both plant pathogens and insects. The impact of this aspect of HT crops on insect and pathogen management has not been adequately studied in order to understand all the implications for IPM.

An interesting IPM-related aspect of the herbicides used with some HT crops is that they may also be helpful in managing plant pathogens. For example, glufosinate controls some pathogens in certain HT crops (10). Glyphosate is also toxic to some microbes and may also play such a role. More study of the impact of HT crops and associated herbicides on the management of insects and pathogens is needed.

Based on currently available technology for weed control, the primary factors farmers use to select a weed control strategy are performance and cost. If HT crops combined with the associated herbicide provide better weed control and cost less than conventional herbicide applications, farmers will embrace the technology. A good example is the rapid market acceptance of glyphosate-tolerant soybeans. The spectrum of weeds in soybeans is a good match to the spectrum of weeds controlled by glyphosate. Soybeans can tolerate some weed competition during early development, which provides application flexibility. The cost of weed management with glyphosate in HT soybeans is less than that of conventional herbicide programs, even when the technology fee and added seed cost are included.

Problems and Issues Associated with Using HT Crops

While there are many benefits to using HT crop technology in agriculture, there are also problems and issues that need to be considered when selecting HT crops and the associated herbicide over conventional herbicide programs. In contrast to genetically engineered insect- and disease-tolerant crops, where the genes that have been introduced are responsible for production of biopesticides, crops genetically engineered to produce weed-killing allelochemicals have not been pursued by industry. Instead, for HT crops, the HT transgene allows substitution of one herbicide for another rather than a gene for a chemical.

At this time, each HT crop has only one herbicide-tolerant trait per seed type; thus, the farmer must purchase a specific seed and herbicide combination if he/she wants to use the technology. If several herbicide tolerances were available within a single seed type, maximum weed control flexibility would be available to the farmer after the crop is planted. However, it is unlikely that multiple herbicide tolerance traits within a single seed type are going to be commercially available in the near future.

A major issue with one HT trait per seed type is the large inventory that seed companies are required to stock. Most crop seeds are available from a seed company in multiple cultivars and maturity zones. Each time a HT trait is introduced, the already large inventory of seed types stocked must be doubled if the HT trait is to be widely available. If a single seed company offers several different types of herbicide tolerances, the inventory that must be managed increases further. While many seed companies now offer HT crops, or are planning on offering HT crops in the near future, most view HT crops as a defensive strategy for their business (i.e., the competition offers this technology, so we must also offer it). Many seed companies will offer glyphosate and glufosinate HT crops, and some will also offer sulfonylurea-, imidazolinone-, bromoxynil-, and sethoxydim-tolerant crops. Few seed companies are interested in increasing their portfolio of HT crops, and some are even focusing on offering non-engineered seed to farmers so they can maintain flexibility in where to market their crops. While increased market share may be possible for those seed companies offering HT crops, the premium received from selling seeds containing the HT trait may not justify the added expense to the seed producer. This has been the view of independent seed companies (6). However, with the large-scale acquisition of seed companies by crop protection/life science corporations, the combination of seed sales, herbicide sales, and technology fee must be considered.

A few additional herbicide types have been evaluated for use in HT crops (e.g., growth regulators, carotenoid synthesis inhibitors, and photosynthesis inhibitors; see ref. 4), but crops having tolerance to these herbicide types are not currently on the market and are expected to have only a minor impact on future markets. There is little commercial interest in HT crops tolerant to inexpensive, non-proprietary herbicides, such as 2,4-D, even though they would greatly decrease the cost of weed management. Press releases from companies reveal new HT crops in development (e.g., corn tolerant to protoporphyrinogen oxidase (PPO) inhibitors from Novartis, and isoxazole tolerance in soybeans from Rhône-Poulenc), but the market impact of these traits will not be known for several years. New combinations of HT crops and herbicides will have to be superior to available technology in terms of having minimal environmental impact and a high degree of efficacy in controlling a wide range of weeds. The technology will need to be applicable to one or more major crops (e.g., corn, soybeans, cotton, or small grains) and will need to be price competitive with current weed control alternatives in these crops.

Because HT crops have only been commercially available for a few years, the possible impact on yield of adding particular HT traits to crops is not fully resolved. There are two factors associated with fitness of the HT crop. First is the possibility of a "yield penalty" being associated with the HT crop cultivar or hybrid. In this case, there is a loss of yield due to adding the trait to a cultivar or hybrid when compared to an isoline without the trait. An insufficient number of side-by-side comparisons have been reported in order to fully assess this possibility. Second is the possibility of a "yield drag" being associated with commercially available seeds. While plant breeders are interested in including the HT trait in their very best germplasm, time is required to complete this task. If plant breeders are able to increase the yield of a crop an average of 2% per year, and it takes 3 years to introduce a HT trait into a specific genetic background, the potential yield loss is 6% when compared to the best seed lines available without the trait. As the ability to rapidly introduce the HT trait into the best elite lines is achieved, the yield drag will become insignificant, provided there are no problems with pleiotrophic effects of the gene or incomplete gene expression. Moreover, less-than-full tolerance at recommended herbicide application rates could be a problem under some environmental conditions. In 1998, a significant number of farmers in several southern states had symptoms of incomplete tolerance with glyphosate-tolerant cotton. The cause of these symptoms was not unequivocally determined, but recent studies in controlled environments with several HT crops have demonstrated that environmental factors can influence expression of tolerance transgenes, resulting in incomplete protection of the crop from the herbicide.

To date, herbicides associated with HT crops are applied postemergence. Late postemergence application is best for weed control, because most of the weeds that can compete with the crop will have germinated. However, depending on the level of competition between weed and crop, early competition can cause crop stunting and yield loss. Early-season herbicide application is best for minimizing competition between the weed and the crop, but later-emerging weeds, if not controlled, can compete with the crop. Because of the variable weather in the spring, there is some risk that if a preemergence herbicide is not used to manage the first significant flush of weeds, application of the postemergence herbicide may have to be delayed to the point where weed competition with the crop causes yield loss. Weed-crop competition can be managed in several ways. Two or more applications of the herbicide associated with the HT crop can be applied during the growing season to make sure weed competition with the crop does not occur. Alternatively, a pre-emergence, crop-selective herbicide having residual activity can be applied at planting to reduce early-season weed competition, or a residual crop-selective herbicide can be included in the first application to the HT crop to increase the time weeds are controlled during crop establishment.

Perhaps the most significant issue related to frequent use of HT crops and associated herbicides is the increased selection pressure for resistant weeds when the same herbicide is used over multiple years with no rotation to herbicides having different mechanisms of action. The risk of selecting resistant weeds in a field population depends on the frequency of resistance in the population. Based on history, the risk of selecting weeds resistant to acetolactate synthase (ALS) or acetyl coenzyme A carboxylase (ACCase) inhibitors is higher than selecting resistance to the herbicides that inhibit EPSPS (glyphosate) or glutamine synthetase (glufosinate). The potential for selection of resistant weeds from a population can be minimized by mixing herbicides, or rotating herbicides over time, that have different mechanisms of action but control a similar spectrum of weeds. Even if the risk of selecting resistant weeds is low, use of a single herbicide year after year can cause weed populations to shift to those that are difficult to control with that herbicide. No herbicide controls every weed in every situation; thus, those species not completely controlled will produce seed, and the population will increase in subsequent years. This risk can be managed by mixing herbicides to assure effective weed control of all species or by rotating herbicides between years to those that control a different spectrum of weeds.

One use for postemergence herbicides is the control of volunteer crops as a weed. For example, it is sometimes necessary to control corn that has germinated in a soybean field from seed not harvested the previous year. One herbicide used for this task is the grass-specific herbicide sethoxydim. If sethoxydim-tolerant corn was planted in the previous year, this herbicide option for control of the volunteer corn is not available. If glyphosate-tolerant soybeans are being grown, glyphosate can be used to control the volunteer corn, unless the corn grown the previous year was also glyphosate tolerant.

There has been general concern related to gene flow of the HT trait from the crop into weeds. This is a potential problem if the weed and crop are interfertile. The selection pressure of the herbicide increases the probability of survival of interspecific crosses. Even if interfertile, the crop and weed must be in proximity, they must flower at the same time, and the crop must produce adequate pollen to be competitive with the weed pollen. In North America, there is potential for cross-pollination between canola and mustard weeds, sorghum and johnsongrass, sugarbeet and wild beet, and rice and red rice. Outcrossing is not a significant problem with the major agronomic crops such as corn, soybeans, wheat, and cotton. The potential for outcrossing can be minimized by targeting the HT trait to the chloroplast, where it will become a maternally inherited trait and not subject to significant movement by pollen. Other precautions are focusing HT on self-pollinating crops, growing HT crops at locations where wild relatives that are interfertile with the crop do not exist, and not using HT indiscriminately as selectable markers if the HT trait is not going to be used in a weed control program.

Introgression of HT transgenes into weedy relatives is unlikely to have any adverse ecological impact outside of areas where the herbicide is used. However, transgenes for insect or pathogen resistance that could improve fitness in a natural ecosystem might be of more concern when incorporated into HT crops, because use of the herbicide could improve the chance of survival of interspecific hybrids and backcrosses of these hybrids with wild relatives.

Research is occurring on using chemically inducible promoters for expression of the HT trait. Thus, if there is outcrossing, the weed can still be controlled by the herbicide if the promoter-inducing chemical is not included with the applied herbicide. Another approach is the engineering of additional genes into the HT crop, which will prevent normal growth and development of second-generation seed (e.g., see U.S. Patent 5,723,765 "Control of Plant Gene Expression," which many have coined as "Terminator Technology"). If outcrossing to weeds does occur, the seed produced will not be viable. This technology is controversial because it prevents the farmer from using saved crop seed for planting the next season. However, saving seed from hybrid crops (e.g., corn) is currently not possible due to loss of hybrid vigor in subsequent generations, and the use of HT crops for some herbicides (e.g., glyphosate) currently requires a contract signed by the farmer stating that seed will not be saved for planting the following year. It is clear that companies utilizing biotechnology traits in their commercial products will demand tighter control over these products to recoup their investment.

One of the most significant problems for weed control in the USA is the lack of availability of herbicides for minor crops. Many have stated that HT crops will solve this problem, but depending on the Reference Dose assigned to the herbicide by the EPA, the number of registrations possible for that herbicide may be limited, particularly if the residue assigned to a specific crop is high. The residues in all crops on which a particular herbicide is registered must not exceed the total residue assigned by the EPA (see reference 9 for details). Many minor crops are assigned rather high residue levels because they are consumed early in their life cycle, which limits dilution of the herbicide or its metabolites as the plant grows and the majority of the crop biomass is harvested for consumption (e.g., lettuce). The decision for the manufacturer is clear if the Reference Dose will only allow a registration for use on lettuce (about 0.1 million ha) or soybeans (more than 28 million ha). Implementation of the Food Quality Protection Act (FQPA) may increase the pressure on minor crop registrations because the accumulation of all possible exposures and all possible risks must be considered in granting additional registrations (filling the "risk cup").

Early in the development of HT crops, there was concern about injury to adjacent crops from spray drift during postemergence application to crops having the HT trait. However, adjacent crop injury due to drift is no more or less of an issue than for conventional herbicides applied to crops where the adjacent crop is sensitive (corn growing adjacent to soybeans). There was also concern about application of the herbicide to a field that was not planted with the

Table 2. Location, as a percent of total, where HT (and insect-tolerant [Bt gene]) crops were planted in 1998 (33 million ha planted worldwide). Data from Wood Mackenzie (11). Used by permission.

USA	82.4%
Canada	9.4%
Latin America	8.0%
Other	0.2%

appropriate HT crop. Drift and applications to the wrong field have been reported, but the frequency of these mishaps has been low. Another issue is who will accept liability if crop injury occurs, the herbicide manufacturer or the seed company. For the most part, the tolerance trait introduced by seed companies has come from the company manufacturing the herbicide; thus, the majority of the complaints related to injury have been directed at the herbicide manufacturer.

Grower Acceptance of HT Crop Technology

In 1998, there were approximately 33 million ha of HT crops grown in the world, with the USA representing over 82% of the total hectares planted (Table 2). In most instances, the reduced acceptance of HT crops outside the USA is due to slow regulatory and consumer acceptance of the use of biotechnology. Restrictions on export of harvested crops containing the HT trait to some countries constitute a substantial problem resulting from this slow acceptance outside the USA. It is estimated (11) that by 2003, there will be about 51 million ha of HT crops planted in the USA, with glyphosate and glufosinate tolerance (Table 3) and corn and soybeans (Table 4) dominating the market.

HT crops are becoming a routine part of crop production in the United States. The impact on two of the most important crops in the USA, corn and soybeans, will be substantial, with HT crops representing 50% to 75% of the market share. Acceptance of HT

Table 3. Millions of HT crop hectares planted in the USA in 1998 and 2003 (estimated) by herbicide type. Data from Wood Mackenzie (11). Used by permission.

	1998	2003
glyphosate	11.1	32.0
glufosinate	2.7	15.4
imidazolinones	2.4	0.8
sulfonylureas	2.4	1.6
sethoxydim	0.8	0.4
bromoxynil	0.4	0.6

Table 4. Millions of HT crop hectares planted in the USA in 1998 and 2003 (estimated) by crop and trait. Data from Wood Mackenzie (11). Used by permission.

Crop	HT Product	1998	2003
Soybeans:	Roundup Ready®	9.3	17.8
	Liberty Link®	none	4.0
	STS®	2.4	1.6
Corn:	Roundup Ready®	0.4	12.1
	Liberty Link®	2.7	11.3
	Poast Protected®	0.8	0.4
	IMI (IR/IT)®	2.4	0.8
Cotton:	Roundup Ready®	1.4	2.0
	BXN®	0.4	0.6

Roundup Ready = glyphosate tolerant; Liberty Link = glufosinate tolerant; STS = sulfonylurea tolerant; Poast Protected = sethoxydim tolerant; IMI (IR/IT) = imidazolinone tolerant; BXN = bromoxynil tolerant.

technology in other parts of the world will continue to remain behind the USA for the foreseeable future. In addition, use of HT in crops other than corn and soybeans will lag behind for technology, regulatory, and market reasons. HT crops will be an important component in weed control; however, it is important that the technology be used wisely. HT technology should constitute one component of an overall weed management program to avoid reducing the effectiveness of the technology through selection of resistant weeds or by enhancing shifts in weed populations toward those that are more difficult to control.

Acknowledgment

The market data were obtained from Wood Mackenzie (11) and published with permission. Wood Mackenzie is located at Kintore House, 74-77 Queen Street, Edinburgh EH2 4NS, United Kingdom. (Telephone: 44 131 225 8525; Fax: 44 131 243 4435)

Literature Cited

1. Bradshaw, L. D., Padgette, S. R., Kimball, S. L., and Wells, B. H. 1997. Perspectives on glyphosate resistance. Weed Technol. 11:189-198.
2. De Block, M., Botterman, J., Vandewiele, M., Dockx, J., Thoen, C., Gosselé, V., Movva, N., Thompson, C., van Montagu, M., and Leeman, J. 1987. Engineering herbicide resistance in plants by expression of a detoxifying enzyme. EMBO J. 6:2513-2518.
3. Dekker, J., and Duke, S. O. 1995. Herbicide-resistant field crops. Adv. Agron. 54: 69-116.
4. Duke, S. O., ed. 1996. Herbicide-Resistant Crops. Agricultural, Environmental, Economic, Regulatory, and Technical Aspects. Lewis Publishers, Boca Raton, FL.
5. Duke, S. O. 1998. Herbicide-resistant crops – their influence on weed science. J. Weed Sci. Technol. (Zasso-Kenkyu, Japan) 43:94-100.
6. Duvick, D. N. 1996. Seed company perspectives. Pages 253-262 in: Herbicide-Resistant Crops. Agricultural, Environmental,

Economic, Regulatory, and Technical Aspects. S. O. Duke, ed. Lewis Publishers, Boca Raton, FL.

7. Dyer, W. E., Hess, F. D., Holt, J. S., and Duke, S. O. 1993. Potential benefits and risks of herbicide-resistant crops produced by biotechnology. Hort. Rev. 15:367-408.

8. Gressel, J. 1993. Advances in achieving the needs for biotechnologically derived herbicide resistant crops. Plant Breeding Rev. 11:155-198.

9. Hess, F. D. 1996. Herbicide-resistant crops: Perspectives from a herbicide manufacturer. Pages 263-270 in: Herbicide-Resistant Crops. Agricultural, Environmental, Economic, Regulatory, and Technical Aspects. S. O. Duke, ed. Lewis Publishers, Boca Raton, FL.

10. Liu, C.-A., Zhong, H., Vargas, J., Penner, D., and Sticklen, M. 1998. Prevention of fungal diseases in transgenic, bialaphos- and glufosinate-resistant creeping bentgrass (*Agrostis palustris*). Weed Sci. 46:139-156.

11. McDougall, J., Phillips, M., Galloway, F., and Mathisen, F. 1998. Biotechnology in Crop Protection and Production. Agrochemical Monitor No. 153. Wood Mackenzie, Edinburgh, UK.

12. McLean, G. D., and Evans, G., eds. 1995. Herbicide-Resistant Crops and Pastures in Australian Farming Systems. Bureau of Resource Sciences, Canberra, Australia.

13. Stalker, D. M., McBride, K. E., and Malyj, L. D. 1988. Herbicide resistance in transgenic plants expressing bacterial detoxification genes. Science 242:419-423.

14. Stetter, J. 1993. Trends in the future development of pest and weed control – an industrial point of view. Regul. Toxicol. Pharmacol. 17:346-370.

15. Thayer, A. M. 1999. Transforming agriculture. Chem. Eng. News 77 (16): 21-35.

Implications of Pest-Resistant/Herbicide-Tolerant Plants for IPM

David C. Bridges
Crop & Soil Sciences Department
Univ. of Georgia, Griffin, GA 30223 USA

Advances in agricultural mechanization that were associated with the industrial revolution at the turn of the century were the last real semblance of revolution in agriculture. While several important milestones are recognizable, such as cultivar improvement, selective pesticides, improved fertilizer amendments, and conservation tillage, these advancements have largely been evolutionary, not revolutionary. To the contrary, biotechnology has resulted in unprecedented changes in agriculture in recent years, especially in the area of pest management.

During the past 10 years, there has been a steady growth in the number of biotechnology products emerging from both the private and public sectors. The U.S. Department of Agriculture is charged with permitting and issuing notice for regulated technologies. USDA indicates that approximately 6,000 permits/notices have been processed to date (Fig. 1). Non-regulated technologies are not included but continue to provide many new agricultural products and technologies, too. So, these USDA numbers are conservative.

Of the nearly 6,000 permits/notices processed by USDA, approximately 70% of all the permits to date deal with pest management. Herbicide-tolerant crops and insect resistance traits lead the pack, comprising nearly 50% of the regulated technologies to date. Clearly, biotechnology will provide important new opportunities for agriculture in areas other than pest management, but the focal topic of this chapter is pest-resistant and herbicide-tolerant crops.

It is anticipated that we are at the brink of the next revolutionary period in agriculture. The question is, Will the benefits of these

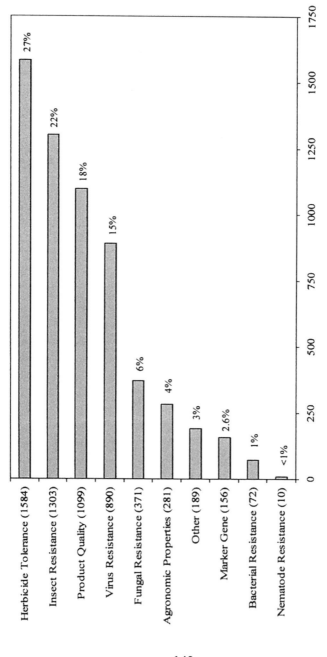

Figure. 1. Total number of USDA permits and notices for regulated organisms, adapted from USDA.

technological "advancements" be realized to their fullest potential? Will these technologies and the resultant products be integrated into sustainable pest management practices, or will they be spoiled? Is integrated pest management a part of future pest management, and, if so, will the tools of biotechnology be compatible with IPM? Will pest management be more sustainable, economical, and environmentally friendly with these tools? Will the onslaught of biotechnology-derived pest management tools have an adverse effect on the research and development efforts associated with traditional agrichemical, seed, and pest management enterprises?

While some would suggest that these questions are rhetorical, answers to these and many other questions are important to the development of sound pest management strategies for the future. Pondering these questions, let alone answering them, is a daunting task.

Depending on the pest management discipline, the issues may be framed somewhat differently. It is important to note that all of the pest resistance and herbicide tolerance technologies that are being commercialized are not the product of biotechnology. Regardless, there are a number of crosscutting issues that need to be addressed. Throughout the remainder of this article, I will raise several serious questions that may be difficult, if not impossible, to answer. Nonetheless, they need to be considered as we enter this era of increased agricultural sophistication.

New Technologies and IPM

WHAT IS IPM?

To determine the potential impact that pest-resistant and herbicide-tolerant crops may have on IPM, we must reach some agreement on the question, "What is IPM?" Furthermore, we must establish a framework within which to conduct an impact assessment.

What is IPM? Simplistically, IPM is an approach to pest management that considers economics and sustainability using all available tactics and utilizing information about the pest(s) and the crop and/or site.

What are the goals of IPM? It has been suggested that IPM should be efficacious, economically viable, ecologically sustainable, and societally acceptable. In recent years, the mainstay approaches used in IPM have been:

- Monitoring - The identification of the pest(s) and assessment of pest populations and their spatial characteristics at the site are essential to IPM.
- Formulation of strategy - Most IPM approaches focus on the maintenance or suppression of pest populations to predetermined levels that will result in a predictable and acceptable outcome.
- Impact determination - The development of strategy and assessment of viability require the determination of biological, ecological, and economic impact. These effects are often codified in thresholds.
- Deployment of multiple tactics - IPM programs employ a diverse set of tactics, which is usually considered to be preferable to the use of a single tactic.
- Follow-up monitoring - Once an IPM strategy has been developed and deployed, follow-up monitoring is usually required to determine success and to facilitate the development of a follow-up strategy.

ASSESSMENT FRAMEWORK

Setting the previously stated boundaries regarding what IPM includes and having identified some of the newly emerging technologies, a framework must be established within which one can evaluate how these technologies will possibly affect IPM.

One may gain some perspective of the future by assessing, at least qualitatively, the impact that these technologies may have on four important aspects of pest management:

- The achievement of IPM goals
- The utilization of practices consistent with IPM principles
- Grower perception of the importance of IPM
- The richness and diversity of cropping and pest management practices that are available to agriculture

144

Assuming that among the goals of IPM are efficacy, economic viability, ecological sustainability, and societal acceptability, then a complete impact assessment should include a determination of the impact of the technologies on achievement of these goals.

Efficacy. It is reasonably safe to assume that these technologies will be efficacious, at least in the short term, or otherwise they probably will not be introduced. For farmers to adopt these technologies, they must provide efficacy that is equal to or greater than that of traditional pest control tools. Some concerns have been expressed regarding the longevity of some of the technologies relative to the development of pest resistance. If they fail to be efficacious over time, they will be eliminated from the market.

Economic viability. As with efficacy, these pest management practices must provide a positive return on investment or they will not be sustainable in the market. Furthermore, they must be compatible with other pest management practices. They could be efficacious on some pests, but could enhance populations of and damage caused by secondary pests, resulting in an overall increase in pest management costs.

Ecological sustainability. Are these pest management practices ecologically sustainable? Are they based on sound ecological principles? Will they work over the long term? Among the principle concerns are pest resistance, pest population shifts, changes in pest behavior, pest acclimation, and adaptation. For example, does incorporation of transgenic toxin genes from *Bacillus thuringiensis* (Bt) into a variety of food and fiber crops pose a serious threat of development of Bt-resistant insects due to persistent selection pressure associated with temporally and spatially diverse use? Will the use of glyphosate on an array of herbicide-tolerant crops, several of which may be rotational crops to one another, result in significant narrowing of weed control tactics to the extent that either resistant populations are selected or that relatively insensitive species will predominate? The jury is out! In fact, many believe the jury (scientific community) has not heard enough evidence to render a verdict.

Societal acceptability. Questions regarding societal acceptability are common. Fortunately, scientists are not often charged with assessing the social science aspects of pest management. However, fair questions remain unanswered. The road of societal acceptability is well traveled by science. Medicine and animal agriculture are also facing questions regarding standards of societal acceptability. The public will sort it out.

How will these technologies affect the utilization of practices that are consistent with IPM principles? Let us look at a few examples.

Monitoring. Monitoring is an integral part of most IPM programs. How will new pest management technologies affect the willingness of farmers to monitor? The introduction of Bt cotton (*Gossypium hirsutum*) had a negative impact on monitoring, at least initially. The expectation of performance was very high. Many growers felt that they could reduce their scouting efforts. In fact, some farmers thought that reduced scouting costs might actually help offset technology costs. This was based on an unrealistic presumption of performance. While performance has generally been good, the reality has begun to set in, and farmers realize that Bt cotton is an important new technology that can help them manage insects, but that it is not a "silver bullet" solution to all problems. Scouting and monitoring remain important aspects of southeastern cotton production after the introduction of Bt cotton, primarily because the Bt toxin is not equally effective on all cotton insect pests.

With herbicide-tolerant crops, the technologies have had a variety of effects on perceptions about the need for monitoring. Changes were almost unrecognizable with the introduction of bromoxynil-tolerant cotton (BXN®). Farmers saw BXN cotton as another important weed control tool. The performance expectation was realistic, and farmers understood that it was a selective herbicide that would control a limited number of dicot weeds. The use of bromoxynil in BXN cotton was not expected to replace the use of all other herbicides in a traditional weed control program. Farmers anticipated the need for continued monitoring and scouting to determine if species present could be controlled with bromoxynil and to quantify their densities.

To the contrary, farmer performance expectations with glyphosate-tolerant (Roundup Ready®) cotton were so high that many farmers believed that there was no need to monitor. Many farmers unrealistically expected glyphosate to control all weeds. In fact, the popular marketing phrase "Freedom from Weeds" is not perceived as a dream by many, but rather as an achievable reality. Farmers quickly learned that glyphosate is a very "broad-spectrum" herbicide, but that it is not "non-selective." Depending on application rate and time of application, some weed species are controlled better than others. For example, morning glory species (*Ipomoea spp.*), hemp sesbania (*Sesbania exaltata*), Florida pusley (*Richardia scabra*), and others can be difficult to control. Therefore, correctly identifying weed species is critical if farmers are to rely solely on glyphosate for weed control.

The importance of monitoring to IPM cannot be overemphasized. Ensuring that the appropriate control measure is applied, and only when pest populations dictate such, necessitates the continued use of monitoring in IPM. For example, within one year of the introduction of Bt cotton in the southeastern United States, entomologists and crop consultants were faced with the question of how to scout for stinkbugs in cotton. Prior to the introduction of Bt cotton, mid- to late-season insecticide applications for worms provided stinkbug suppression or control. With Bt cotton, these insecticide applications were often not needed, and damaging populations of stink bug developed.

Deployment of multiple tactics. Multitactic control strategies are at the heart of IPM (5). Research has repeatedly demonstrated that a diverse approach to pest control is almost always more effective, particularly in the long term, than are single tactic approaches, regardless of whether one is talking about weeds, insects, or disease-causing organisms. Can pest-resistant/herbicide-tolerant crops fit into integrated pest control systems, or will they hamper integrated control efforts? The answer probably depends largely on how reliant farmers become on a specific technology.

The development of disease resistance mechanisms, particularly through biotechnology, has lagged behind the development of commercial offerings for weed and insect control. However, host plant resistance and cultural practices have been an integral part of disease management for so long that the changes that have occurred to date have been perceived as more evolutionary than revolutionary. Disease resistance is already an integral part of many integrated disease management programs. Therefore, novel discoveries in disease resistance have to date simply provided additional tools via host plant resistance.

Many insect control programs involve multiple tactics, including host plant resistance, enhancing or preserving natural enemies or beneficial insects, biological control, and insecticides. How will the introduction of insect-resistant crops affect the use of multiple-tactic approaches for insect control? A good case study might be the introduction of Bt cotton. With cotton, the impact on multiple-tactic use has been strongly expectation driven. Where insect resistance to pyrethroid insecticides has been a problem in previous years, the introduction of Bt cotton has been perceived as another tool. In these areas, it has quickly become a part of an integrated insect control system. However, where insect resistance to pyrethroid insecticides is not an issue, Bt cotton is perceived as a replacement "silver bullet" cure, with little or no need for integration. It simply means that the farmer has to spray less.

Weed populations are very diverse. Unlike economically damaging insects and diseases, numerous species of weeds typically occur is almost every field and in a vast array of combinations. So, single-tactic approaches seldom provide adequate control. With exception of a few very broad-spectrum herbicides, one would not expect to achieve acceptable weed control using a single herbicide on a herbicide-tolerant crop. The benefit associated with the use of diverse tactics is discussed later.

FARMER PERCEPTION

Insect resistance. Farmer attitudes about pest-resistant and herbicide-tolerant crops vary widely. With the Bt cotton grower, perception depends largely on what insect control practices were in place when Bt cotton was introduced. For some farmers, Bt cotton significantly enhances their existing insect IPM program by allowing the development of beneficial insect populations, because spraying with synthetic insecticides is reduced. These farmers may perceive Bt cotton as an important component of their IPM program. Other farmers (those who have not been focused on integrated insect management) may consider Bt cotton as a simple replacement for existing insecticides.

Disease resistance. As mentioned earlier, host plant resistance has been an integral component of disease control for many years. The availability of a disease-resistant plant often dramatically changes the overall approach to disease control. Therefore, farmers often perceive introductions of new disease resistance mechanisms as a "fix" that largely eliminates the need for other control practices.

148

As a result, farmers often do not perceive the need for an integrated approach.

Herbicide-tolerant crops. Weed control in most row crops involves the use of cultural, chemical, and mechanical tactics. Cultural practices include crop rotation, planting weed-free seed, and using stale seedbed techniques. Mechanical practices include preplant tillage and seedbed preparation, postemergence crop cultivation, mowing, and sometimes hand weeding or hoeing. A vast array of chemical weed control practices is used in row crops. For the most part, herbicide-tolerant crops simply allow farmers to use a herbicide that they would otherwise not be able to use. For example, bromoxynil has been used for weed control in cereal grains and corn (*Zea mays*) in the United States for years, but only with the development of BXN cotton could farmers use bromoxynil on cotton. Likewise, triazine-resistant canola (*Brassica napus*) varieties were developed in the 1980s to allow for the use of triazine herbicides on these otherwise sensitive plants (1,2). These examples did not result in revolutionary shifts in the way that farmers thought about weed control. However, with the introduction of glyphosate-tolerant crops (Roundup Ready®), farmer attitudes did begin to change. Prior to the introduction of these glyphosate-tolerant crops, glyphosate was used for "non-selective" weed control. Therefore, experience indicated that with glyphosate-tolerant crops, one could possibly control all vegetation except the tolerant crop. Obviously, this is not true, but weed control with glyphosate is often good enough to substantially change farmer perceptions about the need for integrated control and to substantially change his entire approach to weed management. This will be discussed in greater detail in the following section.

RICHNESS AND DIVERSITY OF PEST CONTROL TACTICS

How will these tools affect the diversity of pest control tactics that are used? Deploying a diverse set of pest control tactics is at the heart of IPM, and doing so provides several benefits. First, tactical diversity provides for stability in pest populations by damping oscillations in pest populations. Tactical diversity also reduces the chance for catastrophic failure. It contributes to diversity within pest populations, and it ensures long-term availability of tactics.

The importance of diverse tactics is evident in the way we manage pests. Chemical tactics are effective and more or less economical, and they provide a range of tools. It is for these reasons that they are so widely used. Mechanical tactics are effective with some pests. With weeds, tillage is an integral part of many production systems. Cultural tactics are important in all of pest control. They alter the pest/environment relationship. They are often preventative rather than curative. Crop rotation, along with the rotation of pest management tactics that go along with it, provides for an important resolution phase with respect to pest populations. Biological control may employ host plant resistance, which makes use of important genetic diversity within the crop plant. Or biological control may involve conservation of beneficials. In either case, the pest/environment relationship is altered.

Pest resistance. Diversity of tactics is critically important in the management of pest resistance, interspecific shifts, and intraspecific shifts. In the war against pests, preservation of pest control tactics requires that a single soldier not become the army. We cannot afford to become over-reliant on a single tactic. It won't work. This has been demonstrated time and time again (4). It has often been said that every good pest control tactic ultimately selects for its own failure. We should now know that this is true.

Interspecific shifts. Again, relying on the war analogy, there is rarely one soldier in the enemy's foxhole. If one pest is removed, most assuredly, another will fill its niche. We know this is true. Previously mentioned was the fact that stinkbugs are now often a problem in Bt cotton during late season because of the reduced number of insecticide sprays being used in this cotton. Several weed scientists have already demonstrated that significant changes have already occurred with only two years of total reliance on glyphosate (H.D. Coble, personal communication).

Intraspecific shifts. Genetic diversity within a pest species (population) is almost always greater than that in the crop. Relying broadly on a single tactic will lead to selection for less sensitive populations of the species and against more sensitive populations. Again, this is fact. Gould (3) eloquently illustrated this in a review article several years ago.

Selection pressure. What leads to pest resistance, interspecific shifts, and intraspecific shifts? Selection pressure is the answer! Will diversity of pest control tactics decrease with these technologies? If it does, will selection pressure increase, leading to resistance, interspecific shifts, and intraspecific shifts? Let's look at two examples. With the introduction of Bt crops, a single tactic will be used over a large area for the control of a common group of pests. Selection force is great, and duration is long. A single site/mode of action will be used across several commodities, i.e., corn and cotton. The trait is intrinsically linked to narrowed genetics in the crop, and a novel insect control tactic has not been developed. Rather, an existing use has been vastly expanded.

The same is true for glyphosate-tolerant crops. A single tactic (herbicide) could be used on 10 or more crops within the next 5 to 10 years if the current pace is continued. The tactic is intrinsically linked to a narrowed crop genetic base. And, again, a novel tactic has not been developed. The use of an existing one has simply been expanded.

Are these technologies sustainable? Clearly, selection force and duration is tremendous. What do you think? The question should not be one of whether problems will occur, but rather, when will they occur? How did we get here? During the past 20 years, weed scientists have been criticized for not fully embracing IPM. It has been suggested that the continued prophylactic use of preemergence herbicides is not responsible, that selection pressure is too great. So, how is the current situation different? What is likely to happen when more than 90% of the hectares of cotton are planted to Bt cotton? Effectively, the cotton was "sprayed" the day it was planted, and it has been "sprayed" every day since, at least metaphorically. The selection pressure is intense. It is forceful and persistent. Can resistance be managed? Many people think that pest resistance can be prevented or mitigated but rarely managed (4).

The current "management" plan for Bt involves the concept of refugia to provide safe haven for susceptible insects, which will result in mating of putative resistant insects with susceptible insects and thus provide resolution within the population. Is this sound? Maybe theoretically, but in reality, it is very difficult to apply. For refugia to significantly delay the development of resistance, the coincidental occurrence of several factors is required. I make it a habit to never bet against nature!

The same problems are lurking in the background for herbicide-tolerant crops, if we become too reliant on a single herbicide. It is already common to use glyphosate several times in a crop within a single growing season. Worse yet, it is being used sequentially on rotational crops. Is it really possible to use glyphosate for weed control in corn, cotton, and soybeans year after year in the same field? Again, the selection pressure for resistance and weed species shifts is tremendous. A diverse set of weed control tactics could prevent what will be inevitable if we become too reliant on a single tactic. With respect to management options, diversity of tactics is a good investment.

Conclusion

The time for reflection has come. Diversity is at the heart of IPM. The "I" in IPM stands for "integration," implying more than one approach. Are we practicing what we preach to our students and the public? Aren't insect-resistant and herbicide-tolerant plants just tools? Have we become silver bullet chasers? There are no silver bullets! History has proven that both pests and Superman are faster than a speeding bullet.

Are expectations realistic? Are we compromising crop genetic diversity and crop performance in an attempt to grasp the ever-elusive silver bullet? All Roundup Ready® Soybeans available today were initially derived from a single line of soybean tissue. All Roundup Ready cotton was derived from just two lines. This represents a significant narrowing of crop genetic diversity, the potential implications of which are not good. In 1972, the U.S. white corn crop was devastated by corn blight. A contributing factor was that a single male-sterile line was used to produce all commercially available hybrids. Could it happen again?

Ask yourself these questions: Do these technologies improve efficacy, increase profitability, and enhance environmental safety? Are they ecologically sustainable? Do they improve the societal acceptability of pest management? Do they fit well within an integrated pest management program? Will they result in fewer options rather than greater options?

Pest-resistant and herbicide-tolerant crops can result in more efficient, effective, and safer pest control, if their use is balanced with other pest control tactics. However, if they are perceived and used as single-tactic, magical solutions, their future is more questionable, and IPM will not have been served well by the introduction.

Literature Cited

1. Beversdorf, W. D., and Hume, D. J. 1984. OAC Triton spring rapeseed. Can. J. Plant Sci. 64:1007-1009.
2. Beversdorf, W. D., Hume, D. J., and Donnelly-Vanderloo, J. J. 1988. Agronomic performance of triazine-resistant and susceptible reciprocal hybrids of canola. Crop Sci. 28:932-934.
3. Gould, F. 1991. The evolutionary potential of crop pests. American Scientist 79:496-507.
4. Hoy, M. 1998. Myths, models, and mitigation of resistance to pesticides. Phil. Trans. R. Soc. Lond. 353:1787-1795.
5. Stern, V. M., Smith, R. F., van den Bosch, R., and Hagen, K. S. 1959. The integration of chemical and biological control of the spotted alfalfa aphid. The integrated control concept. Hilgardia 29:81-101.

Regulation of Plant-Pesticides:
Current Status

Janet L. Andersen[1] and Elizabeth Milewski[2]
[1]Office of Pesticide Programs and [2]Office of Prevention,
Pesticides, and Toxic Substances, United States Environmental
Protection Agency, Washington, DC 20460 USA

Confusion and misinformation about the U.S. Environmental
Protection Agency's regulation of plant-pesticides have led to
misunderstanding and mistrust. We welcome the opportunity to
participate in this conference and explain what the EPA actually has
proposed, what we are doing now in regulating plant-pesticides, and what
to look ahead to in the future. First, it needs to be made clear that the
U.S. government, including EPA, believes that biotechnology can provide
many benefits, such as less reliance on toxic chemical pesticides, lower
worker exposure, and reduced ecological impacts. However, products of
biotechnology are not without potential risks. It is EPA's intention to
manage and mitigate the risks from all pesticides, including those
pesticides produced through biotechnology.

EPA's Regulation of Plant-Pesticides

In 1986, the federal government produced a document called the
"Coordinated Framework for Biotechnology," in which it was made clear
that pesticidal substances produced through biotechnology would be
regulated by EPA under the pesticide laws. Under these laws, EPA
regulates chemical, biochemical, microbial, and plant-pesticides, no
matter how they are made or their mode of action.

The principal laws for regulating pesticides are the Federal
Insecticide, Fungicide, and Rodenticide Act (FIFRA) and the Federal
Food, Drug, and Cosmetic Act (FFDCA). Both of these laws were

amended by the Food Quality Protection Act of 1996. FIFRA defines a pesticide as:

(1) any substance or mixture of substances intended for preventing, destroying, repelling or mitigating any pest, (2) any substance or mixture of substances intended for use as a plant regulator, defoliant, or desiccant, and (3) any nitrogen stabilizer.

FIFRA also defines a pest as:

(1) any insect, rodent, nematode, fungus, weed, or (2) any other form of terrestrial or aquatic plant or animal life or virus, bacteria, or other microorganisms on or in living man or other living animals which the Administrator declares to be a pest.

FIFRA is also an "intent" law, which means that if someone is using a substance with the intention of controlling a pest, that substance is considered a pesticide, whether it is a substance produced through a chemical manufacturing process, a fermentor, or a living plant. Also, FIFRA clearly states that it is either the sale or the distribution of the pesticide that requires the substance to be regulated.

With this background, let's turn to what EPA is actually proposing to do and not do under the proposed Plant-Pesticide Rule. A word of explanation is needed first. This presentation can only be about the **proposed** plant-pesticide rule, because the final rule has not yet been published. EPA realizes that this procedure may make it appear that EPA does not plan to take into account any of the excellent ideas and comments given to the Agency during the comment period. However, in addition to the comments submitted to EPA, the Agency also held three open, public meetings with independent scientific advisory committees, including the Scientific Advisory Panel, as required by FIFRA. In the final rule, EPA will explain how it considered and used every comment made to the Agency during the comment period. Because the comments received during the comment period were exhaustive, EPA believes that our response to those comments also will address all the concerns raised by late-submitted comments. We will now turn to clarifying EPA's proposal for plant-pesticides.

First, EPA is NOT proposing to regulate whole plants. EPA is regulating the pesticidal substance produced and used by a living plant and the genetic material necessary for the production of such a pesticidal substance. Herbicide resistance is not considered a pesticidal trait. In fact, most of the proposed rule is about exemptions from regulation.

There have been several reports calling on EPA to exempt or exclude a number of things from regulation. EPA agrees that many categories of plant-pesticides could be exempted. Indeed, what EPA proposes to exempt and what others have suggested should be exempt are remarkably similar. For example, the Institute of Food Technologists listed five categories for exclusion:

1. Naturally occurring and heritable traits derived from plants of the same or sexually compatible species. EPA advanced a similar exemption in the proposed plant-pesticide rule.

2. Traits that are new to the plant species and its sexually compatible relatives that result in changes in physical structure or form (i.e., leaves with hairs to prevent or discourage insect attack). EPA agrees; this exemption is in the proposed plant-pesticide rule.

3. Viral coat proteins. EPA agrees; this exemption is in the proposed plant-pesticide rule.

4. Traits involved in defense mechanisms expressed as a cascade of biochemical and genetic events triggered by incompatibility between the pest and the plant (i.e., hypersensitive reaction). EPA agrees that the hypersensitive reaction should not be regulated and has included it as part of the proposed exemption.

5. Traits responsible for pest defense effects that are widely known and common within the plant, animal, and microbial kingdoms, and are not characteristic of pesticides, such as enzymes. There is validity in this category, although some compounds that might be included in this category could pose risks to humans or the environment. A number of enzymes are already regulated as biochemical pesticides. In addition, what actually is included in this category is not well described. However, EPA staff will be glad to provide a determination of whether a particular situation comes under the proposed regulation. EPA is willing to consider additional exemptions after the proposed rule is final. We anticipate that a number of exemptions will be granted in the future. EPA plans an extensive outreach program once the rule is final to provide better guidance on what types of products are exempt and what must come to the Agency for approval before commercialization.

The Institute of Food Technologists' report supports EPA regulation of "pest-defense substances that act as pesticides when extracted from their hosts and tested *in vitro* and in the environment, such as nicotine, scorpion toxin, spider venom, and crystalline Bt endotoxin." EPA currently regulates these types of substances when extracted from their hosts, and believes that EPA should consider the health and safety of new exposures to such substances, such as when plants that would not ordinarily express these substances are engineered to do so. The regulation of nicotine serves as a good example of how EPA would proceed.

Tobacco plants naturally produce nicotine, an insecticidal substance. If someone intends for the nicotine in the tobacco plant to act as an insecticide, then, under FIFRA, the nicotine is a pesticide. Similarly, if someone alters the tobacco so that it produces higher amounts of nicotine, either through conventional breeding or through genetic engineering, and the nicotine in the plant(s) is intended to provide protection against an insect, the nicotine is a pesticide. However, if EPA's proposal is enacted, EPA would not regulate the nicotine produced in the tobacco. EPA is proposing to exempt such cases from regulation, because humans and wildlife are already exposed to nicotine produced by tobacco, whether or not someone intends the nicotine to be acting as a pesticide. Similarly, if tobacco plants are altered to have the ability to produce additional amounts of nicotine, and the plants are then sold to kill insects, the nicotine would be classified as a plant-pesticide. However, EPA is proposing to exempt changes of this type from regulation under the proposed rule. If the ability to produce additional nicotine were transferred into ornamental tobacco, EPA would not regulate this, because ornamental tobacco plants are sexually compatible with the agronomic varieties. However, when nicotine is extracted from tobacco and used as an insecticide, EPA has regulated nicotine as a pesticide under FIFRA, and also under FFDCA if the nicotine is used on food crops such as apples. If the ability to make nicotine were transferred from tobacco to apples for use as a pesticide, EPA would regulate nicotine as a plant-pesticide because of the novel exposure, especially to children.

There have been two other common misconceptions reported about EPA's proposed regulation that should be clarified. First, it has been stated that EPA would require every plant variety that contains a plant-pesticide to be registered. EPA does not register varieties for each crop for conventional chemical pesticides, microbial pesticides, or biochemical pesticides, and the Agency does not require plant-pesticides to be registered by variety. For example, EPA has registered Bt potatoes, and the company, NatureMark, has put the approved plant-pesticide into three varieties without coming back to EPA for approval. There are probably hundreds of corn varieties that now produce one of the seven different Bt plant-pesticide substances approved for use in field corn.

Second, some scientists have expressed concern that registration of plant-pesticides will restrict their research. EPA's approach does not affect plant-pesticides until they are used on 4.05 ha or more. This is the point at which an experimental use permit is needed. The USDA Animal Plant Health Inspection Service's (APHIS) regulation under the Plant Pest Act covers small-scale research for plant-pesticides. It has been EPA's experience with many biopesticides that, when testing reaches the 4 ha level, researchers typically have either turned the product totally over to a pesticide company or are working in cooperation with a pesticide company that has experience registering pesticides and intends to market the product through its distribution networks. There are a handful of registrations held by universities, as well as a few more held by the U.S. Forest Service or USDA. EPA has special ways to assist other Federal agencies with registrations. The Biopesticides and Pollution Prevention Division (BPPD) at EPA also has extensive experience working with small pesticide companies and the few universities that register products, because most, if not all, of these are biological pesticides. We recognize that for plant-pesticides, some companies that have not had experience in the past working with EPA (e.g., certain seed companies) will now be dealing with us, and that EPA needs to make special efforts to assist these companies in understanding and complying with pesticide regulations. BPPD is working with several small entities developing plant-pesticides.

Finally, there have been concerns raised regarding the term "plant-pesticides." This was one of the comments that came to the Agency during the official comment period. EPA is willing to consider other terms to refer to these products, although we have been offered few, if any, alternative names. We have considered some other terms within the Agency. However, we do not want to select another name for these products that would be unacceptable to any of the stakeholder groups. EPA intends to publish as part of its final rule package a Federal Register notice to begin the process of finding a more acceptable alternative to the term "plant-pesticides."

Implementing the Registration of Plant-Pesticides

Since the mid-1980s, EPA has been working with companies and individuals that have been developing plant-pesticide products. In developing our approach for an appropriate risk assessment for these products, EPA has held many public meetings and workshops with its Biotechnology Advisory Committee, with the FIFRA Scientific Advisory Panel, with the Office of Pesticide Programs' Pesticide Dialog Committee, and with all interested stakeholders at a number of public hearings and workshops. Through this process, EPA has developed a risk assessment approach that we have applied to the products going through the registration process.

There are established guidelines for testing chemical pesticides, microbial pesticides, and biochemical pesticides published in the Code of Federal Regulation and available from the Agency's Web site. EPA believes that the unique aspects of plant-pesticides dictate that there should be special testing guidelines for these products. For example, worker exposure is a major risk issue for many chemical pesticides, but because the plant-pesticide may be confined within the plant, the potential for worker exposure is much reduced. On the other hand, out-crossing from crop plants to other crop varieties or wild or weedy relatives by spread of pollen may present a risk from a plant-pesticide, where it never would be a concern from a conventional pesticide.

The Agency will not begin the process of establishing testing guidelines for plant-pesticides until the exemptions to the proposed rule are implemented. In establishing these testing guidelines, EPA will propose the tests it believes are appropriate, indicating the circumstances for which each study is either required, conditionally required, or not required. These proposed guidelines will be widely available for public comment and will be reviewed in a public meeting of the Scientific Advisory Panel. Once in place, changes to guidelines can also be made through a process that includes an opportunity for public participation and comment. In addition, anyone applying for registration of a product (called a "registrant" by EPA) can provide a scientific rationale and request a waiver for doing any particular study. Public literature may also fill a data requirement.

Until these guidelines are in place, the Biopesticides and Pollution Prevention Division has been handling the data requirements on a case-by-case basis. The potential registrant typically comes in for a meeting with BPPD staff, at which time we decide upon the appropriate data requirements to support either the Experimental Use Permit (EUP) or full commercial approval, called a Section 3 registration. The studies done under the EUP are used to support the later application for full registration.

For the plant-pesticide products EPA has reviewed to date, we have reviewed data in four categories: product characterization, toxicology, non-target organism effects, and exposure and environmental fate. Product characterization includes reviewing the source of the gene and the expression system, the nature of the pesticidal substance produced, modifications to the introduced trait as compared to that trait in nature, and the biology of the recipient plant. For toxicology, an acute oral toxicity test of the pesticidal substances on mice is required. At times, it has not been possible to make enough of the substance in the plant itself, so EPA has allowed the exact same protein to be produced by bacteria and used for the testing. It should be noted that, to date, all of the plant-pesticides reviewed by EPA are proteins and the genes required to make these proteins within the plant. For these proteins, EPA also required an *in vitro* digestibility test in which the amount of time it takes for the protein to degrade in gastric and intestinal fluids is determined. EPA also considers the amino acid homology to known toxins and allergens. Determination of whether a novel protein is likely to be an allergen is one of the major challenges for the federal agencies. EPA and FDA are working on this issue together.

For ecological effects, EPA examines the exposure and toxicity of the plant-pesticide to non-target organisms. These tests are unique to the crop and pests involved. For example, with Bt potatoes, a test of potential effects of the CryIIIA protein to ladybird beetles showed that there were no adverse effects to these predators of Colorado potato beetles. For Bt corn, tests were conducted on the potential effects on fish, because field corn is manufactured into commercial fish food. No effects were observed in the tests. EPA also has evaluated the degradation rates of the proteins in soil and plant residues.

If adverse effects or potential adverse effects have been observed in the testing, second or higher tiers of testing have been required to allow EPA to evaluate the risks. EPA routinely consults with USDA and FDA on reviews of these plant-pesticides. Data requirements for resistance management will not be routine for plant-pesticides. Resistance management requirements have only been placed on the Bt plant-pesticides.

Since March 1995, EPA has registered eight plant-pesticide active ingredients representing 10 pesticide products (Table 1). Nine of these products were for Bt delta endotoxins. The crops have included potatoes, cotton, field corn, sweet corn, and popcorn. Late in 1998, EPA registered

Table 1. Registered Plant-Pesticide Products

Active Ingredient	Crop(s)	Trade Name	Company	Date Registered
Bt Cry 3A	Potato	NewLeaf	Monsanto	1995
Bt Cry 1Ab	Field Corn	NatureGuard Corn Borer Control Protein	Mycogen	1995
Bt Cry 1Ab	Field Corn Popcorn	NK Brand Bt Corn with KnockOut and Hybrid Popcorn with KnockOut	Novartis	1995 1998
Bt Cry 1Ac	Cotton	Bollguard	Monsanto	1995
Bt Cry 1Ab	Field Corn	Yieldguard	Monsanto	1996
Bt Cry 1Ab	Field corn	NK Brand Bt Corn with YieldGuard	Novartis	1996
Bt Cry 1Ab	Sweet Corn	Attribute	Novartis	1998
Bt Cry 1Ac	Field Corn	DEKALBt	Dekalb	1997
Bt Cry 9c	Field Corn	Starlink	AgrEvo	1998
PLRV virus resistance gene	Potato	NewLeaf Plus	Monsanto	1998

a Monsanto replicase gene for potato leaf roll virus (PLRV). The PLRV resistance will be combined with the Bt cryIIIA delta endotoxin for control of Colorado potato beetle in the NatureMark NewLeaf Plus® potatoes available for the 1999 growing season.

In addition, tolerance exemptions have been approved for these registrations and for seven viral coat proteins. There are now tolerance

161

exemptions in place for the Bt delta endotoxins Cry1Ac and Cry1Ab for all food commodities. This means that these proteins may be used at any concentration in any food commodity. We encourage registrants to apply for tolerance exemptions under FFDCA as broadly as possible, such as for all food commodities, as a way to help expedite future registrations, especially for minor crops.

EPA sees many advantages to the use of plant-pesticides for reducing reliance on chemical pesticides and reducing dietary, worker, and environmental exposure to chemical pesticides. For example, figures from Monsanto indicate that since Bt cotton was introduced for commercial production in 1996, the amount of chemical insecticides used on cotton has been reduced by almost 4 million liters.

Yet, there are potential risks from the use of certain plant-pesticides, and EPA is required by law to regulate all pesticidal substances in order to prevent unreasonable adverse effects from occurring. We recognize that people working in this field may not be clear on whether their product falls under our regulatory umbrella. Staff in the Biopesticides and Pollution Prevention Division are willing to work with anyone interested in developing a plant-pesticide for commercial use. EPA is willing to provide written determinations on whether a product would be exempted by the proposed rule for anyone requesting them, and we are always willing to meet one-on-one with anyone who would like to discuss a particular product or group of products under development. We can help a company to determine what kinds of data might be required and where waiver requests are likely to be accepted and instances in which a particular donor of a new trait may require less testing than another donor. Please feel free to contact the Biopesticides and Pollution Prevention Division at EPA.

Genetically Engineered Pathogens of Insects for IPM: Concepts and Status

Brian A. Federici
Department of Entomology and
Interdepartmental Graduate Program in Genetics
University of California - Riverside
Riverside, CA 92521 USA

Pathogens of insects have long been considered as control agents that could be important components of IPM programs. The most intensively studied and evaluated for use in IPM are bacteria, especially *Bacillus thuringiensis* (Bt), the nuclear polyhedrosis viruses (NPVs), imperfect fungi, including *Metarhizium anisopliae* and *Beauveria bassiana*, and microsporidia (protozoa), such as *Nosema locustae*. Of these, only Bt has achieved a moderate level of commercial success.

Aside from high efficacy, the principal reason for Bt's success is its relative ease of mass production. This enables products based on Bt to be produced and sold at prices competitive with many synthetic chemical insecticides. In addition, Bt is relatively easy to formulate, distribute, and apply using methods similar to those used for chemical insecticides, and it has good shelf life. Even so, Bt products account for less than 2% of insecticide sales in the United States. With respect to other pathogens, the combined sales of viruses, fungi, and protozoans are less than 0.1% of insecticide sales. Although sales of these agents are increasing at an annual rate of around 25%, the number of hectares treated is so small that pathogens will not constitute a substantial proportion of the insect control agents used in IPM for many years to come, unless their efficacy is improved considerably. Recombinant DNA technology, i.e., genetic engineering, has provided a variety of methods for improving pathogen efficacy. Thus, it is my purpose in this presentation to review the key concepts and techniques being used to enhance pathogen efficacy, evaluate the current status of the field, and identify the major hurdles that must be overcome to have pathogens used more widely in IPM. I will begin with an overview of registered, wild-type pathogens, identify their key limitations, and show how these are being overcome with recombinant DNA technology. I close with a view of possible future developments.

Pathogens Registered for Insect Control

The wild-type, i.e., non-recombinant, pathogens registered as microbial insecticides by the U.S. Environmental Protection Agency are summarized in Table 1. From this table, it can be seen that bacterial insecticides, with approximately 2.4 million annual hectare treatments in the United States, are the most successful. In addition, the number of bacteria-based products registered attests to the diversity of uses that have been developed for bacterial insecticides.

Table 1. Pathogens registered as insecticides in the USA

Type	Number of Species	Products	Hectares Treated Annually
Bacteria	9	>50	2,428,000
Viruses	8	3	<20,000
Fungi	3	3	<20,000
Protozoa	1	1	0

Most of these formulations are products based on *Bacillus thuringiensis* subsp. *kurstaki*, which is used to control a variety of lepidopteran pests of vegetable and field crops, ornamentals, and forests (Table 2). Other Bt subspecies used for pest control include *B. t.* subsp. *aizawai*, used especially for control of armyworms; *B. t.* subsp. *israelensis*, used to control mosquito and blackfly larvae; and the tenebrionis strain of *B. t.* subsp. *morrisoni*, used to control certain beetle pests. Products based on these bacteria account for greater than

164

95% of Bt insecticide sales. Other bacteria used in pest control are *B. popilliae,* used for control of scarab larvae, and *B. sphaericus,* used in mosquito control. As noted previously, these products have been successful because, in general, they are easy to produce and use and have good efficacy. An advantage of these products in comparison to synthetic chemical insecticides is that they are generally specific for their main target group, e.g., mosquitoes or lepidopterans, and thus preserve natural enemy populations, thereby often reducing the need for repeated pesticide applications.

Several viruses, especially baculoviruses of the nuclear polyhedrosis virus (NPV) type, have also been registered for use as insecticides (Table 2). Those viruses that have received the most attention are the NPV of the corn earworm, *Helicoverpa zea*, which has been used primarily to control *H. zea* and the tobacco budworm, *Heliothis virescens*, on cotton, and the NPV of the beet armyworm, *Spodoptera exigua*, used for control of this pest on vegetable crops (tomatoes in the USA) and horticultural crops (chrysanthemums in the Netherlands). Two other viruses that are used occasionally and still under development are the NPV of gypsy moth, *Lymantria dispar*, used to control gypsy moth outbreaks in deciduous forests in the eastern USA, and the granulosis virus (GV) of the codling moth, *Cydia pomonella*, used in the northwestern USA and California to control codling moth larvae on apples and walnuts. Production of all these viruses is quite limited at present (4,6).

In other countries, wild-type viruses are used more extensively, but their use is still minor in comparison to the use of chemical insecticides and *Bacillus thuringiensis*. The best-documented use of a virus as an insecticide is the use of the NPV of the velvetbean caterpillar, *Anticarsia gemmatalis*, to control this pest on soybeans in Brazil (11). This virus is used annually on more than 0.89 million hectares to control this important pest, and is the centerpiece of the velvetbean caterpillar IPM program. This program has been successful because the virus only needs to be applied once per season, it is highly efficacious, and the Brazilian government subsidizes the cost of virus production.

Two fungal pathogens, *Beauveria bassiana* and *Metarhizium anisopliae*, have been studied widely for use against a variety of insect pests, including beetle larvae, whiteflies, leafhoppers, and grasshoppers, but their use remains extremely limited. Fungi have an advantage over bacterial and viral pathogens, which must be eaten to be active, because fungi can penetrate through the cuticle, and thus have "contact" activity.

At one time, the only microsporidian produced for use in commercial agriculture was *Nosema locustae*, used to control a variety of grasshopper species in rangeland. The product was not sufficiently efficacious and is no longer produced for this market.

Limitations of Wild-type Pathogens as Insecticides

The reason that insect pathogens are not used more routinely as components of IPM programs is that they all have limitations in comparison to chemical insecticides. In the case of *B. thuringiensis*, each naturally occurring subspecies is typically active against only one type of insect. Moreover, most products are often only highly effective against certain members of the pest target group. For example, *B. t.* subsp. *kurstaki* is active against lepidopterous insects but not against beetles or mosquitoes. Against its target group, *B. t.* subsp. *kurstaki* is much more effective against the cabbage looper (*Trichoplusia ni*) than it is against the beet armyworm (*S. exigua*). Thus, in situations where a complex of insect pests attacks a particular crop, broad-spectrum chemical insecticides become the control agent of choice. Fortunately, there are many situations where only one or a few target species dominate (e.g., in field crops, forests, or mosquito control), and these are the situations where bacterial insecticides are favored. Another limitation of Bt is that pest species of low sensitivity to the endotoxin proteins produced by this bacterium, though paralyzed, may not die until 24-48 hours after treatment.

Viruses are also very restricted in their target spectra, even more so than Bt subspecies; this is a key limitation to their more widespread use in IPM programs. In fact, many viruses are virtually species-specific, such as the *S. exigua* NPV and *L. dispar* NPV. Development and production of these viruses therefore necessitates that their host be a very important pest for which other cost-effective control technologies do not exist or are not used. For example, in the eastern United States, public pressure has restricted the use of chemical insecticides to treat gypsy moth populations, and thus, either Bt or the gypsy moth NPV is used to control this pest. Another key limitation of viruses is that after infection, the virus typically requires 5-10 days to kill the host, during which time the host continues to feed and cause

Table 2. Principal pathogens registered for insect control

Species	Active Ingredient	Targets
Bacteria		
Bacillus thuringiensis	Protein endotoxins	Various
subsp. *kurstaki*	Protein endotoxins	Caterpillars
subsp. *israelensis*	Protein endotoxins	Mosquitoes
subsp. *morrisoni*	Protein endotoxins	Beetles
Virions		
Helicoverpa zea NPV	Virion	Cotton bollworm Tobacco budworm
Spodoptera exigua NPV	Virion	Beet armyworm
Autographa californica NPV	Virion	Caterpillars
Fungi		
Beauvaria bassiana	Conidium	Beetles Whiteflies Grasshoppers
Metarhizium anisopliae	Spore	Grasshoppers

damage. Finally, cost-effective methods for virus mass production *in vitro* (cell culture) are not available. Consequently, the viruses must be produced in caterpillars, requiring mass rearing of millions of larvae to treat large areas (5). The velvetbean caterpillar IPM program mentioned earlier solved this problem by producing large amounts of the virus using caterpillars that occurred in field populations. But this is not a practical production method in most agricultural situations in the USA. Methods were developed decades ago for the commercial production of large quantities of virus in caterpillars (8). However, large companies in the USA have been reluctant to use these methods. Recently, Thermo-Trilogy of Columbia, MD, has re-established this method for commercial production of the *S. exigua*, *H. zea*, and *Autographa californica* (alfalfa looper) NPVs (6).

A major limitation of the fungi *B. bassiana* and *M. anisopliae* is also limited production efficiency, although Mycotech Inc., of Butte, MT, has developed a production process that may eventually be able to supply *B. bassiana* in amounts sufficient to treat large areas. They currently market products based on *B. bassiana* for use against whiteflies and other homopterous pests in glasshouses. With regard to the microsporidia, because these cause chronic diseases and kill slowly, they are no longer envisioned for use as microbial insecticides. They are still being developed for IPM programs, but more as population management tools for long-term control of forest pests and ants.

Improved Genetically Engineered Pathogens

Recombinant DNA technology has been used to develop more effective bacterial and viral insecticides, and preliminary studies indicate these methods can also be used to enhance the infectivity of fungi. The major developments for each group are summarized below.

BACTERIA

The focus of most of the engineering of bacteria has been *Bacillus thuringiensis*. This bacterium kills insects primarily through the action of protein toxins. The genes encoding these toxins are located on plasmids, and more than 100 of these genes have been cloned and sequenced, providing a rich source of material for genetic improvement of bacterial insecticides (3).

The first genetically manipulated products were not developed using recombinant DNA technology, but rather by transferring plasmids from one strain to another. The concept was to combine toxins with different insect spectra that did not occur naturally into a single strain, referred to as a transconjugate (or transconjugant). Ecogen Inc., of Langhorne, PA, pioneered this strategy and developed, for example, a product with the trade name Foil®, which has as its active ingredient a combination of toxins active against both beetles and lepidopteran larvae (2). This strategy allowed products to be marketed with novel host spectra that were developed without the use of recombinant DNA technology, thereby facilitating registration more quickly than would occur with true recombinant products.

After the development of transconjugates, different strategies were used to produce true recombinant bacterial insecticides. The two main concepts were to engineer bacteria to produce Bt Cry toxins, either alone or in combination with either *Psuedomonas fluorescens* or *B. thuringiensis*. Mycogen Corporation (recently acquired by Dow Agrosciences) of San Diego, CA, pioneered the work on *P. fluorescens*, and Ecogen pioneered the development of recombinant Bts.

The rationale for using *P. fluorescens* as a host cell was that it grew well, it did not produce a spore, and the cell wall could be chemically fixed around the Cry toxin crystal to enhance residual activity by protecting the toxin from UV inactivation. Mycogen developed and marketed three products based on this technology, MVP® and MATTCH® for control of lepidopteran pests, and MTRAK® for control of beetles, especially the Colorado potato beetle (Table 3). MTRAK contains only a single toxin, whereas MVP contains two toxins produced in a single cell active against lepidopterans. MATTCH is a combination of two different engineered strains, each producing a single toxin.

The recombinant Bt strains developed by Ecogen contain a greater range of toxins, and were developed using a strategy whereby the toxin genes were engineered into Bt plasmids, after which foreign DNA was eliminated (2). The complexity of the toxin mixtures and ratios in the strains developed by Ecogen demonstrates the utility of recombinant DNA for designing improved bacteria for use as insecticides (Table 3).

Table 3. Registered recombinant bacterial insecticides[a,b]

Product	Company	Bacterial host	Active	Target
MVP	Mycogen/Dow	*P. fluorescens*	Cry1Ac, Cry1Ab	Lep.
MTRAK	Mycogen/Dow	*P. fluorescens*	Cry3A	Coleop.
MATTCH	Mycogen/Dow	*P. fluorescens*	Cry1Ac, Cry1C	Lep.
CRYMAX	Ecogen	*B. thuringiensis*	Cry1Ac. Cry1C, Cry2A	Lep.
Lepinox	Ecogen	*B. thuringiensis*	Cry1Aa, Cry1Ac-Cry1F, Cry2A	Lep.
Raven	Ecogen	*B. thuringiensis*	Cry1Aa, Cry3A, Cry3Ba	Lep. Coleop.

a/ Lep. = Lepidoptera; Coleop. = Coleoptera
b/ Modified from Baum et al. (2).

Field performance of these recombinant products has been good to excellent, but it remains to be seen whether they will be more useful and sell better than bacterial insecticides based on wild-type strains, which still dominate the Bt market.

VIRUSES

As noted earlier, two of the key limitations of NPVs are that they kill slowly compared to chemical insecticides and Bt, and the host range of most is very narrow. To increase the speed of paralysis and time until death, two tactics have been employed, eliminating viral genes (deletion mutants) that delay death, and adding genes to speed up paralysis and death (10,14).

Research on NPVs has shown that most contain a gene which encodes ecdysone glucosyl-UDP-transferase (EGT), an enzyme that glucosylates the molting hormone ecdysone. Glucosylation of ecdysone prevents molting, which benefits the virus by prolonging the life of the caterpillar so that it can eat more, resulting in greater virus production. Deletion of this gene from the virus reduces the time from infection to death by about 25%.

To further increase the speed of kill, genes encoding primarily toxins, especially insect-specific neurotoxins, have been added to NPVs. And to increase the target spectrum of activity, the model virus used for most engineering studies has been the *Autographa californica* NPV (AcNPV), a virus capable of infecting a wide range of caterpillar species.

By combining these concepts and strategies, several different recombinant AcMNPVs have been produced (Table 4). Most of these recombinant viruses produce only a single neurotoxin. The toxins that have proven to be the most effective to date have been those from scorpions and the straw itch mite, *Pyemotes triticii*. In the laboratory, EGT deletion mutant AcNPVs that also produce a neurotoxin have reduced the time from infection until cessation of feeding by as much as 50%. Preliminary field trials at rates of application of 10^{12} polyhedra per hectare have shown that the recombinant AcNPV producing the AaIT scorpion toxin provided control of the *H. zea* and *H. virescens* for four weeks, with damage reduction being equivalent to that obtained with the chemical insecticide esfenvalerate (1). For a review of the recent literature on the development and evaluation of recombinant baculoviruses, see Treacy (14).

While the above results are very promising, it must be noted that cost-effective methods for the mass production of recombinant NPVs remain to be perfected. Recombinant viruses that produce neurotoxins are difficult to produce in larvae because they result in rapid paralysis and lower virus yields than are obtained with wild-type viruses. The recombinant viruses can be produced *in vitro* (cell culture), but large-scale production of amounts that would be required to compete with chemical insecticides and Bt is not yet practical.

FUNGI

Fungi have received little attention with regard to engineering

171

them to improve their insecticidal properties. The feasibility of the concept has been demonstrated, though, in *M. anisopliae,* which was engineered to produce more protease by addition of a protease gene (12). The engineered strain had increased virulence because of improved cuticle-penetrating properties. While this result is positive, engineered fungi will no doubt receive a high level of scrutiny for registration because of their naturally broad host range.

Table 4. Recombinant viral insecticide strategies[a]

Mutation	Gene	Function	Source	Virus
Gene deletion	EGT	ecdysone inactivation	AcMNPV LdMNPV	AcMNPV
Gene additions	AaIT	sodium channel agonist	Scorpion	AcMNPV
	LghIT2	neuromuscular depressant	Scorpion	AcMNPV HzSNPV
	JHE	enzyme toxin	Lepidopteran	AcMNPV
	TxP-1	presynaptic channel blocker	Mite	AcMNPV
	mu-A4	sodium channel activation	Spider	AcMNPV
	P2	fungal protease	*Metarhizium*	AcMNPV
	cry	midgut lysis	*Bacillus*	AcMNPV

a/ Modified from Treacy (14).

Competing technologies and future prospects

While there have been significant advances during the past decade in our ability to improve the insecticidal properties of insect pathogens, other pest control technologies have also advanced and will continue to provide strong competition for microbial insecticides based on wild-type and recombinant pathogens. For example, several new types of potent and more-specific chemical insecticides are now on the market, e.g., imidocloprid and the spinosins. Perhaps an even greater threat to the future success of microbial insecticides is insect-resistant transgenic crops. At present, three types of insect-resistant transgenic crops, all of which have been engineered to produce Bt Cry proteins, are on the market. These are: Bt potatoes, which produce the Cry3A protein for resistance to the Colorado potato beetle (*Leptinotarsa decemlineata*); Bt cotton, which produces the Cry1Ac protein for resistance primarily to the tobacco budworm and pink bollworm (*Pectinophora gossypiella*); and Bt corn, which produces either the Cry1Ab or Cry9 protein for resistance to the European corn borer (*Ostrinia nubilalis*) or the Cry1Ac protein for resistance to the corn earworm (*H. zea*). In the first two years, these crops have proven to be very effective overall. The threat of insects developing resistance to these crops is real, but several resistance management strategies are being developed to delay or avoid its development (13; 7). Should these strategies prove effective, many other crops will be engineered to produce Bt and other insecticidal proteins to protect them from insect damage.

During the next decade, the future success of microbial insecticides will depend on a variety of factors, including the success of insect-resistant transgenic crops (and the extent to which they penetrate the market in the USA and foreign countries) and the continuing availability of other types of cost-competitive microbial and chemical insecticides. Even if transgenic crops prove to be an effective long-term technology, many markets for microbial insecticides will continue to exist, such as forestry, ornamentals, vegetables, and fruit trees, where transgenic crops either do not exist or are not appropriate due to the diversity of the pest complex. In these markets, the primary competition for individual types of microbial insecticides will be chemicals or other microbials. For example, NPV and Bt products often target the same pests in the same markets, namely the gypsy moth and spruce budworm in forests, and noctuid pests in vegetable and field crops. The viruses, whether conventional or engineered, will only be successful in these markets if methods for mass producing them on a large scale are developed. If this is not done, products based on wild-type and engineered Bts will continue to dominate the microbial insecticide market.

Engineered fungal insecticides are only in the earliest stages of development. However, even if improved strains of fungi, such as *B. bassiana* and *M. anisopliae,* are developed and approved for use by the EPA, existing production problems will have to be overcome to make them competitive with new chemical insecticides.

Summary and Conclusions

In this presentation, I have provided a glimpse of how recombinant DNA technology has been and is being used to improve the efficacy of insecticides based on insect pathogens. Methods have been developed during the past decade that allow insect-pathogenic bacteria, viruses, and fungi to be manipulated genetically with relative ease. In addition, the genes for numerous proteins, primarily toxins, have been cloned that can be added to these pathogens to improve their efficacy. In the case of bacteria, especially *B. thuringiensis*, the focus is now on improving the strains used in products by recombining genes from different bacteria and other organisms to increase specific toxicity and expand the spectrum of activity against key insect pests of major commodity and vegetable crops. The focus on viruses is similar, in that the goal is to manipulate NPVs (especially the AcMNPVs) by the addition of neurotoxin genes to the point where they paralyze and kill a broad spectrum of target lepidopterans within 24-48 hours. The primary challenge associated with Bt is to develop strains that are cost-effective against the development of all instars of difficult-to-control pests, such as those of the *Spodoptera* and *Heliothis/Helicoverpa* complexes. With NPVs, a major obstacle to the use of recombinant viruses is the lack of a cost-effective cell-culture technology for mass production on a commercial scale. The advent of other novel technologies, such as insect-resistant transgenic crops and new types of chemical insecticides, makes it imperative that these obstacles be overcome if pathogens are to be used more routinely in IPM programs, thereby significantly increasing their share of the insecticide market. If these obstacles cannot be overcome, pathogens will remain useful for insect control but will continue to account for only a small percentage of

174

the insecticide market.

Literature Cited

1. All, J. N., and Treacy, M. F. 1997. Improved control of *Heliothis virescens* and *Helicoverpa zea* with a recombinant form of *Autographa californica* nuclear polyhedrosis virus and interaction with Bollgard® cotton. Pages 1294-1296 in: Proc. Beltwide Cotton Conf.
2. Baum, J. A., Johnson, T. B., and Carlton, B. C. 1998. *Bacillus thuringiensis*, natural and recombinant bioinsecticide products. Pages 189-209 in: Methods in Biotechnology, Vol. 5: Biopesticides: Use and Delivery. F. R. Hall and J. J. Menn, eds. Humana Press Inc., Totowa, NJ.
3. Crickmore, N., Zeigler, D. R., Feitelson, J., Schnepf, E., Van Rie, J., Lereclus, D., Baum, J., and Dean, D. H. 1998. Revision of the nomenclature for *Bacillus thuringiensis* pesticidal crystal proteins. Micribiol. Mol. Biol. Rev. 62:807-813.
4. Entwistle, P. F. 1998. A world survey of virus control of insect pests. Pages 189-200 in: Insect viruses and pest management. F. R. Hunter-Fujita, P. F. Entwistle, H. F. Evans, and N. E. Crook, eds. John Wily & Sons, Chicester.
5. Federici, B. A. 1998. Naturally occurring baculoviruses for insect pest control. Pages 301-320 in: Methods in Biotechnology, Vol 5: Biopesticides: Use and Delivery. F. R. Halland, and J. J. Menn, eds. Humana Press Inc., Totowa, NJ.
6. Federici, B. A. 1999. A perspective on pathogens as biological agents for insect pests. Pages 517-550 in: Handbook of Biological Control. T. W. Fisher, T. S. Bellows, L. E. Caltagirone, D. L. Dahlsten, C. Huffaker, and G. Gordh, eds. Academic Press, San Diego, CA.
7. Gould, F. 1998. Sustainability of transgenic insecticidal cultivars: integrating pest genetics and ecology. Ann. Rev. Entomol. 43: 701-726.
8. Ignoffo, C. M. 1973. Development of a viral insecticide: concept to commercialization. Exp. Parasitol. 33:380-406.
9. Jenkins, J. J. 1998. Transgenic plants expressing toxins from *Bacillus thuringiensis*. Pages 211-232 in: Methods in Biotechnology, Vol 5: Biopesticides: Use and Delivery. F. R. Hall, and J. J. Menn, eds. Humana Press Inc., Totowa, NJ.

10. Miller, L. K. 1998. The Baculoviruses. Pelnum Press.
11. Moscardi, F. 1999. Assessment of the application of baculoviruses for control of lepidoptera. Ann. Rev. Entomol. 44: 257-289.
12. St. Leger, R. J., Lokesh, J., Bidochka, M. J., and Roberts, D. W. 1996. Construction of an improved mycoinsecticide overexpressing a toxic protease. Proc. Natl. Acad. Sci. USA 93:6349-6354.
13. Tabashnik, B. E. 1994. Evolution of resistance to *Bacillus thuringiensis*. Ann. Rev. Entomol. 39:47-79.
14. Treacy, M. F. 1998. Recombinant baculoviruses. Pages 321-367 in: Methods in Biotechnology, Vol. 5: Biopesticides: Use and Delivery. F. R. Hall and J. J. Menn, eds. Humana Press Inc., Totowa, NJ.

Success and Failure of Bt Products:
Colorado Potato Beetle -- A Case Study

David N. Ferro
Department of Entomology, University of Massachusetts
Amherst, MA 01003 USA

Biotechnology has ushered in a new era of insect pest management. Currently, the cornerstone of this technology for insect control is based on toxins produced by *Bacillus thuringiensis* (Bt) Berliner. Bt has been formulated for application as a foliar spray, and Bt genes that encode for endotoxins have been incorporated into the genome of plants. For integrated pest management programs designed to become less dependent on broad-spectrum, synthetic insecticides, Bt products offer the IPM practitioner the option to target specific insect pests without disrupting the biology of natural enemies within the cropping system. The Bt-based insecticides developed since the mid-1980s were formulated from genetically modified micro-organisms or from new fermentation processes that produced higher concentrations of the toxin(s). Bt products were shown to be as effective at controlling pests as their synthetic counterparts; however, these products never gained a major share of the market, except where synthetic insecticides had failed. The question that many of us are asking ourselves today is why have Bt products failed to gain market share? I will use the Colorado potato beetle, *Leptinotarsa decem-lineata* (Say) and the potato cropping system as one example of why Bt products failed to garner significant grower support.

It is necessary to provide adequate historical information on the pest status of the Colorado potato beetle, on its biology, and on control tactics developed to manage this important pest. In the United States, potato has a comparatively simple pest complex, including several species of pathogens and insects; however, the Colorado potato beetle is the most destructive insect pest of potato in North

America (7). The beetle has developed resistance to most categories of insecticides currently registered for its control (14). Resistance is widespread but is most pronounced in the northeastern United States (14,20) and to a lesser extent in Virginia (26) and North Carolina (16), with isolated cases as far west as Michigan (17), and in the Maritime Provinces of Canada, New Brunswick (3), and Prince Edward Island (24). Because of widespread resistance by the beetle to a wide range of insecticides (Table 1), it became very costly for growers to control this pest (Table 2).

During the late 1970s, aldicarb was introduced as a soil-applied systemic insecticide to control the beetle, and proved to be most effective. However, within a few years aldicarb was detected in groundwater (21), which led to its being banned from use a few years after being introduced. However, even if aldicarb had not been banned, the beetle would have soon escaped its control (14). Biochemical evidence from beetle populations from Long Island, NY,

Table 1. Abbreviated chronology of Colorado potato beetle resistance to insecticides on Long Island, New York.

| | Year | | |
| | introduced | 1st failure | |
Insecticide	introduced	1st failure	Chemical group
carbaryl	1957	1958	carbamate
azinphosmethyl	1959	1964	organophosphate
phosmet	1973	1973	organophosphate
phorate	1973	1973	organophosphate
carbofuran	1974	1976	carbamate
oxamyl	1978	1978	carbamate
fenvalerate	1979	1981	pyrethroid
permethrin	1979	1981	pyrethroid
fenvalerate + PBO	1982	1983	pyrethroid + synergist
imidacloprid	1995	1998	nitroquanidine

After Forgash (14)

178

Table 2. Number of insecticide applications and costs per hectare for controlling the Colorado potato beetle in the northeastern United States, 1991.

State	No. of applications	Cost $ (US)
Massachusetts		
Bio-intensive IPM	5 Btt	$254
IPM Growers (nonrotated fields)	1 esfenvalerate + PBO; 6 Btt; 1 oxamyl + endosulfan	$423
Conventional growers	3 esfenvalerate + PBO; 4 oxamyl + endosulfan; 1 Btt	$512
Maine		
Southern Maine	1 esfenvalerate + PBO; 1 permethrin + PBO; 2 cryolite; 2 oxamyl; 1 azinphosmethyl + endosulfan	$348
Northern Maine (Aroostock County)	2 esfenvalerate; 1 azinphosmethyl + endosulfan; 1 oxamyl	$126
New Jersey	1 esfenvalerate; 1 cryolite; 2 oxamyl; 1 totenone	$368
New York (Long Island)	1 esfenvalerate + PBO + endosulfan. 1 esfenvalerate + PBO + permethrin; 1 esfenvalerate + PBO + parathion; 1 esfenvalerate + PBO + cryolite; 1 esfenvalerate + PBO + azinphosmethyl; 1 esfenvalerate + PBO + azinphosmethyl + endosulfan; 1 Btt 2 cryolite 1 rotenone + PBO + endosulfan; 1 rotenone + PBO + parathion + permethrin; 1 phosmet + permethrin + PBO + azinphosmethyl	$988

showed a dramatic change in levels of susceptibility. Once aldicarb had been banned, the synthetic pyrethroid insecticides permethrin and esfenvalerate were quickly registered for control of Colorado potato beetle on potato under the specific "exemption from tolerance" provision (Section 18) of FIFRA. However, even before these products gained full registration, the beetle had developed resistance to the pyrethroids. Within a few years, these insecticides failed, and growers had to resort to using rotenone, a very expensive botanical insecticide. Rotenone did not provide high levels of control, and growers were faced with extremely large populations of overwintered beetles colonizing their fields.

This situation was the impetus for a flurry of activity to develop innovative approaches to managing the beetle. These included the use of plastic-lined trenches for capturing beetles as they colonized potato fields from overwintering sites, propane flamers to kill beetles within the potato crop, vacuum collectors to remove adult and larval potato beetles from potato foliage, and trap cropping to control colonizing adults (6). Although some of these tactics were very effective at reducing beetle populations, each tactic required the growers to invest in new equipment and educate themselves in a new way to control this major pest, and most important, these tactics required a great deal of the grower's time. New management tactics were needed.

Ferro et al.(10) showed that potato plants could tolerate high levels of defoliation early and late in the season with no reductions in yields. Action thresholds were developed based on this study. Growers could reduce the number of applications by 25% if they only sprayed when beetle densities reached these levels. Lashomb and Ng (18) showed that fields rotated into potato from another crop were colonized later and at lower densities than non-rotated fields. The success of the beetle as a pest of potato is largely determined by its remarkably diverse and flexible life history (29). Migrations, closely connected with diapause, feeding, and reproduction, allow this insect to employ "bet-hedging" reproductive strategies (27), distributing its offspring in both space (within and between fields) and time (within and between seasons). As a result, the risk of catastrophic losses of offspring due to insecticides or crop rotation is diminished (22,27). Furthermore, the beetle's mating behavior is strongly directed toward maximizing genetic variability of its progeny. After summer-generation beetles accumulate at least 34 degree-days (DD) following eclosion (1), both males and females perform multiple copulations with different partners (25). Multiple matings are necessary for females to realize their full reproductive potential (2).

Colorado potato beetles overwinter in the soil as adults, with the majority of them aggregating in woody areas adjacent to fields in which they have spent the previous summer (28). After diapause is induced by a short-day photoperiod (5), the beetles engage in a low altitude flight directed toward tall vegetation. Upon arrival at overwintering sites, they immediately burrow into the soil to diapause (27), and their flight muscles undergo significant degeneration (23). The refractory phase of diapause, during which the beetles do not react to changes in environmental conditions, lasts for approximately 3 months. After that, the beetles respond to elevation of temperature above 10°C by emerging from the soil (5). The beetles usually accumulate 50-250 DD before they appear on the soil surface (12). Males and females terminate diapause simultaneously and require only 60-80 DD after emergence before they are able to mate (12).

After emergence from the soil, overwintered Colorado potato beetles colonize potato fields both by flight and by walking (27). Beetle flight is strongly encouraged by the absence of food (4,11,30). If the fields are rotated, the beetles are able to fly up to several kilometers to find a new host habitat (11,30). Mating starts before beetles disperse, with at least half of the population mating within the overwintering sites (12). However, post-diapause females need not mate in the spring to produce viable offspring; enough sperm is carried over the winter from the previous fall/summer matings to produce 80% of potential offspring (11).

Crop rotation had long been the main line of attack for battling the beetle. An alternative to crop rotation for growers who were unable to rotate their fields was to plant their potato crop later so that the plants did not emerge until most of the beetles had taken flight. This tactic was highly successful for dealing with high densities of overwintered beetles. However, these practices were largely abandoned beginning in the late 1940s, once growers were able to control the beetle with highly effective insecticides. We were able to reduce the number of insecticide applications by 50% by having growers rotate their fields, and by having them better time the application of insecticides based on action threshold levels for the different beetle life stages. Because rotated fields were colonized later in the season than non-rotated fields, most of the beetles emerged after the day-length had decreased below the critical photoperiod for diapause induction; hence, they laid few eggs. For this reason, there was no need for growers to treat fields late in the season unless the larval population exceeded the action threshold. Prior to learning of the importance of diapause induction in the beetle, growers were applying insecticides to control summer generation beetles, assuming that they were going to lay eggs and produce a late-season, damaging larval population.

We also noticed that rotated fields were colonized early in the season by the ladybird beetle, *Coleomegilla maculata*, which fed on Colorado potato beetle eggs. Hazzard et al. (15) showed that *C. maculata* caused about 50% generational mortality of potato beetle eggs in a rotated field in Great Barrington, MA. About this time, we also found that potato beetle larvae were attacked by two species of parasitic flies, *Myiopharus aberrans* and *M. doryphorae*. Lopez et al (19) found up to 75% of larvae in organically grown potatoes to be parasitized by these flies. These findings convinced us that biological control agents could suppress beetle populations, especially if densities of beetles could be reduced. However, at this time, there was no selective insecticide available to reduce beetle densities without killing its natural enemies.

During the winter of 1984-85, Wendy Gelernter of Mycogen Corporation contacted me about evaluating a new strain of *Bacillus thuringiensis* against the beetle. Under laboratory conditions, the fermentation broth proved to be toxic to potato beetle larvae. The following summer, we evaluated a crude formulation in the field and found the toxin to be effective at reducing larval populations of potato beetle. However, we noticed that larger larvae survived treatment. The following winter, we completed studies showing efficacy of the toxin to be dependent upon weight of the larvae, and, based on these studies, we established the IU and named the bacterium *Bacillus thuringiensis* subsp. *san diego* (8). Later studies showed this bacterium to be the same as a previously described bacterium, *B. thuringiensis* subsp. *tenebrionis* (Btt), which produced the Cry3A delta endotoxin. During the spring of 1986, Mycogen Corporation gained a federal label to market M-One®, a product formulated from

Btt. We completed another study that showed temperatures below 20°C to dramatically reduce effectiveness of M-One (9). Because the toxin was not very persistent in the field, it was critical that growers apply M-One when the larvae were actively feeding. This added another level of complexity to the successful use of this material, requiring a support system to help growers.

Mycogen hired several people in sales to market this new product. I continued to work closely with Mycogen to help them assess different formulations and to advise them on how to use this product. Since the people in sales came from traditional agricultural chemical industries, I was worried that they would market M-One in the same way that they would synthetic insecticides. We knew M-One was highly effective at controlling early instars (1st and 2nd instars), but ineffective at controlling large larvae, so I advised Mycogen to enter the market slowly and to work closely with growers using their product. For M-One to be successful, it would be necessary for potato growers to increase their awareness of the density and age structure of the larval populations in their fields. This required growers to change their approach to controlling the beetle. In Massachusetts, I worked very closely with growers in the University of Massachusetts Potato IPM Program to make sure that the first application of M-One was applied before 3rd instars appeared in the field. Growers had great success in using M-One to combat populations of beetles that were highly resistant to all materials labeled for controlling the beetle. They quickly learned that it was safer to make the first application a little earlier rather than a little later. Because they had to constantly scout their fields and be more vigilant in timing applications, M-One required a more labor-intense approach to beetle control. However, economic savings were quickly realized. The Szawlowski family of Hatfield, MA, having struggled to control the beetle on their 1,000 acres of potato in previous years, was quick to embrace this new insecticide. In 1987, the first year their beetle control program was based on M-One, they saved more than $100,000 in insecticide costs over the previous year! However, this was not the case in other parts of the country.

Extension personnel did not have experience with M-One, and the product was sold to growers without providing the necessary support. M-One failed to provide control in these situations. M-One did not fail; Mycogen Corporation failed in providing the necessary support to make sure growers knew how to properly use this product. Mycogen soon developed a new product (M-Trak) based on their Cell Cap® technology, in which the gene that regulates production of the Cry3A toxin was inserted into *Pseudomonas infestans.* At the end of the fermentation process, the broth was treated in such a way that the cell wall coalesced around the protoplasm containing the toxin, in effect, encapsulating the toxin.

This new formulation was much more persistent in the field than M-One (7). About this time, Novo Nordick of Denmark produced a highly efficacious formulation of the Cry3A toxin, and Ecogen of Pennsylvania also developed a very effective material. So, by the early 1990s, there were several very effective formulations of Btt available. However, even with these new materials and their proven ability to control the beetle, it was very difficult for the industry to overcome the earlier failures. Yet, growers like the Szawlowskis of Massachusetts, who had immediate success with M-One, continued to base their beetle management plan on the Btt products. The success of the Btt products in Massachusetts allowed us to implement a second-phase IPM program for potato.

The University of Massachusetts Potato IPM program implemented the Bio-intensive IPM Program for managing the beetle that was dependent on crop rotation or delayed planting and on judicious use of foliar Btt products. Because of insecticide resistance and inability to control the beetle, growers were no longer able to use the broad-spectrum, synthetic insecticides to control the beetle, and they quickly embraced the Bio-intensive IPM Program. However, this program required growers to change the way they used insecticides. Unlike conventional pesticides, Btt products had to be targeted against the early instars, and the air temperature at the time of application was critical (9). Larvae stop feeding within 15 minutes after ingesting the Btt toxin, and if it is too cold, they do not consume a lethal dose of toxin before feeding is inhibited. So, we encouraged growers to apply the Btt products during the daylight hours, when larval consumption is at its maximum. A survey (Table 2) showed growers in the Bio-intensive IPM Program (which was conducted only in Massachusetts) spent an average of $255/ha to control the beetle compared to $512/ha for non-IPM growers. Another survey from 1995 showed that 83% of the potato growers in 1994 used Btt products to control the beetle. The Btt products had gained a major share of the insecticide market for controlling the beetle; however, this was short-lived. A new broad-spectrum, synthetic insecticide, imidacloprid (Admire®, a soil-applied systemic, and Provado®, a foliar material), gained full registration by EPA for potatoes during the spring of 1995.

Imidacloprid represented new chemistry to which the beetle had not previously been exposed. This insecticide was the most effective chemical for controlling the beetle since aldicarb. In a mark-recapture study, we showed that beetles collected from overwintering sites and then placed on potato plants that had been treated with imidacloprid moved less than one meter before dying. Another study showed that, on average, each imidacloprid-treated potato plant in the outside row closest to overwintering sites had 142 dead beetles, the third row had 10, and by the fifth row, there were only four dead beetles per plant. When imidacloprid became available as a soil-applied, systemic insecticide, growers quickly abandoned the Bio-intensive IPM Program, an effective but labor-intensive program. Within the first year of imidacloprid's registration (1995), 77% of the growers in Massachusetts were using it and had abandoned the Btt products.

Imidacloprid could be applied once at planting and provide control of the beetle through most of the growing season. This greatly simplified beetle management, and unlike the Btt products, imidacloprid provided virtually complete control of colonizing beetles. Before imidacloprid became registered for use on potatoes, many potato entomologists urged EPA to require a resistance management plan for imidacloprid. Because of the beetle's past history of becoming resistant to every insecticide registered for its control, most potato entomologists felt that it was only a matter of time before the beetle would become resistant to imidacloprid. They predicted that if growers used imidacloprid at planting to control the same population of beetles in consecutive years, the potato beetle would become resistant to imidacloprid within 4 years. During August 1998, we collected several hundred potato beetle egg masses from a potato field that had been treated with Imidacloprid for 4 consecutive years. Larvae (2nd instars) from this population were 34 times more tolerant to imidacloprid than a susceptible population collected from the University of Massachusetts Vegetable Research Farm, South Deerfield, MA. It seems that there is once again a place for the Btt products in the management of potato beetle control. However, it will be interesting to see if companies will respond and produce more Btt products. The recent and pending registration of several new chemical insecticides that are highly effective in controlling Colorado potato beetle may deter the rejuvenation of Btt products.

The Btt products failed for several reasons. Companies that introduced the Btt products failed to appreciate the need for providing support to growers to help them properly use these materials. For this reason, there were failures that should not have occurred. Also, these products required new approaches to timing of applications and required growers to closely monitor beetle populations; this was very labor intensive compared to using conventional insecticides. With the registration of imidacloprid, growers had an insecticide that was much more effective and easier to use than the Btt products. Because growers could obtain season-long control by applying imidacloprid at planting, it was cost effective and required less time in the field scouting for beetles. Even though growers were warned repeatedly about the high risk that the potato beetle would develop resistance if imidacloprid were used year after year as a soil-applied systemic insecticide, they chose convenience over sustainability.

Literature Cited

1. Alyokhin, A. V., and Ferro, D. N. 1999. Dispersal and reproduction of the summer-generation Colorado potato beetle (Coleoptera: Chrysomelidae). Environ. Entomol. (in press).
2. Boiteau, G. 1988. Sperm utilization and post-copulatory female-guarding in the Colorado potato beetle, *Leptinotarsa decemlineata*. Entomol. Exp. Appl. 47:183-187.
3. Boiteau, G., Parry, R. H., and Harris, C. R. 1987. Insecticide resistance in New Brunswick populations of the Colorado potato beetle (Coleoptera: Chrysomelidae). Can. Entomol. 119:459-463.
4. Caprio, M., and Grafius, E. 1990. Effects of light, temperature and feeding status on flight initiation in postdiapause Colorado potato beetle. Environ Entomol. 19:281-285.
5. De Kort, C. A. D. 1990. Thirty-five years of diapause research with the Colorado potato beetle. Entomol. Exp. Appl. 56:1-13.
6. Ferro, D. N. 1996. Mechanical and physical control of the Colorado potato beetle and aphids. Pages 53-67 in: Potato Insect

Pest Control: Development of a Sustainable Approach. R. M. Duchesne and G. Boiteau, eds. Agriculture and Agri-Food Canada.

7. Ferro, D. N., and Boiteau, G. 1993. Management of insect pests. Pages 103-115 in: Potato Health Management. R. C. Rowe, ed. APS Press, St. Paul, MN, USA.

8. Ferro, D. N., and Gelernter, W. D. 1989. Toxicity of a new strain of Bacillus thuringiensis to Colorado potato beetle (Coleoptera: Chrysomelidae). J. Econ. Entomol. 82:750-755.

9. Ferro, D. N., and Lyon, S. M. 1991. Colorado potato beetle (Coleoptera: Chrysomelidae) larval mortality: operative effects of *Bacillus thuringiensis* subsp. *tenebrionis*. J. Econ. Entomol. 84: 806-809.

10. Ferro, D. N., Morzuch, B. J., and Margolies, D. 1983. Crop loss assessment of the Colorado potato beetle (Coleoptera: Chrysomelidae) on potatoes in western Massachusetts. J. Econ. Entomol. 76:349-356.

11. Ferro, D. N., Tuttle, A. F., and Weber D. C. 1991. Ovipositional and flight behavior of overwintered Colorado potato beetle (Coleoptera: Chrysomelidae). Environ. Entomol. 20:1309-1314.

12. Ferro, D. N., Alyokhin, A. V., and Tobin, D. B. 1999. Reproductive and dispersal behavior of the overwintered Colorado potato beetle (Coleoptera: Chrysomelidae). Entomol. Expt. Appl. (in press).

13. Ferro, D. N., Yuan, Q.-C., Slocombe, A., and Tuttle, A. F. 1993. Residual activity of insecticides under field conditions for controlling the Colorado potato beetle (Coleoptera: Chrysomelidae). J. Econ. Entomol. 86:511-516.

14. Forgash, A. G. 1985. Insecticide resistance in the Colorado potato beetle. Pages 33-52 in: Proc. Symposium on the Colorado potato beetle. D. N. Ferro and R. H. Voss, eds. XVIIth International Congress of Entomology. Res. Bull. 704, Mass. Agric. Exp. Stn. Circ. 347.

15. Hazzard, R. V., Ferro, D. N., Van Driesche, R. G., and Tuttle, A. F. 1991. Mortality of eggs of Colorado potato beetle (Coleoptera: Chrysomelidae) from predation by *Coleomegilla maculata* (Coleoptera: Coccinelidae). Environ. Entomol. 20:841-848.

16. Heim, D.C., Kennedy, G. G., and Van Duyn, J. W. 1990. Survey of insecticide resistance among North Carolina Colorado

potato beetle (Coleoptera: Chrysomelidae) populations. J. Econ. Entomol. 83:1229-1235.

17. Ioannidis, P. M., Grafius, E., and Whalon, M. E. 1991. Patterns of insecticide resistance to azinphosmethyl, carbofuran, and permethrin in the Colorado potato beetle (Coleoptera: Chrysomelidae). J. Econ. Entomol. 84:1417-1423.

18. Lashomb, J. H., and Ng, Y. 1984. Colonization by Colorado potato beetles, *Leptinotarsa decemlineata* (Say) (Coleoptera: Chrysomelidae) in rotated and non-rotated potato fields. Environ. Entomol. 13:1352-1356.

19. Lopez, E. R., Ferro, D. N., and Van Driesche, R. G. 1993. Direct measurement of host and parasitoid recruitment for assessment of total losses due to parasitism in the Colorado potato beetle *Leptinotarsa decemlineata* (Say) (Coleoptera: Chrysomelidae) and *Myiopharus doryphorae* (Riley) (Diptera: Tachinidae). Biol. Control 3:85-92.

20. Roush, R. T., Hoy, C. W., Ferro, D. N., and Tingey, W. M. 1990. Insecticide resistance in the Colorado potato beetle (Coleoptera: Chrysomelidae): influence of crop rotation and insecticide use. J. Econ. Entomol. 83:315-319.

21. Sherman, S., and Mlay. 1986. Pesticides in ground-water: background document. Office of Ground-Water Protection. USEPA, WH-550G.

22. Solbreck, C. 1978. Migration, diapause, and direct development as alternative life histories in a seed bug, *Neacoryphus bicruci*. Pages 195-217 in: Evolution of insect migration and diapause. H. Dingle, ed.. Springer, New York.

23. Stegwee, D., Kimmel, E. C., de Boer, J. A., and Henstra, S. 1963. Hormonal control of reversible degeneration of flight muscle in the Colorado potato beetle, *Leptinotarsa decemlineata* Say (Coleoptera). J. Cell Biol. 19:519-527.

24. Stewart, J. G., Kennedy, G. G., and Sturz, A. V. 1997. Incidence of insecticide resistance in populations of Colorado potato beetle, *Leptinotarsa decemlineata* (Say) (Coleoptera: Chrysomelidae), on Prince Edward Island. Can. Entomol. 129:21-26.

25. Szentesi, A. 1985. Behavioral aspects of female guarding and inter-male conflict in the Colorado potato beetle, Pages 127-137 in: Proc. Symposium on the Colorado potato beetle. XVIIth

International Congress of Entomology. D. N. Ferro and R. H. Voss, eds. Res. Bull. 704, Mass. Agric. Exp. Stn. Circ. 347.

26. Tisler, A. M., and Zehnder.,G. W. 1990. Insecticide resistance in the Colorado potato beetle (Coleoptera: Chrysomelidae) on the Eastern Shore of Virginia. J. Econ. Entomol. 83:666-671.

27. Voss, R. H., and Ferro, D. N. 1990. Phenology of flight and walking by Colorado potato beetle (Coleoptera: Chrysomelidae) adults in western Massachusetts. Environ. Entomol. 19:117-122.

28. Weber, D. C., and Ferro, D. N. 1993. Distribution of over-wintering Colorado potato beetle in and near Massachusetts potato fields. Entomol. Exp. Appl. 66:191-196.

29. Weber, D. C., and Ferro, D. N. 1994. Colorado potato beetle: diverse life history poses challenge to management, Pages 54-70. in: Advances in Potato Pest Biology and Management. G. W. Zehnder, R. K. Jansson, M. L. Powelson, and K. V. Raman, eds.. APS Press, St. Paul.

30. Weber, D. C., and Ferro, D. N. 1996. Flight and fecundity of Colorado potato beetles fed on different diets. Ann. Entomol. Soc. Am. 89:297-306.

Reassessing Autocidal Pest Control

Fred Gould[1] and Paul Schliekelman[2]

Departments of Entomology[1] and Biomathematics[2],
North Carolina State University, Box 7634, Raleigh, NC 27695 USA

The development of insecticidal, transgenic crops has proven valuable for control of key insect pests of major crops (44, other chapters of this volume). However, there is a question about whether it will be profitable to engineer minor crops for resistance to specialized pests, especially if these pests are not susceptible to previously developed forms of insecticidal proteins or if the crop is not easily transformed. For example, it is hard to imagine roses, poinsettia, oranges, or peaches engineered for homopteran or dipteran active toxins in the near future. There is also a large array of pests that do not feed on plants (e.g., urban and medical/veterinary pests), and therefore cannot be controlled by engineered plants.

As transgenic insecticidal row crops keep eroding the market for new, conventional insecticides, producers of minor crops may be relying on an ever-decreasing arsenal of effective and safe pest control tools. Because of grower needs, and because pesticide poisoning of workers in minor crops is a significant health problem (15), there is a need to develop appropriate new technology for these farming systems. In this chapter, we will examine the use of genetically engineered pests as one potential new technology.

The concept of genetically modifying pests by use of classical genetic manipulations dates back at least to the 1940s. Serebovsky (52) and Vanderplank (58) independently suggested that chromosomal abnormalities and hybrid sterility could be used for insect control. The most heralded success in using genetic manipulation for pest control is the eradication of the screwworm from infested areas of the United States (5), Mexico (38) and Libya (59) through release of irradiated, sterile

190

insects. The genetically damaged males released in these programs mated successfully with native females, which subsequently laid eggs that could not develop properly. Repeated releases finally resulted in population extinction.

The theoretical and empirical understanding of sterile-release techniques matured in the 1970s and 1980s (2,13,18,48,62). More sophisticated genetic manipulation approaches, such as using conditionally lethal genes (13) and chromosomal translocations (2), also received significant attention during that time. Conditional lethals could spread into populations during favorable times but could induce a heavy genetic load (i.e., reduce the population's overall fitness) if they inhibited proper diapause initiation or hindered survival at high temperatures. Release of strains with single or double translocations could impose a genetic load on a native population while replacing native genes with those of the released strains. At least theoretically, a population of malaria-vectoring mosquitoes or plant pathogen-vectoring leafhoppers could be replaced by artificially developed, non-vectoring strains that had one or two translocations (11).

A massive infrastructure was needed in any attempt to control pests with these genetic techniques. Because of this, and perhaps due to the limited number of successes, government support for this general area of pest management declined in the late 1980s and 1990s (but see 29,37, 39). Nevertheless, these early theoretical and empirical studies determined some of the limits and potential of this species-specific control technology.

One of the ideas behind our conference on "New Technologies in IPM" was to reexamine some of the issues dealt with at the "Concepts of Pest Management" symposium held at NCSU in 1970 (49), in light of modern technologies. At the 1970 symposium, Max Whitten, who was a leader in the field of genetic control, reviewed the literature in this area (60). While he was optimistic about this new approach, he did lament the lack of knowledge about the genetics of important pests. He also emphasized the limitation that for each pest we would have to find a species-specific, genetic "Achilles' heel." It is worth assessing whether breakthroughs in genetic technologies have diminished the concerns voiced by Whitten, and if so, whether these breakthroughs should again make us optimistic about genetic pest control approaches.

Since the first successful transfer of genes to *Drosophila* from other species, there has been general speculation about the potential for using gene transfer technology to improve natural enemies (26) and other

191

beneficial insects (e.g., honey bees), while diminishing the survival and reproductive capacity of pest populations (12,32). An often-discussed approach for improving natural enemies involves engineering a natural enemy strain to be pesticide resistant by adding a gene that detoxified a targeted class of insecticides. Once the engineered strain was released, it would be expected to quickly establish itself in agroecosystems that were sprayed with those insecticides. In this case the most difficult task would probably be the transformation itself. Establishing a strain in the field that has one gene conferring to it higher fitness than any native strain in the habitat is unlikely to present any extremely challenging problems.

In contrast, establishing a pest strain in the field that is carrying fitness-decreasing genes is a more difficult task. Natural selection is typically acting against such an event. However, a number of ingenious schemes that had potential were developed by Max Whitten and his colleagues (see 13,61,62 for reviews). Even more elaborate theoretical approaches and analyses have been developed since the advent of genetic engineering (31).

However, until recently, advances in theory were about as far as we could go with these new approaches, because the only insect that could be reliably transformed was *Drosophila*. In the last few years, this picture has been changing. In 1998, two labs succeeded in reliably transforming the mosquito, *Aedes aegypti* (10,27). Although we are still far from transforming insects as routinely as we transform crop plants, the road to such routine insect transformation has been laid out (1). It is, therefore, time to more rigorously assess what would be the best use of these transformation techniques if we could routinely use them with pest insects. If we can proactively assess where, and for what insects, genetic control approaches would be most valuable, we might be able to offer useful guidance to molecular biologists who choose to invest time and money in control projects.

Our lab has begun some theoretical studies for this purpose. Our goal is to compare techniques such as the introduction of transposable elements, conditional lethals, meiotic drive, sex ratio distortion, and the classical sterile insect release approach to determine which of these would be most useful for specific applications. In most of our work, we chose to use efficacy of the sterile release strategy as a general yardstick by which to assess other techniques.

Conditional Lethals

In examining the literature on release of conditionally lethal traits, we realized that almost all of the theoretical studies focused on having a single gene (at one locus) introduced into the population (17,48,53). In the few studies that dealt with more than one gene (33,34), the assumption seemed to be that multiple, additively acting genes were needed in order to confer a high enough level of conditional lethality.

Restricting theoretical studies to these conditions made sense in the pre-transformation era. Indeed, one of the limiting factors in using the conditional lethal approach was the lack of any appropriate single (or multiple) genes in the target species. Once we are able to transform insect pests, we will no longer be limited by a lack of appropriate genes. Indeed, Max Whitten's concern that we will need to discover new genes for each pest should be completely alleviated.

Research with engineered plants and *Drosophila* has shown that many promoter sequences (i.e., sequences that control whether or not a protein coding gene is active in a specific cell at a specific time) are available that cause coding-gene expression only under specific environmental conditions. Heat shock promoters are a classic example (4, 64). As indicated by their name, these promoters turn on expression of nearby coding genes when the organism is exposed to high temperature. If a heat shock promoter is attached to a gene that disrupts the development or reproduction of an insect, and this genetic construction is engineered into an insect, the disruptive gene product would (and perhaps in response to other stresses) be produced at higher-than-normal temperatures. If insects with such a genetic construct were released in a greenhouse, they could develop and reproduce normally until the thermostat in the greenhouse was turned up extremely high. At the high temperature, the coding sequence would be turned on and would produce a disruptive protein.

Fortunately, many environment-sensitive promoters maintain their specificity even when transferred to an unrelated organism or placed together with novel protein coding genes (4). The protein coding genes can also operate in novel genomic backgrounds (e.g., Bt genes from bacteria can be expressed in plants). The lack of organism specificity of these DNA sequences means that once a useful construct is developed, it should be useful in many pests. An array of promoters could be developed that switched on coding gene expression under a number of different internal and external environmental conditions (e.g., when diapause was triggered, at high/low gut pH, when feeding on a specific plant species). One could even develop genetic systems with environmentally induced, irreversible genetic changes, as has been proposed for "terminator" seeds that become infertile (Pat.# 5,723,765).

If a single conditional lethal construct were engineered into a pest strain, and relatively high numbers of the homozygous, engineered strain were released at a time when the resident field or greenhouse pest population numbers were low, the conditional lethal trait would become established, at least in heterozygous form, in most of the native genomes. Under appropriate environmental conditions, the insects with at least one copy of the conditionally lethal gene would die. The details of how specific pest, gene, and environmental characteristics impact on the effectiveness of a conditional lethal release are explored in a number of studies (16,17,36,40,53).

Because these papers focused on release strains with a single gene (or multiple genes with additive effects), we set out to determine how much more efficient a release would be if an engineered pest strain were used that had multiple copies of a single transgenic construct, where even a single copy caused 100% mortality of insects exposed to the appropriate environmental conditions (Shliekelman and Gould, in review). Further, we asked what was the optimal number of copies of the transgene that should be added to a release strain in a defined situation. Some of the parameters we explored were as follows:
1. Number of conditional lethal insertions in the release strain
2. Number of generations between the time of release and the time at which the conditional lethals are triggered.
3. Ratio of released to native pest individuals.
4. Release of only males vs release of both sexes.
5. Fitness cost to pest individuals that carry 1-20 insertions, before environmental conditions trigger the lethal effects.
6. General form of population regulation for the pest species.
7. Extent of inbreeding depression in the released strain.

Each of these factors had an impact on effectiveness of the release, and many of the factors interacted with each other in influencing efficacy. Although some restricting assumptions were used in the modeling effort (e.g., random mating, spatially homogeneous environment, complete absence or presence of density dependence), the results of the model are informative. For example, Figure 1 A & B illustrate how the decision

194

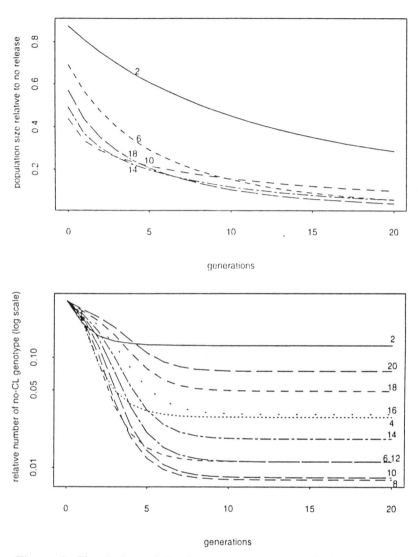

Figure 1. Simulation of the impact on pest population dynamics caused by release of pests engineered to contain multiple copies of a single, dominant, conditional lethal gene. The simulation assumes that each copy of the gene has independent assortment from all of the other copies of the conditional lethal gene, and that each copy reduces fitness by 5% until the conditional lethality is triggered. All insects carrying even one copy of the conditional lethal gene die once

the lethality is triggered. It is assumed that only males are released. The release ratio is 2:1 (released to native insects), and there is no density dependence. A) The impact of releasing insects with varying numbers of copies of the conditional lethal on the relative population size of the target pest even before the lethality is triggered (The simulation allows for up to 20 generations before lethality is triggered. The relative decrease in population size is caused by the 5% fitness cost per gene copy. B) The number of pest insects in the population that are not carrying any conditional lethal (CL) genes in the case when there has been a release, compared to the total number of pests in a population when there has been no release. The relative number of no-CL pests is important because this is the number of insects that survive when the conditional lethality is triggered. (Number of copies per released insect is given by the number next to the respective population trajectory.)

about the number of insertions to use depends on the generation in which the lethal trait is triggered, as well as the fitness cost imposed by each inserted copy of the gene in generations before lethality is triggered. It can be seen from examining Fig. 1B that if lethality is triggered after two versus four generations, a different number of inserted copies results in optimal control (i.e., number of individuals that have no conditional lethal alleles [No-CL]).

When the use of multiple insertions is compared to the traditional idea of using single insertions, it becomes clear that the use of multiple insertions can be much more efficient, even when there is a 5% fitness cost to each inserted copy. We further compared the efficacy of multiple insertions to the efficacy of the sterile insect release technique as described by Knipling (35). Unless there is a very high non-conditional fitness cost, the multiple conditional lethal approach is typically over 10 times more efficient per insect released. One disadvantage of the conditional release compared to the sterile release is the lag time in activation. However, if local or widespread eradication is the goal, this lag time should not be too problematic.

Efforts are needed to test our model predictions with real insects in laboratories and greenhouses to see if our predictions hold true. At this point, tests with population cages of transformed *Drosophila* are the only readily available test system. There is also a need to develop models that can deal with spatially and temporally structured and regulated insect populations.

The very important issue of "gene silencing" must be addressed before embarking on any molecular-genetic project aimed at inserting multiple copies of a gene. Gene silencing is the decreased, or lack of, expression of a gene that results from the presence of multiple copies of that specific gene in an individual. While there is less information on gene silencing in animals (6,19) than there is for plants (23,52), the data available for *Drosophila* and mice indicate that the phenomenon of gene silencing has somewhat different attributes in different taxa. Having a few gene copies in *Drosophila* sometimes results in no gene silencing at all (55). In cases where there is gene silencing, the silencing is typically only partial. For example, Pal-Bhadra et al. (46) found that even with six copies of an introduced ADH gene, there was still significant gene expression. Assessment of their procedures (46, and J.A. Birchler personal communication) indicates that gene silencing was not permanent, so it had to be reinitiated each generation. Work with mice also found that gene silencing depended on the number of gene copies in the generation studied, not the number in previous generations (19). These findings are important for the release of pest strains with multiple copies of conditional lethals. Even if the high copy number in the released generation caused complete gene silencing, this wouldn't doom the release. By the time the conditional lethal was triggered, most of the insects would have only one or two copies of the conditional lethal gene. Hopefully, advances in molecular genetics over the next few years will enable us to avoid gene silencing, even when there is high copy number (28).

Translocations and Transposable Elements

In the older literature on genetic control of pests, the idea of releasing pest strains with one or more translocations was explored as a means of introducing a new trait into a population of pests. In an insect with a translocation, a piece of one chromosome has "broken off" and become attached to another chromosome. An insect strain with a translocation can be viable because it has all of the species' DNA code. In the F2 generation of a cross between a translocated strain and a normal strain, a problem arises because one-fourth of the offspring lack the piece of DNA that was involved in the translocation. Therefore, the fitness of the normal and translocated strains is decreased by their intermating. The proportion of mixed matings is highest for the strain that is least abundant. That results in the less abundant strain having lower fitness than the more abundant strain. The strain (or more specifically, the chromosome type) that is initially less abundant is therefore generally expected to go extinct (11,13,61). If a translocated strain was released into the wild at a higher density than the native strain, the translocated chromosome would replace the normal chromosome, as long as it had approximately the same fitness. If the translocated chromosome was carrying a new trait, that trait would be substituted into the general population. Potentially useful traits would be the inability to vector an animal or plant disease or hyper-susceptibility to an insecticide. The problem with using this approach is that even if the translocated strain is just as fit as the wild strain, a large number of insects would need to be released in order to assure success. This approach has been attempted in some field trials with limited success (2).

In the early 1990s, entomologists and molecular biologists realized that transposable elements such as the P element in *Drosophila* could be used in a manner similar to, but more efficient than, the use of translocations. This spurred new theoretical and empirical studies (12,41, 50). While replacement of a native chromosomal type by insects with translocations required release of large numbers of laboratory-reared insects, it seemed theoretically possible to release a relatively small number of insects with transposable elements (about 1% to 10% of the native population) in order to substitute a new genetic trait into a population (20,31,50,56).

Transposable elements are pieces of DNA that are inserted into the normal chromosomes of a species but have the ability to excise themselves and move to a different place on the chromosome or to a different chromosome. They can be considered an infectious disease of chromosomes because once one copy of a transposable element enters a cell, it can increase in number and infect other chromosomes within the cell, including germline cells. A sexual, diploid insect with one copy of a typical gene only transmits that gene to one-half of its offspring. Because the transposable elements can multiply within the insect, an insect which starts off as an embryo with one copy of a transposable element may transmit this element to more than one-half of its offspring. If this happens generation after generation, all of the insects in a population end up carrying the transposable element. Laboratory experiments have shown that if 10% of the *Drosophila* in a cage start out carrying a transposable element, more than 90% can be carrying it within 10 generations (21,22).

It is possible to engineer a useful gene into the middle of a transposable *Drosophila* P element. When a population is "infected" with such a loaded P element, the new, useful gene is expected, at least theoretically, to become fixed in the population. If the new gene coded for inability to vector a disease organism, the whole population would become refractory. Indeed, there is considerable effort underway to find genes for refractoriness in mosquitoes.

There is a high cost, and potentially a high benefit, associated with developing and releasing a mosquito strain that could minimize the threat of human diseases. This approach has, therefore, attracted much discussion (14,47,54). One of the major concerns with releasing a strain that has a loaded transposable element is the finding that larger transposable elements do not transpose as rapidly as smaller elements (any loaded transposable element is large). One *Drosophila* laboratory study found that a P element loaded with an ADH reporter gene was spread through a population from a starting frequency of 1% or 10% (41). However, another *Drosophila* study (7) found that the reporter gene was not completely spread into the population. In this second case, the evidence indicated that some of the P elements lost the reporter gene, and these unloaded elements spread through the population. Once a P element spreads into a population, that population becomes resistant to accommodating more P elements. If the first attempt to insert a useful gene using a specific transposable element fails to establish the desired gene but does spread the transposable element into the population, it could be impossible to do the release again. It is therefore critical to develop a reasonably fail-proof system before any release is conducted. There will always be the problem that engineered transposable elements may not behave the same in the field as they do in the laboratory. Some researchers have therefore proposed that the first attempts to introduce loaded transposable elements should involve less important disease vectors where a failure will not be as detrimental.

One other criticism of using transposable elements is the potential for these elements to spread between species. There is evidence that the *Drosophila* P element spread from *Drosophila willistoni* to *Drosophila melanogaster* during the past 50 years (30). It is now practically impossible to find wild *Drosophila melanogaster* populations without P elements. Furthermore, there is evidence that a mite may have served as the vector for transferring the transposable element (25). Transposable elements, by themselves, can decrease an individual's fitness if they insert within an important DNA sequence. If the loaded gene is used to decrease fitness of the target species, it could also decrease the fitness of other species. It is also possible for an active transposable element to activate "dormant" transposable elements that reside in other species. Because of the widespread taxonomic distribution of some transposable elements, it is not clear how far an introduced element could move taxonomically (43, 51). Although the potential for environmental disruption posed by cross-species transfer may be relatively small, the risk may be perceived as much larger by the public.

Meiotic Drive and Other Segregation Distortion Mechanisms

In typical sexual reproduction, haploid gametes are formed from diploid cells during meiosis. While the diploid cells have two copies of each of the organism's chromosomes, the haploid gametes have only one copy. Typically, each of the two copies of the chromosome from the diploid cell has a 50% chance of winding up in a gamete that produces offspring. If a gene arises on a chromosome that gives that chromosome even a 51% chance of being passed on through the meiotic and fertilization process, that chromosome (or at least the parts of the chromosome that are tightly linked to the "driving" gene) will become fixed in the population. Evidence of meiotic drive is most easily seen when the genes and chromosomes are involved in sex ratio determination. If the X chromosome out competes the Y chromosome in *Drosophila* or mice, more than 50% of the fertilized eggs have XX genotypes and become females. This results in distortion of the sex ratio (3,42,45). Because individual natural selection favors a 1:1 sex ratio, modifier genes evolve which negate the action of sex ratio distorting genes (e.g., 8,9). This phenomenon has been studied in a number of organisms, but most attention has been focused on mice and *Drosophila*. While the result of segregation distortion in the XY system is opposed by natural selection, the result of segregation distortion in autosomes may be neutral in terms of natural selection. Some evidence for such autosome segregation distortion comes from hybrid crosses of genetically differentiated plant populations where the chromosomes from one population are over-represented in the segregating F2 generation (24,63).

This phenomenon attracted attention of applied insect geneticists, who saw the opportunity to link their favorite gene to a segregation distorter gene as a way to spread it into a population (61). This idea, like many other ideas for genetically manipulating pest populations, did not get past the pencil and paper stage. Molecular biologists have learned a lot about segregation distortion since this possibility was discussed in the 1970s and 1980s. Segregation distortion genes have been mapped in plants and animals (42,63). As we learn more about segregation distortion genes, we may be able to build transgenic strains that have constructs that express a segregation distorter gene and a gene coding for non-vectoring of a disease organism. Depending on the strength of the segregation distorter, such a construct may be more useful in pest management than transposable elements. We do not know at this point whether such segregation distorter genes will be species specific, or if some distorter genes will be active in a variety of taxa.

Summary

We have outlined how advances in molecular genetics could change the feasibility of using genetic manipulation of insect genetics as a pest control tool. It seems clear that new genetic tools will enable us to make more useful modifications of pests, with overall greater efficiency. What is not clear is where and when these engineered pests will be most useful economically and environmentally.

For most of the modifications discussed that result in acute autocidal control, there is still a need to release a high ratio of engineered pests compared to those in the natural population. This means that it will be difficult to implement releases for a pest that always exists at high densities and moves long distances (e.g., cotton bollworm), unless it is sponsored by a large government program. However, for pests that can cause high levels of damage at low densities (e.g., high-value fruit-infesting pests), or in situations where pests have very low densities at the start of each season (e.g., greenhouse pests), establishing a large enough release may be more feasible. In the case of pests whose mobility is culturally limited (e.g., greenhouse pests) or is limited by pest biology (e.g., Colorado potato beetle, corn billbugs), it could be economically reasonable for individual growers to purchase engineered insects for release.

Many of the constraints that are associated with engineered pests are similar to those for inundative biocontrol releases. The autocidal approach should be considered when the required biocontrol agent does not exist or is difficult to produce in high numbers. Indeed, autocidal control is in some ways a highly species-specific form of biological control

The use of insects with transposable elements or genes carried next to segregation distorter loci does not require release of high ratios of engineered insects because, at least theoretically, these genes will push themselves into populations over time. Although these genes may not offer suppression of the pest densities, they will be quite useful if they inhibit pest transmission of plant diseases.

There is no doubt that some of the public response to the release of genetically engineered insects will be negative, even if the released insects carry genes that cause debilitation. We must therefore take extra care to explain this technology to the public and to avoid approaches that involve any real risks.

We are just approaching an era where genetic manipulation of all organisms will be possible. We need to start defining the potential uses of this technology if we are to gain maximum benefits from it.

Acknowledgments

We would like to thank the organizing committee for inviting us to present this information. Discussions with Steve Ellner and his laboratory group, as well as with Sara Oppenheim, Amy Sheck, and Nick Storer, improved the manuscript. This work was in part supported by the Keck Foundation Program in Behavioral Biology and the North Carolina Agricultural Research Service.

Literature Cited

1. Ashburner, M., Hoy, M. A., and Peloquin, J. J. 1998. Prospects for the genetic transformation of arthropods. Insect Molec. Biol. 7:201-213.
2. Asman, S. M., McDonald, P. T., and Prout, T. 1981. Field studies of genetic control systems for mosquitoes. Annu. Rev. Entomol. 26:289-343.
3. Atlan, A., Mercot, H., Landre, C., Montchamp-Moreau, C.,

Ashburner, M., Hoy, M. A., and Peloquin, J. J.. 1997. The sex-ratio trait in *Drosophila simulans*: geographical distribution of distortion and resistance. Evolution 51:1886-1895.

4. Bachiller, D., Dubouled, M. A., and Morata, G. 1994. Conservation of a functional hierarchy between mammalian and insect hox/hom genes. Embo J. 13:1930-1941.

5. Baumhover, A. H. 1966. Eradication of the screwworm fly as agent of myiasis. J. Am. Med. Assoc. 196:240-248.

6. Birchler, J. A., Pal-Bhadra, M., and Bhadra, U. 1999. Less from more: cosuppression of transposable elements. Nature Genet. 21:148-149.

7. Carareto, C. M. A., Kim, W., Wojciechowski, M. F., Martin, F., O'Grady, P., Prokchorova, A. V., Alla, V., Silva, J. C., and Kidwell, M. G. 1997. Testing transposable elements as genetic drive mechanisms using *Drosophila* P element constructs as a model system. Genetica 101:13-33.

8. Cavalho, A. B., Sampaio, M. C., Varandas, F. R., and Klaczko, L. B. 1998. An experimental demonstration of Fisher's principle: evolution of sexual proportion by natural selection. Genetics 148:719-731.

9. Cazemajor, M., Landre, C., and Montchamp-Moreau, C. 1997. The sex-ratio trait in *Drosophila simulans*: genetic analysis of distortion and suppression. Genetics 147:635-642.

10. Coates, C. J., Jasinskiene, N., Miyashiro, L., and James, A. A. 1998. Mariner transposition and transformation of the yellow fever mosquito, *Aedes aegypti*. Proc. Natl. Acad. Sci. USA 95:3748-3751.

11. Curtis, C. F. 1968. Introducing vector incompetence. Nature 218: 368-369.

12. Curtis, C. F. 1992. Selfish genes in mosquitoes. Nature 357: 450.

13. Davidson, G. 1974. The Genetic Control of Insect Pests. Academic Press, London

14. Eichner, M., and Kiszewski, A. 1995. Ribeiro and Kidwell's transposal model. J. Medical Ent. 32:1-4.

15. EPA. 1992. Worker protection standard, hazard information, hand labor tasks on cut flowers and ferns exception; final rule, and proposed rules. Fed. Regist. 57:38102-38176.

16. Fitz-Earle, M., and Suzuki, D. T. 1975. Conditional mutations for the control of insect populations. Pages 365-374 in: Sterility Principle for Insect Control. Int. Atomic Energy Agenc., Vienna.

17. Fitz-Earle, M., Holm, D. G., and Suzuki, D. T. 1975. Population

control of caged native fruitflies in the field by compound autosomes and temperature-sensitive mutants. Theor. Appl. Genet. 46:25-32.

18. Foster, G. G., Vogt, W. G., Woodburn, T. L., and Smith, P. H. 1988. Computer simulation of genetic control. Comparison of sterile males and field-female killing systems. Theor. Appl. Genet. 76:870-879.

19. Garrick, D., Fiering, S., Martin, D. L. K., and Whitelaw, E. 1998. Repeat-induced gene silencing in mammals. Nature Genet. 18:56-59.

20. Ginzburg, L. R., Bingham, P. M., and Yoo, S. 1997. On the theory of speciation induced by transposable elements. Genetics 107:331-341.

21. Good, A. G., and Hickey, D. A. 1987. Hybrid dysgenesis in *Drosophila melanogaster*: the elimination of P elements through repeated backcrossing to an M-type strain. Genome 29:195-200.

22. Good, A. G., Meister, G. A., Brock, H. W., Grigliatti, T. A., and Hickey, D. A. 1989. Rapid spread of transposable P elements in experimental populations of *Drosophila melanogaster*. Genetics 122:387-396.

23. Grant, S. R. 1999. Dissecting the mechanisms of posttranscriptional gene silencing: divide and conquer. Cell 96:303-306.

24. Harushima, Y., Kurata, N., Yano, M., Nagamura, Y., Sasaki, T., Minobe, Y., and Nakagahra, M. 1996. Detection of segregation distortions in an *indica-japonica* rice cross using a high resolution molecular map. Theor. Appl. Genet. 92:145-150.

25. Houck, M. A., Clark, J. B., Peterson, K. R., and Kidwell, M. G. Possible horizontal transfer of *Drosophila* genes by the mite *Proctolaelaps regalis*. Science 253:1125-1129.

26. Hoy, M. A. 1993. Transgenic beneficial arthropods for pest management programs: an assessment of their practicalities and risks. Pages 357-369 in: Pest Management: Biologically Based Technologies. R. D. Lumsden and J. L. Vaughn, eds. Am. Chem. Soc. Conf. Proc. Series, Washington, DC, USA.

27. Jasinskiene, N., Coates, C. J., Benedict, M. Q., Cornel, A. J., Rafferty, C. S., James, A. A., and Collins, F. H. 1998. Stable transformation of the yellow fever mosquito, *Aedes aegypti* with the *Hermes* element from the housefly. Proc. Natl. Acad. Sci. USA 95: 3743-3747.

28. Kasschau, K. D., and Carrington, J. C. 1998. A counterdefensive strategy of plant viruses: suppression of posttranscriptional gene

silencing. Cell 95:461-470.

29. Kerremans, P., and Franz, G. 1995. Isolation and cyto-genetic analyses of genetic sexing strains for the medfly, *Ceratitis capitata*. Theor. Appl. Genet. 9:255-261.

30. Kidwell, M. G. 1983. Evolution of hybrid dysgenesis determinants in *Drosophila melanogaster*. Proc. Natl. Acad. Sci. USA 80: 1655-1659.

31. Kidwell, M. G., and Ribeiro, J. M. C. 1992. Can transposable elements be used to drive disease refractoriness genes into vector populations? Parasitol. Today 8:325-329.

32. Kidwell, M. G., and Wattam, A. R. 1998. An important step forward in the genetic manipulation of mosquito vectors of human disease. Proc. Natl. Acad. Sci. USA 95:3349-3350.

33. Klassen, W., Knipling, E. F., and McGuire, J. U. Jr. 1970a. The potential for insect-population suppression by dominant conditional lethal traits. Ann. Entomol. Soc. Am. 63:238-255.

34. Klassen, W., Creech, J. F., and Bell, R. A. 1970b. The potential for genetic suppression of insect populations by their adaptations to climate. USDA-ARS Misc. Publ. No. 11788.

35. Knipling, E. F. 1955. Possibilities of insect control or eradication through the use of sexually sterile males. J. Econ. Entomol. 48:459-462.

36. Knipling, E. F. 1960. The use of insects for their own destruction. J. Econ. Ent. 53:415-420.

37. Knipling, E. F. 1998. Sterile insect and parasite augmentation techniques: unexploited solutions for many insect pest problems. Fla. Ent. 81:134-160.

38. Krasfur, E. S. 1998. Sterile insect technique for suppressing and eradicating insect population: 55 years and counting. J. Agric. Ent. 15:303-317.

39. Krasfur, E. S., Whitten, C. J., and Novy, J. E. 1987. Screwworm eradication in North and Central America. Parasitol. Today 31:131-137.

40. LaChance, L. E., and Knipling, E. F. 1962. Control of insects through genetic manipulations. Ann. Entommol. Soc. Am. 55:515-520.

41. Meister, G. A., and Grigliatti, T. A. 1994. Rapid spread of a P element/Adh gene construct through experimental populations of *Drosophila melanogaster*. Genome 36:1169-1175.

42. Montagutelli, X., Turner, R., and Nadeau, J. H. 1996. Epistatic control of non-mendelian inheritance in mouse interspecific crosses. Genetics 143:1739-1752.

43. O'Brachta, D. A., Warren, W. D., Saville, K. J., and Atkinson, P. W. 1994. Interplasmid transposition of *Drosophila hobo* elements in noon-drosopholid insects. Molec. Gen. Genet. 244:9-14.

44. Ostlie, K. R., Hutchison, W. D., and Hellmich, R. L., eds. 1997. Bt Corn and European Corn Borer. North Central Reg. Ext. Publ. Univ. Minn. Ext. Serv., St. Paul, Minnesota, USA.

45. Owusu-Daaku, K. O., Wood, R. J., and Butler, R. D. 1997. Variation in Y chromosome meiotic drive in *Aedes aegypti* (Diptera: Culicidae): a potential genetic approach to mosquito control. Bull. Ent. Res. 87:617-623.

46. Pal-Bhadra, M., Bhadra, U., and Birchler, J. A. 1997. Cosuppression in *Drosophila*: gene silencing of alcohol dehydrogenase by white-Adh transgenes is polycomb dependent. Cell 90:479-490.

47. Pfeifer, T. A., and Grigliatti, T. A. 1996. Future perspectives on insect pest management: engineering the pest. J. Invert. Pathol. 67:109-119.

48. Prout, T. 1978. The joint effects of the release of sterile males and immigration of fertilized females on a density regulated population. Theor. Popul. Biol. 13:40-71.

49. Rabb, R. L., and Guthrie, F. E. 1970. Concepts of Pest Management. North Carolina State University, Raleigh, North Carolina, USA.

50. Ribeiro, J. M. C., and Kidwell, M. G. 1994. Transposable elements as population drive mechanisms: specification of critical parameter values. J. Med. Entomol. 31:10-16.

51. Robertson, H. 1993. The mariner transposable element is widespread insects. Nature 362:241-245.

52. Serebovsky, A. S. 1940. On the possibility of a new method for the control of insect pests. Zool. Zh. 19:618-630.

53. Smith, R. H. 1971. Induced conditional lethal mutations for the control of insect populations. Pages 453-465 in: Sterility Principle for Insect Control or Eradication. Int. Atomic Energy Agency., Vienna.

54. Spielman, A. 1994. Why entomological antimalaria research should not focus on transgenic mosquitoes. Parasitol. Today 10:374-376.

55. Spradling, A. C., and Rubin, G. M. 1983. The effect of chromosomal position on the expression of the *Drosophila xanthine* dehydrogenase gene. Cell 34:47-57.

56. Uyenoyama, M. K. 1985. Quantitative models of hybrid dysgenesis: rapid evolution under transposition, extrachromosomal inheritance, and fertility selection. Theor. Popul. Biol. 27:176-201.

57. Vaucheret, H., Beclin, C., Elmayan, T., Feuerbach, F., Godon, C., Morel, J. B., Mourrain, P., Palauqui, J. C., and Vernhettes, S. 1998. Transgene-induced gene silencing in plants. Plant J. 16:651-659.

58. Vanderplank, F. L. 1944. Hybridization between *Glossina* species and suggested new method for control of certain species of tsetse. Nature 154:607-608.

59. Vargas, T. M., Hursey, B. S., and Cunningham, E. P. 1994. Eradication of the screwworm from Libya using the sterile insect technique. Parasitol. Today 10:199-122.

60. Whitten, M. J. 1970. Genetics of pests in their management. Pages 119-137 in: Concepts of Pest Management. R. L. Rabb and F. E. Guthrie, eds, North Carolina State Univ., Raleigh, North Carolina, USA.

61. Whitten , M. J. 1985. The conceptual basis for genetic control. Pages 465-528 in: Comprehensive Insect Physiology, Biochemistry, and Pharmacology. G. A. Kerkut and L. I. Gilbert, eds. Vol. 12. Pergamon, New York.

62. Whitten, M. J., and Foster, G. G. 1975. Genetical methods of pest control. Annu. Rev. Entomol. 20:461-476.

63. Xu, Y., Zhu, L., Xiao, J., Huang, N., and McCouch, S. R. 1997. Chromosomal regions associated with segregation distortion of molecular markers in F-2, backcross, doubled haploid, and recombinant inbred populations in rice (*Oryza sativa* L.). Molec. Gen. Genet. 253:535-545.

64. Zhoa, Y. G., and Eggleston, P. 1999. Comparative analysis of promoters for transient gene expression in cultured mosquito cells. Insect Molec. Biol. 8:31-38.

Section 4

Biological Control

Biological control has been a fundamental element of the IPM concept since the concept was initially formulated more than 30 years ago. In the intervening years, there have been significant advances in population theory, microbial ecology, and pesticide technology, as well as in the theory and practice of biological control. Although classical biological control (the importation and release of exotic species to effect control of insect or weed pests) has experienced numerous successes during this period, perhaps the greatest advances have been made in the modification of pest control and production practices to enhance the suppression of insect pests and plant pathogens by naturally occurring beneficial organisms, and in applying knowledge from microbial ecology to develop improved biological controls for plant pathogens. Additionally, advances in molecular biology and changes in pesticide regulations have facilitated the development and commercialization of a number of biological control technologies for insect, weed, and disease management in IPM.

Marjorie Hoy begins this section with a discussion of the challenges facing biological control of insects in the 21st century and of possible approaches to meet those challenges. Joyce Loper and Virginia Stockwell then present a brief historical overview of the development of biological control of plant pathogens. They use biological control of fire blight in pome fruits to illustrate the characteristics of biological control products that enhance or limit their adoption by producers and then discuss strategies to optimize biological control of plant pathogens. Next, Douglas Landis et al. discuss the potential for using habitat modification to enhance biological control of insect pests. James Ligon et al. then describe the use of recombinant DNA technology to produce genetically improved strains of *Pseudomonas fluorescens* having enhanced activity as a biological control of plant pathogens. This is followed by Raghavan Charudattan's description of the current status of biological control of weeds using bioherbicides. He identifies constraints on and opportunities for the use of plant pathogens in integrated weed management. In the final chapter of this section, J. C. Meneley analyzes the current market for commercial biological control products and describes the challenges that exist to commercialization of biological control technologies for IPM.

Current Status of Biological Control of Insects

Marjorie A. Hoy
Department of Entomology and Nematology,
University of Florida, Gainesville, FL 32611-0620 USA

Biological control of arthropod pests and weeds is a method of pest management that employs parasitoids, predators, pathogens, and entomo-pathogenic nematodes (= natural enemies) to reduce pest populations. Many insects or mites are not pests because natural enemies keep them suppressed with no assistance from humans. This naturally occurring biological control often is discovered only after it has been disrupted and arthropod populations have increased dramatically to become "new" pests. Biological control also includes an applied technology through which humans attempt to restore, enhance, or mimic a natural phenomenon using three basic approaches: classical biological control, augmentation, or conservation.

Classical biological control involves importing and establishing natural enemies to provide long-term control of foreign and, occasionally, native pests (14,23,46). Interest in classical biological control in the USA developed after the spectacularly successful introduction and establish-ment in 1888 of the Vedalia ladybeetle, *Rodolia cardinalis*, and an endoparasitic fly, *Cryptochetum iceryae*, to control the cottony cushion scale, *Icerya purchasi*, throughout California citrus groves. These natural enemies have provided effective control ever since, unless disrupted by applications of pesticides, and they have been distributed to more than 25 additional countries, where they have provided excellent control (7). Classical biological control has received the most attention in the USA, but interest in conservation and augmentation is increasing (2,26,33).

Augmentation involves efforts to increase populations or beneficial effects of natural enemies, most often employing periodic releases of natural enemies, which provide either immediate or delayed effects on pest populations. Augmentation also involves environmental manipulation, including provision of alternate or factitious hosts or prey, provision of food or nesting sites, and modification of cropping practices to favor natural enemies.

Conservation involves efforts to preserve and maintain existing natural enemies by modifying crop management techniques or other environmental components. Conservation is achieved most often by altering pesticide use patterns to preserve natural enemies and increase their effectiveness. Other methods involve altering management practices, such as strip cropping or changing the timing of planting or harvesting. The techniques employed in augmentation and conservation overlap to some degree.

For more than 40 years, biological control has been considered by some to be one of the most important components of agricultural integrated pest management (IPM), although chemical control has prevailed as the most commonly used tactic (4,8,23). Interest in making biological control a more important component of IPM, as originally conceived by Stern et al. (37), was the impetus for a workshop in 1984 at the University of Florida, where the following point was made:

> "Many students in recent years have been taught that biological control and plant resistance are cornerstones of IPM. Despite the supposed, and often-quoted, status of biological control as a basic subdiscipline underlying IPM, research to incorporate biological control into agricultural cropping management systems is still relatively uncommon, and funding for such effort is scarce. It is our perception that IPM program managers have often tended to support projects that focused on research designed to time pesticide applications effectively, to model plant/pest interactions, to identify and/or refine economic injury levels, and to predict pest populations. These subjects are certainly important and deserve the attention that they have received. However, funds for biological control -- whether by importation, augmentation, or conservation -- probably constitute 20% or less of the funding allocated to IPM programs." (26)

Fourteen years later, those statements generally remain valid. The irony is that while the stage is set for biological control to become a more important component in agricultural, forestry, and urban IPM systems in the 21st century, its enhanced use could be limited by a variety of social, ecological, legal, and organizational issues to be discussed below.

The Loss of Registered Pesticides Offers a Challenge to Biological Control Workers

LEGISLATIVE CHANGES

The Food Quality Protection Act (FQPA) in the USA has yet to be implemented, but substantial changes likely are in store for pest management. Whether fewer pesticides will be available is unclear; however, without changes induced by the FQPA, demand for insecticide chemical active ingredients is still projected to grow by 4% annually to $2 billion in the year 2000 (8). The loss of traditional registered products, especially for the so-called "small (minor) crops," may shift our focus from a pesticide-centered IPM system to a more biologically based IPM model. The challenge to biological control workers will be to respond to this shift in a timely manner, especially if some crops are left with few or no registered pesticides. This response will be particularly difficult in ornamental and vegetable crops, where cosmetic standards are high.

RESISTANCE TO PESTICIDES

Some pesticides are being lost due to the development of resistance in the pests. The problems of resistance, especially in pests that are particularly prone to develop resistance, such as the Colorado potato beetle, aphids, and spider mites, are unlikely to be resolved by the registration of new classes of pesticides. As noted by toxicologists (8), "This process of continually shifting approaches may ultimately be limited by a finite number of practical targets for pest control." During the past 40 years, each new product class has been hailed as a solution to the resistance problem, yet each has been conquered by resistance in key pests. The solution to resistance will be found only when we can manage pests in a coordinated, multi-tactic manner, using pesticides as solutions of the last resort -- when other tactics fail (23,37).

The development of resistance to Bt toxin genes inserted into transgenic crops also will occur; the proper question is not *if*, but *when*. The options of pyramiding or rotating toxin genes as resistance mitigation methods may be ineffective. Resistance to Bt toxins in the diamondback moth is complex and varies from one population to another, making it impossible to meet the underlying assumptions found in resistance management models (38). Fortunately, pests rarely develop resistance to parasitoids and predators (18).

Challenges to Biological Control for the 21st Century

CLASSICAL BIOLOGICAL CONTROL

The increasing need. There is an increased need for more classical biological control rather than less, as pests continue to invade new geographic regions due to the globalization of agriculture and commerce (12,41). Our borders are increasingly "leaky," and trade and diplomatic issues often determine whether agricultural products are allowed into the USA from foreign countries that have exotic pests not yet established in the USA.

Despite the inspections made on imported products by the U.S. Department of Agriculture Animal and Plant Health Inspection Services (USDA-APHIS) and the quarantines imposed when new pests are detected, inadequate funding and the ever-increasing volume of products imported, both legally and illegally, into the USA limits the effectiveness of APHIS. It appears that invading pests will continue to ravage our agricultural, forest, and urban environments unless improvements can be made in the exclusion, interception, and eradication of newly established pests.

In February 1999, President Clinton signed an Executive Order establishing an Invasive Species Council, which is charged with developing a comprehensive plan to minimize the economic, ecological, and human health effects of invasive species and to determine how to prevent the introduction and spread of additional species. The budget for the fiscal year 2000 proposes an increase of more than $28.8 million for this purpose, including accelerating research on habitat restoration and biologically based IPM tactics. The Council's Management Plan is due within 18 months, and it won't come too soon! We are being inundated with pests.

Just since 1992, pest managers in Florida's citrus have been challenged by the invasion of the Mediterranean fruit fly, the citrus leafminer, the brown citrus aphid, and the citrus psylla (not to mention citrus canker). The response to the Medfly has been to eradicate it using

213

aerial applications of malathion bait spray and releases of sterile insects, but the leafminer, brown citrus aphid, and psylla are targets of classical biological control. This is crucial, because most arthropod pests in citrus are under complete-to-substantial biological control. We have adopted a "proactive" biological control strategy in an effort to shorten the time required to identify potentially effective natural enemies and potential collaborators overseas who can assist us in obtaining them. If, however, a pest invades that cannot be controlled biologically or by other tactics compatible with natural enemies, then our entire citrus IPM program will be jeopardized, indicating we are on a biological control "treadmill."

The biodiversity issue. Classical biological control of arthropod pests can be effective in suppressing newly established species that invade without their natural enemies, but it is under increasing criticism from people concerned about biodiversity and the potential threat to endangered or nontarget species (19,35,36). It is ironic that, at the time when invasive species are increasingly recognized as a serious environmental threat, one of the most effective solutions is under threat itself.

Other challenges include the declining number of qualified taxonomic specialists who can identify both pest and natural enemy species. Without their essential services, classical biological control will be hindered.

The art-into-science challenge. Effective classical biological control depends upon the establishment of one or more foreign natural enemies in a new environment. Estimates of successful establishment rates range from 16% to 34% (20). Depending on your viewpoint, these rates are high or low, but they can, and should, be improved. With additional research on climatic tolerances, host range, genetics of colonization, and release tactics, the results of natural enemy releases could become more predictable. Another challenge is the need to evaluate the effects of natural enemies imported in classical biological control programs on nontarget species.

AUGMENTATIVE BIOLOGICAL CONTROL

Mass rearing. Bigler (5) estimated that approximately 150 species of entomophagous insects and mites are mass reared and released worldwide for biological control, especially in greenhouses. Augmentation of natural enemies of other pests could be more widely adopted. For example, releases of natural enemies of livestock ectoparasites are attractive to livestock producers if costs can be competitive with pesticides and users can be educated effectively (17).

At present, augmentative releases in the USA primarily are conducted against pests of greenhouse crops, nurseries, or field-grown strawberries -- crops with high value and a long history of pests becoming resistant to pesticides (31). I am unaware of any operational programs in large-scale agriculture (such as the vast hectares of cotton, wheat, or corn) that employ augmentative releases.

The relatively low rate of use of augmentative biological control probably is due to a combination of high costs and low quality. For example, the average cost of control using *Trichogramma brassicae* is $102 per hectare (40), which must be more expensive than chemical control. Quality control is often poor, and the number of natural enemies shipped and number received alive or in good health may vary widely from company to company and from time to time (11). Sometimes the incorrect species is shipped (27). For example, approximately 20% of coccinellids shipped were parasitized, and a range of 75-500 females were received when 1000 were ordered (32). Furthermore, it is not unusual for the consumer to receive inadequate information on release rates or timing or user-friendly sampling methods for determining whether the natural enemies are performing as expected. Such problems create credibility issues with consumers.

CONSERVATION OF NATURAL ENEMIES

Compatibility of pesticides and natural enemies. Conservation of natural enemies is limited by inadequate information. When we have adequate information on the compatibility of pesticides with natural enemies, it is often possible to reduce pesticide use by 50% to 80%, with no reduction in crop yield or quality (23,29). We also can reduce pesticide application numbers and rates with improved knowledge of economic injury levels, particularly in the case of indirect pests, where the threshold for control could be higher than commonly assumed. It is unlikely that pesticide use can be eliminated completely, because some key pests cannot be controlled by any other method and, for these situations, learning how to conserve natural enemies will be critically important.

Transgenic crops and natural enemies. The conservation of natural enemies does not just involve knowing whether a product causes acute toxicity to the natural enemy. If the host/prey are eliminated from the system, then natural enemies cannot persist and have to migrate back into the crop from some distances. The use of high doses of Bt toxins in transgenic crops could lead to the disruption of predator/prey or host/parasitoid interactions if crops are grown over large areas. Additional research should be conducted to determine how transgenic crops and natural enemies can be used in a compatible manner. Employing natural enemies in the system would have the added benefit of delaying the development of resistance to genes used in the transgenic plants (44).

POTENTIAL NEW TECHNOLOGIES

Transgenic natural enemies. The development of transgenic technology opens the possibility that genetically improved natural enemies will be developed, including transgenic entomopathogens, entomopathogenic nematodes, parasitoids, and predators (1,13,15,24). A variety of scientific and other issues must be resolved before improved transgenic natural enemies can be deployed in practical pest management programs.

Genes and regulatory sequences must be identified that would potentially improve the effectiveness of transgenic natural enemies, but we currently lag far behind in this endeavor compared to the researchers working to develop transgenic pest arthropods. We need effective guidelines on methods for containing the transgenic strains in the laboratory until they have been approved for release into the environment (28). While we have guidelines on how to assess the potential risks of releasing these agents into the environment for short-term experiments (39), we lack information on potential risk assessment issues associated with the permanent establishment of transgenic natural enemies in the environment (21,22,25).

Molecular taxonomy and ecology. The loss of taxonomic specialists to identify pest and beneficial species is crucial to classical biological control, and can be offset only partially by the use of new molecular tools to identify races, biotypes or cryptic species.

Potential Solutions

CLASSICAL BIOLOGICAL CONTROL

Classical biological control could have a more important role to play in the future, because classical biological control could be extended to more ecosystems (e.g., ephemeral crops and urban environments) if approaches, mechanisms, theory, and principles for effective natural enemies can be developed for these habitats (14,45,46).

Molecular tools for biological control. The development of several molecular methods, such as random amplified polymorphic DNA polymerase chain reaction (RAPD-PCR), allows us to identify cryptic species or biotypes of natural enemies and pests. This could result in improved biological control. Molecular tools will allow us to detect low densities of pathogenic organisms in or on natural enemies imported for classical biological control. For example, we are testing parasitoids of the citrus psylla using the PCR to ensure we do not accidentally introduce greening disease of citrus (Hoy, unpublished).

Molecular tools also will allow us to carry out research in the laboratory and field that previously was difficult or impossible. For example, it may be possible to examine the way in which new biotypes become established and spread in the environment (9). These tools also may allow us to answer questions such as: "How much genetic variability has been lost during laboratory colonization?" "How can we maintain quality in laboratory-reared natural enemies?" "Which strain has become established?" "Has post-release genetic adaptation occurred?" Molecular methods applied to systematics, population genetics, and ecological questions will aid us in our endeavors to make classical biological control less of an art and more of a science.

Effects on nontarget species. Increasingly, we will have to conduct studies on the effects of imported natural enemies on nontarget species. Such studies already have begun (e.g., 3,16,42). This research may be aided by enlisting the help of the general community (6). Engaging the aid of Master Gardeners, 4-H Youth, and similar groups in redistributing natural enemies imported for classical biological control projects and in evaluating their effectiveness and effects on nontarget species has the added benefit of educating more of the public about the benefits of biological control.

Global cooperation. The FAO Code of Conduct for the Import and Release of Exotic Biological Control Agents was ratified by the FAO Council in November 1995 (10). The Code identified the responsibilities of government and importers and exporters of natural enemies, and, while the Code is voluntary, it creates an opportunity to establish and harmonize safety procedures for the introduction of natural enemies so that the exchange of agents and information can be beneficial to all (41).

With the likelihood that testing will be required to ascertain the host specificity of natural enemies of arthropod pests (as part of increased concerns about effects on nontarget species), and molecular tests may be required to confirm that natural enemies are free of plant pathogens, enhanced exchange of information and of certified clean colonies will reduce costs and time delays.

AUGMENTATIVE BIOLOGICAL CONTROL

Augmentative releases of natural enemies require that we have improved diets and quality controls. Significant improvements in artificial diets and quality control programs are needed to allow true mass production of effective natural enemies (27,32).

Labeling. Reliability in outcome, ease of use, cost effectiveness, and consistent availability are critical if augmentation is to be used more often in pest management programs. There are few standards or regulations in the USA, although professional organizations are attempting to develop their own standards. The development of data on release rates, timing, and monitoring methods, as well as assurances of quality and purity, may have to be achieved by requiring labels for natural enemies so that consumers do not lose confidence in this pest management tactic (27). Regulation and registration of natural enemies is not new; in Switzerland, registration of macroorganisms released for biological control became mandatory in 1986 (5).

CONSERVATION OF NATURAL ENEMIES

Currently registered and experimental pesticides should be evaluated to determine whether changes in application methods and dosage can be made. The lethal/sublethal effects on pests and natural enemies must be determined in realistic ways. The results of research on using pesticides selectively are not easily accessible to the end-users. A national database of the impact of pesticides on target and nontarget species (including natural enemies) should be developed so that duplications and deficiencies in research efforts could be avoided and information made available (23).

Because pesticides can disrupt biological control of nontarget pests, information on selectivity should be provided when registering or reregistering pesticides. Ideally, the label for each pesticide should include specific information on the relative toxicity of the product to specific natural enemies in each crop and each area for which the product is registered. Industry and university scientists should cooperate to identify priority crops, selected beneficial species to be tested, preferred test methods, and reporting formats (23).

Educational Challenges

Under most circumstances, pest managers will need to employ diverse tactics, including host plant resistance, biological controls, cultural controls, or biorational controls in order to achieve economic control of all the pests in a particular ecosystem. A fully integrated IPM system is more complex than a program based on a single-tactic approach. How should the pest manager prioritize the tactics to be evaluated and employed? One principle should be that all tactics considered should be examined for their impact on natural enemies. This conceptual basis for organizing an IPM program with biological control as a key tactic was spelled out more than 30 years ago (37).

Farmers have to deal with complex problems during the growing season and need large amounts of information to adopt IPM programs (43). Improved methods are needed to provide this information in an efficient manner, perhaps including the use of "farmer schools" and "participatory learning" to adapt biological control in a site-specific manner (41). Much information will no doubt be delivered over the World Wide Web, making it feasible for the consumer to obtain information from many sources.

Professional crop consultants offer another way to provide large amounts of integrated information to farmers, and their role may increase. Independent crop consultants can monitor for pests, diseases, and other problems, and provide detailed information and recommendations for their management. Ideally, these crop consultants will provide information and advice, but will not sell specific products (including natural enemies) in order to avoid any appearance of conflict of interest.

Cosmetic standards. One of the biggest problems facing pest management is the difficulty in achieving sufficiently high levels of pest control so that cosmetic standards are achieved. Is it impossible to educate the consumer that the most perfect fruit or vegetables are not necessarily the best? Is it impossible to change how foods are graded by the processors?

Implementing multi-tactic pest management programs. Implementing a multi-tactic pest management program requires more of the researcher, cooperative extension specialist, and end-user than relying on single-tactic pest management approaches such as chemical control or transgenic plants (30). With these options, the end-user needs only to read the label on the pesticide product or to plant the seed. To some degree, classical biological control can be just as simple to implement, but to achieve full efficacy of the introduced natural enemies, it may be necessary to conserve them by modifying a variety of cropping practices.

Conclusions

The increased use of biological controls in IPM programs will reduce pesticide use and thus reduce ground water contamination by pesticides, negative impacts on nontarget organisms, pesticide residues on food, and production costs. In many cases, improved control of pests and increased farm worker safety will occur.

Incorporating biological control into agricultural pest management often is an information-intensive operation. In many cases, growers will need to understand complex predator/prey or parasite/host dynamics, or hire pest management specialists to assist them. Effective educational and monitoring programs will be needed to provide training and information.

Augmentative biological control can be expanded, but more information on release rates and timing is required, and reliable sources of high-quality natural enemies need to be established. In some cases, once an exotic natural enemy has been established in a new environment for classical biological control, it may continue to control the pest species indefinitely without additional inputs by growers, as long as toxic pesticides are not used that interfere with their activities. Pesticide selectivity remains a particularly important issue as long as some pest species remain recalcitrant to control by biological, cultural, or other biorational approaches. Increased efforts to use pesticides in a manner compatible with biological control agents could yield relatively rapid and substantial improvement without requiring new technologies.

One impediment to the implementation of biological control in agricultural IPM systems appears to be the belief that biological control agents should be able to "do it all." Biological control often must be combined with several tactics if economic control is to be achieved.

Another factor limiting adoption of biological control is the impression that biological control must be simple, cheap, and perpetual. Because dramatic, complete, inexpensive, and permanent biological control of the cottony cushion scale was achieved in citrus, this prototype is used in considering potential targets for classical biological control. This is unfortunate, because it is likely that we will have to use several tactics *in combination* to achieve effective control of many pest arthropods.

The importance of biological control as a central component of IPM was clearly delineated more than 40 years ago (37) and has been reiterated regularly. People developing IPM programs must have as their first priority the goal of making biological control effective. To do so, other control tactics must be considered in terms of their impact on biological control as well as of their efficacy. Biological control must receive additional resources and attention.

Several studies have indicated that it is feasible to reduce pesticide use in the USA by 35% to 50% without reducing crop yields or significantly increasing food costs (34). Social, political, and environmental forces are converging to trigger a broad-based change in pest management strategy away from one that relies primarily on pesticides. A change to a truly integrated pest management strategy is long overdue.

Acknowledgments

I thank the organizers of the conference for inviting me to contribute. This work was supported in part by the Davies, Fischer, and Eckes Endowment in biological control. This is Florida Agricultural Experiment Station Journal Series R-06858.

Literature Cited

1. Ashburner, M., Hoy, M. A., and Peloquin, J. 1998. Transformation of arthropods--research needs and long term prospects. Insect Molec. Biol. 7:201-213.
2. Barbosa, P. 1998. Conservation Biological Control. Academic Press, San Diego.
3. Bathon, H. 1996. Impact of entomopathogenic nematodes on non target hosts. Biocontrol Sci. Techn. 6:421-434.

221

4. Benbrook, C. M. 1996. Pest Management at the Crossroads. Consumers Union, Yonkers, NY.
5. Bigler, F. 1997. Use and registration of macroorganisms for biological crop protection. Bull. OEPP/EPPO Bulletin 27:95-102.
6. Briese, D. T., and McLaren, D. A. 1997. Community involvement in the distribution and evaluation of biological control agents: Landcare and similar groups in Australia. Biocontr. News Inform. 18:39N-49N.
7. Caltagirone, L. E., and Doutt, R. L. 1989. The history of the Vedalia beetle importation to California and its impact on the development of biological control. Annu. Rev. Entomol. 34:1-16.
8. Casida, J. E., and Quistad, G. B. 1998. Golden age of insecticide research: past, present, or future? Annu. Rev. Entomol. 42:1-16.
9. Edwards, O. R., and Hoy, M. A. 1995. Random amplified polymorphic DNA markers to monitor laboratory-selected, pesticide-resistant *Trioxys pallidus* (Hymenoptera: Aphidiidae) after release into three California walnut orchards. Environ. Entomol. 24:487-496.
10. FAO. 1996. International Standards for Phytosanitary Measures. Code of Conduct for the import and release of exotic biological control agents. Food and Agriculture Organization of the United Nations, Rome.
11. Fernandez, C., and Nentwig, W. 1997. Quality control of the parasitoid *Aphidius colemani* (Hymn., Aphidiidae) used for biological control in greenhouses. J. Appl. Ent. 121:447-456.
12. Frank, J. H., and McCoy, E. D. 1992. The immigration of insects to Florida, with a tabulation of records published since 1970. Florida Entomol. 75:1-28.
13. Gaugler, R., Wilson, M., and Shearer, P. 1997. Field release and environmental fate of a transgenic entomopathogenic nematode. Biological Control 9:75-80.
14. Gilstrap, F. E. 1997. Importation biological control in ephemeral crop habitats. Biological Control 10:23-29.
15. Hashmi, S., Hashmi, G., Glazer, I., and Gaugler, R. 1998. Thermal response of *Heterorhabditis bacteriophora* transformed with the *Caenorhabditis elegans hsp70* encoding gene. J. Exp. Zool. 281: 164-170.
16. Hawkins, B. A., and Marino, P. C. 1997. The colonization of native phytophagous insects in North America by exotic parasitoids. Oecologia 112:566-571.

17. Hogsette, J. A. 1999. Management of ectoparasites with biological control organisms. Intern. J. Parasitol. 29:147-151.

18. Holt, R. D., and Hochberg, M. E. 1997. When is biological control evolutionarily stable (or is it)? Ecology 78:1673-1683.

19. Howarth, F. G. 1991. Environmental impacts of classical biological control. Annu. Rev. Entomol. 36:485-509.

20. Hoy, M. A. 1985. Improving establishment of arthropod natural enemies, pages 151-166. In: Biological Control in Agricultural IPM Systems. M. A. Hoy and D. C. Herzog, eds. Academic, Orlando.

21. Hoy, M. A. 1992. Commentary: Biological control of arthropods: genetic engineering and environmental risks. Biological Control 2: 166-170.

22. Hoy, M. A. 1992. Criteria for release of genetically improved phytoseiids: an examination of the risks associated with release of biological control agents. Exp. Appl. Acarol. 14:393-416.

23. Hoy, M. A. 1993. Biological control in U.S. agriculture: Back to the future. Founders' Memorial Lecture honoring Robert van den Bosch. Amer. Entomol. 39: 140-150.

24. Hoy, M. A. 1994. Transgenic pest and beneficial arthropods for pest management programs: an assessment of their practicality and risks, pages 641-670. In: Pest Management in the Subtropics: Biological Control a Florida Perspective. D. Rosen, F. D. Bennett, and J. L. Capinera, eds. Intercept Ltd., Andover, U.K.

25. Hoy, M. A. 1995. Impact of risk analyses on pest management programs employing transgenic arthropods. Parasitology Today 11: 229-232.

26. Hoy, M. A., and Herzog, D. C., eds. 1985. Biological Control in Agricultural IPM Systems. Academic Press, Orlando.

27. Hoy, M. A., Nowierski, R. M., Johnson, M. W., and Flexner, J. L. 1991. Issues and ethics in commercial releases of arthropod natural enemies. Amer. Entomol. 37:74-75.

28. Hoy, M. A., Gaskalla, R. D., Capinera, J. L., and Keierleber, C. 1997. Laboratory containment of transgenic arthropods. Amer. Entomol. 43: 206-209, 255-256.

29. Hull, L. A., and Beers, E. H. 1985. Ecological selectivity: modifying chemical control practices to preserve natural enemies. Pages 103-122. In: Biological Control in Agricultural IPM Systems. M. A. Hoy, and D. C. Herzog, eds. Academic Press, Orlando.

30. Lewis, W. J., van Lenteren, J. C., Phatak, S. C., and Tumlinson, J. H., III. 1997. A total system approach to sustainable pest management. Proc. Natl. Acad. Sci. USA 94:12243-12248.

31. Obrycki, J. J., Lewis, L. C., and Orr, D. B. 1997. Augmentative releases of entomophagous species in annual cropping systems. Biological Control 10:30-36.

32. O'Neil, R. J., Giles, K. L., Obrycki, J. J., Mahr, D. L., Legaspi, J. C., and Katovich, K. 1998. Evaluation of the quality of four commercially available natural enemies. Biological Control 11:1-8.

33. Parrella, M. P., Heintz, K. M., and Nunney, L. 1992. Biological control through augmentative releases of natural enemies: a strategy whose time has come. Amer. Entomol. 38:172-179.

34. Pimentel, D., McLaughlin, L., Zepp, A., Lakitan, B., Kraus, T., Kleinman, P., Vancini, F., Roach, W. J., Graap, E., Keeton, W. S., and Selig, G. 1991. Environmental and economic effects of reducing pesticide use. BioScience 41:402-409.

35. Ruesink, J. L., Parker, I. M., Groom, M. J., and Kareiva, P. M. 1995. Reducing the risks of nonindigenous species introductions: guilty until proven innocent. BioScience 45:465-477.

36. Simberloff, D., and Stiling, P. 1996. How risky is biological control? Ecology 77:1965-1974.

37. Stern, V. M., Smith, R. F., van den Bosch, R., and Hagen, K. S. 1959. The integrated control concept. Hilgardia 29:81-101.

38. Tabashnik, B. E., Liu, Y. B., Malvar, T., Heckel, D. G., Masson, L., and Ferre, J. 1998. Insect resistance to *Bacillus thuringiensis*: uniform or diverse? Phil. Trans. R. Soc. Lond. B 352:17451-1756.

39. U.S. Department of Agriculture, Animal and Plant Health Inspection Service. Transgenic Arthropod Web site. **http://www.aphis.usda.gov:80/bbep/bp/arthropod/#tgenadoc**

40. van Lenteren, J. C., Roskam, M. M., and Timmer, R. 1997. Commercial mass production and pricing of organisms for biological control of pests in Europe. Biological Control 10:143-149.

41. Waage, J. 1996. "Yes, but does it work in the field?" The challenge of technology transfer in biological control. Entomophaga 41:315-332.

42. Walter, D. E., Azam, G. N., Waite, G., and Hargreaves, J. 1998. Risk assessment of an exotic biocontrol agent: *Phytoseiulus persimilis* (Acari: Phytoseiidae) does not establish in rainforest in southeast Queensland. Austr. J. Ecol. 23:587-592.

43. Wearing, C. H. 1988. Evaluating the IPM implementation process. Annu. Rev. Entomol. 33:17-38.
44. Wearing, C. H., and Hokkanen, H. M. T. 1994. Pest resistance to *Bacillus thuringiensis*: case studies of ecological crop assessment for Bt gene incorporation and strategies of management. Biocontrol Sci. Technol. 4:573-590.
45. Wiedenmann, R. N., and Smith, J. W., Jr. 1997. Attributes of natural enemies in ephemeral crop habitats. Biological Control 10:16-22.
46. Wiedenmann, R. N., and Smith, J. W., Jr. 1997. Novel associations and importation biological control: the need for ecological and physiological equivalencies. Insect Sci. Appl. 17:51-60.

Habitat Management to Enhance
Biological Control in IPM

Douglas A. Landis, Fabián D. Menalled, Jana C. Lee,
Department of Entomology and Center for Integrated Plant
Systems, Michigan State University, E. Lansing, MI 48824 USA

Dora M. Carmona
Facultad de Ciencias Agrarias UNMdP-INTA Balcarce
Balcarce, Buenos Aires, Argentina

Amalia Pérez-Valdéz
Universidad Autonoma Chapingo, Departamento de
Parasitologia Agricola, Chapingo, Mexico

Since the articulation of the integrated control concept by Stern et al. (34), biological control has been viewed as a fundamental part of integrated pest management (IPM) systems. However, in spite of producing some stunning successes, production systems designed to rely on biological control are still the exception rather than the rule. Due to increases in pesticide resistance and loss of pesticide registrations, there is a growing national consensus that sole reliance on chemically based pest management systems is not desirable (26). Habitat management to enhance biological control is one emerging technology that is positioned to help fill the pesticide gap.

Habitat Management Defined

Manipulation of the environment to enhance the survival and the physiological and/or behavioral performance of natural enemies

226

and to increase their effectiveness is the aim of conservation biological control (3). Conservation practices generally focus on reducing mortality, providing supplementary resources, controlling secondary enemies, or manipulating host plant attributes to the benefit of natural enemies (31,37). Habitat management is that subset of conservation biological control practices where habitats are the focus of manipulation. The aim of habitat management is to promote optimal performance of natural enemies by manipulating areas within the agricultural ecosystem to provide needed resources. The area manipulated may be the crop field itself, the land at field margins, or the surrounding agricultural landscape. These manipulations should be designed so that resources are provided at the correct time and in a spatial array that allows the natural enemies to take full advantage of them. Thus, habitat management favors natural enemies over pests by providing the appropriate ecological infrastructure at varying spatial and temporal scales within the agricultural landscape.

The Need for Habitat Management

ECOLOGICAL BASIS FOR HABITAT MANAGEMENT

Integrated pest management as a whole is based on the recognition that the regulation of pest populations is fundamentally an ecological process (33). Two important developments in the ecological sciences have further paved the way for the development of habitat management. First, the concept of disturbance and its impact on the development of ecological communities (29), and second, the discipline of landscape ecology, which examines structure, function, and change in landscapes. These two interacting areas of ecology help us understand the template on which pest management takes place.

Disturbance is a natural part of all ecosystems. Disturbances are defined as relatively discrete events in time that disrupt ecosystem, community, or population structure, and change resources, substrate availability, or the physical environment (29). Some common examples of natural disturbances include herbivore defoliation, storm events that cause sudden loss of vegetation, such as treefalls, and periodic fires or floods. The immediate effect of disturbance is an initial loss of organisms from the affected area. This is followed by gradual re-colonization of the disturbed area from surviving individuals or from individuals dispersing from undisturbed source areas. Repeated disturbance over space and time results in a mosaic of habitats at different successional stages.

The disturbance regime in any given area can be characterized by a variety of variables, such as the frequency of disturbance, the intensity of the disturbance (i.e., relative number of organisms removed), the area affected, the spatial distribution of disturbance within the affected area, and the predictability with which the disturbance events reoccur. The disturbance regime and the environmental characteristics (climate, topography, etc.) of a given area set the template for the types of organisms that are able to live there. For example, in temperate forests, where the disturbance regime is infrequent and generally of low intensity, the community will tend to be made up of long-lived individuals, and the relative species diversity will be high. In contrast, if the same area were disturbed frequently or with high intensity, a very different community would develop consisting of shorter-lived species and exhibiting lower-species diversity.

Agricultural systems can also be examined in terms of their disturbance regimes. If we take the example of temperate annual crop agriculture, the following picture emerges. Initially, land is cleared for agriculture in a disturbance of very high intensity. This typically includes removal of all above-ground vegetation over a large area. Each subsequent season, the soil is disturbed by tillage followed by fertilization and planting. The purpose of these disturbances is to create a habitat at a very early successional stage in order to promote the crop species of interest. Subsequent pesticide applications aimed at weeds, insects, and plant pathogens are intended to ensure that pest organisms are maintained at very low levels. While the frequency and extent of natural disturbances are regulated by subtle environmental variations (e.g., soil type, topography, and weather), in large part, technology allows agricultural disturbances to occur irrespective of these underlying variations. Thus, in annual crop landscapes, disturbances are repeated over very large areas with a high degree of predictability and spatial uniformity.

The result of the disturbance regime induced by modern annual crop agriculture can be seen in the physical structure and ecological interactions of agricultural landscapes (15). These landscapes tend to be dominated by large fields that have very low species diversity.

Many non-crop habitats, such as fencerows, hedgerows, woodlots, and roadside ditch banks, have been progressively reduced or removed. Fragmentation and isolation of non-crop habitats occur along with an increase in the contiguous area under crop production. The central question, then, for habitat management is how well do natural enemies of crop pests function within such landscapes?

NATURAL ENEMIES

It is well known that natural enemies may require certain resources in their environment in addition to the presence of their prey/hosts. These resources may include: overwintering sites, sources of food for non-entomophagous stages, alternate hosts for entomophagous stages, presence of con-specifics for mating, and shelter from adverse conditions (weather, pesticides, etc.). Lack of these resources can limit the effectiveness of natural enemies (30,31). The fact that many agricultural ecosystems lack one or more of these resources can be viewed as a direct result of the frequent, intense, and spatially uniform disturbance regimes imposed on these systems (16).

Non-crop plants and habitats in agricultural systems can provide for many of the needs of natural enemies and can enhance biological control. The role of non-crop plants in providing these resources for arthropod natural enemies was first summarized by van Emden (38) and later reviewed by Altieri and Whitcomb (2) and Altieri and Letourneau (1). These resources may include shelter, alternate prey/hosts, pollen, and nectar. Many studies have observed that adjacent habitats can influence natural enemy communities in crop fields (6,7,39). The challenge is in identifying the specific needs of the natural enemy and then conducting habitat management to meet those needs in a way that is compatible with modern production systems. There is a growing international interest in the study of habitat management (28) as shown by the increasing number of relevant publications appearing in recent years (Fig 1.)

HABITAT MANAGEMENT

There are several well-documented examples of habitat management to enhance natural enemies. The development of the so-called "beetle banks" now in use in several western European countries is an excellent model of habitat management to enhance

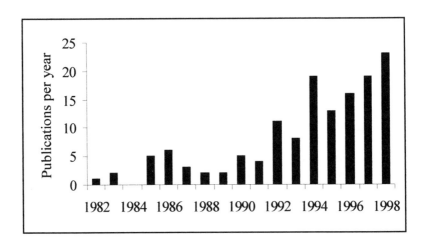

Figure 1. Publications relevant to habitat management since 1982. The above data from a forthcoming review of habitat management (D. Landis, G. Gurr, and S. Wratten, unpublished) do not include all relevant publications but are illustrative of the overall trend.

overwintering of natural enemies and to promote the early colonization of crop habitats (32). Researchers found that many carabid and staphylinid beetles that prey on aphids in cereal crops overwinter in vegetation in hedges along field edges. They reasoned that by recreating similar habitats in crop fields, they might increase the numbers of overwintering beetles and favor their early establishment in fields. By creating raised banks and planting them with tussock-forming grasses that provided favorable overwintering sites, they were able to increase overwintering of these predators and reduce aphid pressure (35,36).

Flowering "weed strips" containing a diversity of native plants have been studied for some time in Germany and Switzerland (27). These strips are established by sowing a seed mixture containing approximately 30 species. These strips undergo succession, with the plant community shifting toward perennial plants over time. The presence of these strips has been shown to contribute to increased activity-density of Carabidae (Coleoptera) (18), spiders (Arenae), Nabidae (Heteroptera), Dolichopodidae (Diptera), and Syrphidae

(Diptera) (11). One finding is that weed strip-management appears to increase the availability of food for carabids and results in increased reproduction (40).

Implementing Habitat Management

HABITAT MANAGEMENT RESEARCH IN MICHIGAN

We have examined the ecological factors influencing the abundance and effectiveness of natural enemies in various annual cropping systems, examining the role of habitat and landscape structure on European corn borer parasitoids (8-10,13), coccinellid communities (5,19), parasitoid communities (14,16,20-22,25), and generalist predator communities (17). During the past 3 years, we have focused attention on both ground beetle communities as generalist consumers of arthropods and weed seeds in agricultural systems (4,20).

We have been examining a system of habitat management for annual crop (primarily corn, soybean, wheat) production systems typical in the north-central United States. The system combines various elements in an attempt to provide for the needs of a diverse assemblage of natural enemies. It is also designed to be flexible to meet the needs of individual producers. The basic component is a refuge strip ca. 3 m wide, containing a cool-season grass (*Dactylis glomerata* L.) or a grass-legume mixture (*D. glomerata* plus red clover *Trifolium prataense* L., white clover, *Trifolium repens* L., and/or sweet clover *Melilotis officinalis* L.) This component provides a sheltered overwintering site for many natural enemies and begins growing early in the spring to provide a source of potential alternate prey/hosts. A mixture of perennial flowering plants may be added to strips to provide pollen and nectar resources. Such strips can be located within fields or at field edges. In addition, in some areas we are establishing high-value woody species, such as red oak and black cherry, in the grass-legume strip, adding long-term value to the farm.

CARABIDS AND REFUGE STRIPS

Pitfall trap studies conducted in 1996-1997 examined the carabid community of refuge strips and adjacent crop area. Carabid beetle activity-density was significantly higher in refuge strips than in the non-refuge crop interfaces (controls) in both years of the study (4). Using a cluster analysis, it was shown that the refuge strips contained a carabid community that was distinct from non-refuge areas or the surrounding crop. Refuge habitats were dominated by *Pterostichus melanarus* Illiger.

While overall carabid activity-density was ca. 2.5 times greater in refuge strips, no significant effect of refuge presence/absence could be detected in the carabid community of the surrounding crop area. It was believed that the relatively small plot size (30 X 30 m^2) may have prevented observation of significant refuge effects, since larger species could disperse between plots with relative ease. A subsequent study in 1998 used 15 X 15 m^2 plastic barriers to surround refuge and non-refuge areas to prevent movement of carabids between plots. This study showed that refuge plots could contribute beetles to the surrounding crop area, since the carabid community in plots with refuges recovered from insecticide disturbance, while the community in plots without refuges remained depleted (J. Lee, unpublished).

REFUGE HABITATS AND FOLIAR ARTHROPODS

Refuge habitats may influence the abundance of foliar arthropods as well as ground dwelling species. We have investigated the effect of refuge strips on the abundance of dispersing foliar arthropods using sticky traps and water pan traps. Two yellow water pan traps and one yellow sticky trap were placed in either refuge habitats (described above) or an adjacent crop field located at least 30 m from a refuge. Twenty-four-hour trap captures were collected on 3, 5, 8 and 10 June, 1999. Samples were returned to the lab, where arthropods were sorted to family level and classified generally as herbivores, predators, or parasitoids. The crop (maize) during this period was between 5-10 cm tall. Significant crop residue remained in this minimum-tillage system from the previous maize crop, and herbicides maintained the crop area nearly weed free. Refuge vegetation reached ca. 1 m tall (*D. glomerata*). None of the perennial flowering species were in bloom at this time.

Sticky traps captured a greater number of herbivores than either predators or parasitoids (Fig 2). There were no significant differences in the number of herbivores captured in crop or refuge habitats. Captures of both predators and parasitoids were greater in

232

refuge than crop habitats. Pan traps captured more parasitoids than herbivores or predators. There were, again, no significant differences in the number of herbivores captured in crop or refuge habitats, while predator and parasitoid captures were significantly greater in refuge habitats (Fig. 2). While refuges harbor herbivores that potentially serve as alternate prey, to date we have not observed any increase in insect pest activity in adjacent crops related to the presence of refuges.

MANIPULATING CARABID ABUNDANCE ALTERS PREDATION RATES

In order to determine if increasing the carabid activity-density results in higher predation rates, we assessed the relationship between carabid beetle abundance and field rates of prey removal (24). To do so, we created plots surrounded by different boundaries that selectively altered carabid communities with minimal habitat alteration and without use of insecticides. Three treatments were established: naturally occurring communities, augmented communities using ingress boundaries, and reduced communities using egress boundaries. Three times during the growing season, a fixed number of prey were placed in plots to evaluate the impact of carabid abundance on predation rates. A combination of vertebrate and invertebrate exclosures allowed us to evaluate prey removal by invertebrates alone. In comparison to the no-boundary treatment, carabids increased by 54% and decreased by 83% in the plots surrounded by ingress and egress boundaries, respectively. Predation rates were positively correlated with activity-density of the four most abundant carabids $(r^2 = 0.70)$.

FILTER STRIPS, CARABID ABUNDANCE AND WEED SEED PREDATION

Some carabids are known to feed on weed seeds, and we assessed the impact of stable habitats such as filter strips on the abundance and diversity of carabid beetles and their impact on weed seed predation (23). In 1997 and 1998, we studied a 9.3 ha crop field and two adjacent 30 m-wide filter strips in Midland County, MI. One filter strip was composed of switchgrass (*Panicum virgatum* L.), the other a legume-grass mixture (*Medicago sativa* L., *Phleum pratense* L., *M. officinalis*). The activity-density, species richness, and diversity of carabid beetles within filter strips and the crop were estimated using pitfall traps, and seed removal by invertebrates was

233

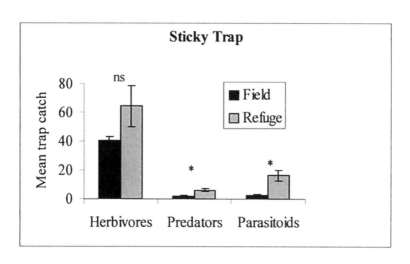

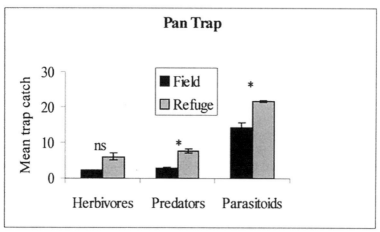

Figure 2. Mean daily capture of arthropods in yellow sticky traps and yellow water pan traps in fields versus refuges containing a mix of grass- legume and perennial flowering plants, E. Lansing, Michigan, USA, 3-10 July, 1999. An * above bars indicates a significant difference (P ≤ 0.05) for field versus refuge comparison from ANOVA, n = 4 replicates.

compared between the center of the crop field and the filter strips. Both filter strips contained significantly more seed predators than the crop field, and seed predators were highly associated with filter strip habitats. Carabid beetles and other invertebrates were responsible for significant removal of weed seeds, which ranged from 0.6%/day in soybean to 4.4%/day in switchgrass filter strips.

Summary and Conclusions

These studies have begun to provide insight into the factors necessary to develop a successful habitat management system for annual crops. We are encouraged that producers are beginning to take notice of these systems. We have several collaborators who have established up to 0.8km-long refuge habitats on their farms. It is clear that much practical knowledge and innovation in this area will occur only after producers begin to experiment and adapt these basic systems. For habitat management to be accepted, it is likely that growers will need to see additional benefits beyond pest management. Fortunately, habitat management for natural enemies can frequently be integrated with other conservation practices on the farm. For example, grass waterways, riparian buffers, cross-wind trap strips, and conservation headlands are soil, water, and wildlife conservation practices already in use that can provide benefits to natural enemies as well (12).

ECOLOGICAL INFRASTRUCTURE IN AGRICULTURAL LANDSCAPES

Agricultural landscapes have become increasingly fragmented and simplified in recent decades to the point where certain ecological services, such as maintenance of biodiversity, soil and water erosion prevention, nutrient cycling, etc., are threatened. The concept of restoring these functions by managing the ecological infrastructure of landscapes appears to have arisen in Europe in the 1990s. Just as society requires and benefits from investment in certain infrastructure (e.g., transportation and educational systems), so to do ecological systems. An ecological infrastructure is required to support the function of biological systems that provide ecological services to society. For example, if forest-nesting birds are to persist in increasingly fragmented agricultural landscapes, they must be provided with appropriate corridors to travel from one breeding habitat patch to another. These wooded corridors may serve additional functions as well, such as windbreaks to reduce soil erosion or as riparian buffers to prevent pesticide and nutrient run-off, maintaining water quality. In some cases, these habitats may also be managed in a way that provides needed resources in the landscape to support natural enemies for biological control. The costs of this ecological infrastructure should be spread across the various segments of society that benefit.

Scientists interested in habitat management to enhance biological control in IPM should consider the benefits of linking their research to others with a similar overall goal of maintaining an appropriate ecological infrastructure in agricultural landscapes. This will require seeking partnerships among disciplines such as land-use planning, economics, landscape ecology, and wildlife ecology, as well as the plant, soil, and water sciences.

Literature Cited

1. Altieri, M. A., and Letourneau, D. K. 1982. Vegetation management and biological control in agroecosystems. Crop Protection 1:405-430.
2. Altieri, M. A., and Whitcomb, W. H. 1979. The potential use of weeds in manipulation of beneficial insects. Hort. Sci. 14:12-18.
3. Barbosa, P. 1998. Conservation Biological Control. Academic Press, San Diego.
4. Carmona, D. M. 1998. Influence of refuge habitats on seasonal activity-density of ground beetles (Coleoptera: Carabidae) and northern field cricket *Gryllus pennsylvanicus* Burmeister (Orthoptera: Gryllidae). M.S. thesis. Michigan State University, E. Lansing.
5. Colunga-Garcia, M., Gage, S. H., and Landis, D. A. 1997. Response of an assemblage of Coccinellidae (Coleoptera) to a diverse agricultural landscape. Environ. Entomol. 26:797-804.
6. Dennis, P., and Fry, G. L. A. 1992. Field margins: can they enhance natural enemy population densities and general arthropod diversity on farms? Agric. Ecosyst. Environ. 40:95-115.
7. Duelli, P., Studer, M., Marchand, I., and Jakob, S. 1990. Population movements of arthropods between natural and cultivated areas. Biological Conservation 54:193-207.
8. Dyer, L. E., and Landis, D. A. 1996. Effects of habitats, temperature, and sugar availability on longevity of *Eriborus*

terebrans (Hymenoptera: Ichneumonidae). Environ. Entomol. 25:1192-1201.

9. Dyer, L. E., and Landis, D. A. 1997a. Diurnal behaviour of *Eriborus terebrans* (Hymenoptera: Ichneumonidae). Environ. Entomol. 26:1385-1392.

10. Dyer, L. E., and Landis, D. A. 1997b. Influence of noncrop habitats on the distribution of *Eriborus terebrans* (Hymenoptera: Ichneumonidae) in cornfields. Environ. Entomol. 26:924-932.

11. Hausmmann, A. 1996. The effects of weed strip-management on pests and beneficial arthropods in winter wheat fields. Zeitschrift fur Pflanzenkrankheiten und Pflanzenschutz 103:70-81.

12. Landis, D. A., and Dyer, L. E. 1998. Conservation buffers and beneficial insects, mites & spiders. Conservation Information Sheet, Agronomy Series. USDA Natural Resources Conservation Service. Lansing, Michigan.

13. Landis, D. A., and Haas, M. J. 1992. Influence of landscape structure on abundance and within-field distribution of *Ostrinia nubilalis* Hübner (Lepidoptera: Pyralidae) larval parasitoids in Michigan. Environ. Entomol. 21:409-416.

14. Landis, D. A., and Marino, P. C. 1999. Conserving parasitoid communities of native pests: Implications for agricultural landscape structure. In: Biological Control of Native or Indigenous Pests. Entomological Society of America. L. Charlet, G. Brewer, eds. Thomas Say Publications. (in press).

15. Landis, D. A., and Marino, P. C. 1999. Landscape structure and extra-field processes: Impact on management of pests and beneficials. Pages 79-104 in: Handbook of Pest Management. J. Ruberson, ed. Marcel Dekker Inc., New York (in press).

16. Landis, D. A., and Menalled, F. D. 1998. Ecological considerations in the conservation of effective parasitoid communities in agricultural systems. Pages 101-121 in: Conservation Biological Control. P. Barbosa, ed. Academic Press, San Diego.

17. Landis, D. A., and van der Werf, W. 1977. Early-season aphid predation impacts establishment and spread of sugar beet yellows virus in the Netherlands. Entomophaga 42:499-516.

18. Lys, J. A., Zimmermann, M., and Nentwig, W. 1994. Increase in activity density and species number of carabid beetles in cereals as a result of strip-management. Entomologia Experimentalis et Applicata 73:1-9.

19. Maredia, K. M., Gage, S. H., Landis, D. A., and Scriber, J. M. 1992. Habitat use patterns by the seven-spotted lady beetle (Coleoptera: Coccinellidae) in a diverse agricultural landscape. Biological Control 2:159-165.
20. Marino, P. C., Gross, K. L., and Landis, D. A. 1997. Post-dispersal weed seed loss in Michigan maize fields. Agric. Ecosyst. Environ. 66:189-196.
21. Marino, P. C., and Landis, D. L. 1996. Effect of landscape structure on parasitoid diversity and parasitism in agroecosystems. Ecol. App. 6:276-284.
22. Marino, P. C., and Landis, D. A. 1999. Parasitoid community structure: implications for biological control in agricultural landscapes. In: Interchanges of Insects between Agricultural and Surrounding Habitats. B, Ekbom, ed. Kluwer, Dordrecht. (in press).
23. Menalled, F. D., Renner, K., and Landis, D. A. 1998. Impact of landscape and field structure on post-dispersal weed seed removal by invertebrates. In: Weed Science Society of America Abstracts. Vol. 38. February 8-12. Chicago, IL.
24. Menalled, F., Lee, J., and Landis D. 1999. Manipulating carabid beetle abundance alters prey removal rates in corn fields. BioControl. 44:1-16.
25. Menalled, F. D., Marino, P. C., Gage, S. H., and Landis, D. A. 1999. Does agricultural landscape structure affect parasitism and parasitoid diversity? Ecol. Applic. (in press).
26. National Research Council. 1996. Ecologically Based Pest Management: New Solutions for a New Century. National Academy Press. Washington, DC.
27. Nentwig, W. 1998. Weedy plant species and their beneficial arthropods: potential for manipulation in field crops. Pages 49-72 in: Enhancing Natural Control of Arthropod Pests through Habitat Management. C. H. Picket and R. L. Bugg, eds. University of California Press, Berkeley.
28. Pickett, C. H., and Bugg, R. L. 1998. Enhancing Biological Control: Habitat Management to Promote Natural Enemies of Agricultural Pests. University of California Press, Berkley.
29. Pickett, S. T. A., and White, P. S. 1985. The Ecology of Natural Disturbance and Patch Dynamics. Academic Press, New York.
30. Powell, W. 1986. Enhancing parasite activity within crops. Pages 314-340 in: Insect Parasitoids. J. K. Waage and D. Greathead, eds. Academic Press, London.

31. Rabb, R. L., Stinner, R. E., and van den Bosch, R. 1976. Conservation and augmentation of natural enemies. Pages 233-254 in: Theory and Practice of Biological Control. C B. Huffaker and P. S. Messenger, eds. Academic Press, New York.
32. Sotherton, N. W. 1995. Beetle banks - helping nature to control pests. Pesticide Outlook 6(6):13-17.
33. Southwood, T. R. E., and Way, M. J. 1970. Ecological background to pest management. Pages 6-29 in: Concepts of Pest Management. Proc. of a Conference held at North Carolina State University, Raleigh, NC, 25-27 March, 1970. R. L. Rabb and F. E Guthrie, eds. North Carolina State University. Raleigh.
34. Stern, V. M., Smith, R. F., van den Bosch, R. and Hagen, K. S. 1959. The integrated control concept. Hilgardia 29:81-101.
35. Thomas, M. B., Sotherton, N. W., Coombes, D. S., and Wratten, S. D. 1992. Habitat factors influencing the distribution of polyphagous predatory insects between field boundaries. Ann. App. Biol. 120:197-202.
36. Thomas, M. B., Wratten, S. D., and Sotherton, N. W. 1992. Creation of "island" habitats in farmland to manipulate populations of beneficial arthropods: predator densities and species composition. J. App. Ecol. 29:524-531.
37. van den Bosch, R., and Telford, A. D. 1964. Environmental modification and biological control. Pages 459-488 in: Biological Control of Pests and Weeds. P. DeBach, ed. Reinhold, New York.
38. van Emden, H. F. 1965. The role of uncultivated land in the biology of crop pests and beneficial insects. Sci. Hort. 17:121-136.
39. Wallin, H. 1985. Spatial and temporal distribution of some abundant carabid beetles (Coleoptera: Carabidae) in cereal and adjacent habitats. Pedobiologia 28:19-34.
40. Zangger, A., Lys, J. A., and Nentwig, W. 1994. Increasing the availability of food and the reproduction of *Poecilus cupreus* in a cereal field by strip-management. Entomologia Experimentalis et Applicata 71:111-120.

Current Status of Biological Control of Plant Disease

Joyce E. Loper and Virginia O. Stockwell

USDA-ARS, Horticultural Crops Research Laboratory; and
Department of Botany and Plant Pathology,
Oregon State University, Corvallis, OR 97331 USA

As a large and diverse subdiscipline of plant pathology, biological control of plant disease has been the subject of many excellent and recent reviews (2,6,9,14,43). In this chapter, we make no attempt to represent the full spectrum of biocontrol systems being studied, but instead, we focus on some current trends in the study and practice of biocontrol.

Two early events were pivotal in setting the direction for biocontrol research in the last half of the 20th century. The first of these occurred in 1963, when plant pathologists studying the ecology of soilborne plant pathogens came together to articulate their common view that epidemics of soilborne diseases could not be explained unless the mediating influences of soil microorganisms were considered (1). Many of the concepts and perspectives articulated in The International Symposium on Factors Determining the Behavior of Plant Pathogens in Soil, held in Berkeley, CA, in 1963, have been central to the field of biological control since that time. For example, the focus of biological control research on soilborne diseases has persisted to the present. Furthermore, the view that biological control should be based in a clear and fundamental understanding of the ecology of the pathogen was articulated clearly in the symposium proceedings (1), and the focus on ecological principles has proved to be a well-founded approach to the development of biological control (8). A second important event occurred in the 1970s, when *Agrobacterium radiobacter* K84, a biological control agent for crown gall, was discovered. *A. radiobacter* K84 can be applied to root surfaces of nursery tree crops, and will protect roots from infection by the crown gall pathogen *Agrobacterium tumefaciens*, a ubiquitous soil inhabitant. Furthermore, strain K84 can be packaged and distributed to

240

growers as a commercial product, and it has been very successful in controlling crown gall on a variety of deciduous hosts around the globe (11,31).

Commercially Available Biological Control Agents

The remarkable success of *A. radiobacter* K84 encouraged scientists to seek out comparable biocontrol agents for a variety of plant diseases, and this effort has dominated the field of biological control for the past 2 decades. The search for new biological control agents has focused on those like strain K84, which can suppress plant disease by inhibiting the pathogen on plant surfaces. One outcome of this effort has been the development of commercial biological control agents for the management of plant disease. At least 37 biocontrol agents are now commercially available (**http://www.barc.usda.gov/psi/bpdl/bioprod.htm**) world-wide. The composition of these products reflects the current focus of biological control research on soilborne diseases, a focus that was already in place when the Berkeley symposium was held in 1963. It also reflects more recent directions that biological control research and development has taken, such as diseases of foliar or floral tissues, post-harvest diseases, and diseases of plants grown in defined substrates.

How well do these commercially available biological agents work? This is a difficult question to answer, because each of these products has its detractors as well as its advocates. Many of the arguments posed by both groups apply to general characteristics of biological control agents, rather than to one product in particular. In the following discussion, we describe BlightBan A506, a commercially available biological control agent for fire blight of pear and apple, and we present features of the product that have contributed to or detracted from its adoption by pome fruit growers.

Biological of Fire Blight with BlightBan A506

Fire blight is an important disease of pome fruits caused by the bacterium *Erwinia amylovora*. In the western United States, *E. amylovora* infects its plant hosts primarily through blossoms. When blossoms first open in the spring, their surfaces are nearly devoid of microorganisms (V. O. Stockwell, unpublished), but open blossoms soon

241

support populations of epiphytic bacteria that are introduced primarily by rain or insects. Under conducive environmental conditions (temperatures exceeding 15.6° C with adequate moisture), *E. amylovora* can establish populations in excess of 10^6 cfu/blossom (17). The stigmatic surfaces are the primary site of colonization by *E. amylovora,* and when the population size of the pathogen exceeds 10^5 cfu/blossom, infection is likely to occur through the hypanthium. Once the pathogen infects floral tissues, it can spread into the supporting peduncle and twig, resulting in necrosis that appears blackened and burned (hence, the name fire blight). Infection can continue to spread, threatening fruit-bearing wood, scaffold branches, and even the entire tree.

CONVENTIONAL FIRE BLIGHT MANAGEMENT

Since the 1950s, antibiotic sprays that suppress populations of the pathogen in blossoms have been a mainstay of fire blight management programs. Both streptomycin and oxytetracycline are registered for management of fire blight, and, historically, streptomycin has been the more effective of the two. During the past 2 decades, however, streptomycin-resistant strains of the pathogen have become prevalent in many pome fruit-growing regions of the United States. Therefore, growers have relied increasingly upon oxytetracycline in recent years. Because oxytetracycline is bacteriostatic rather than bactericidal, and less stable than streptomycin on aerial plant surfaces, it is generally not as effective as streptomycin for fire blight management. Commonly, timing of antibiotic sprays is directed by a variety of fire blight prediction models (17). The exact criteria for prediction of fire blight infections vary among the models, but each model predicts that infection periods occur after environmental conditions are conducive to growth of the pathogen.

Sanitation and fertility management are other important components of fire blight management practices (17). Pruning trees to eliminate branches displaying symptoms of fire blight is commonly done to prevent further spread of the bacterium within the tree and to reduce pathogen inoculum for subsequent infections of shoots or blossoms.

APPLICATION OF BLIGHTBAN A506

BlightBan A506 is distributed as a pellet formulation comprised of freeze-dried bacterial cells of the biocontrol agent *Pseudomonas fluorescens* A506, which suppresses fire blight through pre-emptive

242

exclusion of the pathogen (22,47). A506 grows on blossom surfaces, utilizing chemical resources and physical colonization sites that would otherwise be available to support the growth of *E. amylovora*.

According to label recommendations, BlightBan A506 should be sprayed onto blossoms soon after blossoms open (i.e., 30% bloom), and again later in the bloom period. If applied as recommended, the biocontrol agent can reduce the proportion of blossoms supporting populations of the pathogen exceeding 10^5 cfu/blossom (18,32). In nine field trials conducted in California, A506 decreased the severity of fire blight (measured as the number of fire blight strikes per tree) by an average of 60% (26). Therefore, like virtually all biological control agents, BlightBan A506 suppresses but does not eliminate disease.

ADOPTION OF BLIGHTBAN A506 BY POME FRUIT GROWERS

BlightBan A506 was first available commercially in 1997. In 1998, approximately 25% of the hectares planted to pear in California received the recommended two applications of BlightBan A506, as estimated from sales volumes of the product (Steve Kelly, Plant Health Technologies, personal communication). Several factors have contributed to the adoption of BlightBan A506 by growers: **1) Growers lack effective chemicals for disease management**. Populations of *E. amylovora* present in many pome fruit-growing areas are resistant to streptomycin; oxytetracycline is registered for pear but not apple; oxytetracycline is less effective than streptomycin; and registration of new antibiotics for management of bacterial diseases of plants is not expected. **2) Application of the biological control agent can decrease labor costs.** Because applications of BlightBan A506 reduce the number of fire blight strikes by an average of 60%, its use can reduce labor costs incurred by growers later in the season, when strikes need to be pruned out to maintain the health of the tree and reduce inoculum for the following season. **3) Disease does not need to be eliminated, only suppressed.** BlightBan A506, like other biological control agents, suppresses disease but does not eliminate it. Because growers can also use fertility management and sanitation to manage fire blight, biological control provides an acceptable complement to other components of a disease management program. **4) Biological control can be integrated into the disease management program**. A506 is resistant to streptomycin and tolerant of oxytetracycline in culture, and strategies to apply BlightBan A506 in combination with antibiotics have been described (26,32). In large-scale field trials in California, for example, combined applications of A506 and streptomycin suppressed fire blight better than applications of streptomycin alone (26). Johnson and Stockwell (17) propose to further develop disease-existing forecasting models to include recommendations for applications of BlightBan A506, in addition to antibiotic sprays. Such models could recommend antibiotic sprays only under environmental conditions especially conducive to fire blight epidemics, thereby extending the usefulness of antibiotics by reducing unnecessary applications and consequent increases in the prevalence of antibiotic-resistant strains of the pathogen. **5) BlightBan A506 suppresses other bacterial diseases of pome fruits.** BlightBan A506 is registered for the suppression of frost injury incited by ice nucleation active strains of epiphytic bacteria, including *Pseudomonas syringae*, and russeting of pear, caused in part by indole-3-acetic acid-producing strains of epiphytic bacteria (22-24). Therefore, applications of BlightBan A506 provide other potential benefits that enhance its value to growers.

The distributors of BlightBan A506 have also encountered obstacles to its adoption by pome fruit growers, including the following: **1) Fire blight is suppressed by an average of 60% by BlightBan A506.** Growers commonly observe fire blight strikes in orchards that have been sprayed with BlightBan A506. Such observations are consistent with experimental evidence indicating that fire blight is suppressed by an average of 60% by A506 (26). Nevertheless, in the absence of side-by-side comparisons of disease severity on treated and non-treated trees in controlled field experiments, a lack of complete disease control can result in the perception that the biological control agent is not effective. Consequently, growers can become dissatisfied with the product. It is likely that similar perceptions limit the adoption of many biological control agents by agricultural producers. **2) The level of disease suppression varies somewhat from site to site and year to year.** A506 significantly decreased the number of fire blight infections in five of nine field trials in California (26). It is important to note, however, that streptomycin or oxytetracycline also failed to suppress fire blight in the four field plots in which A506 was ineffective. Population sizes established by A506 on blossoms and biological control effected by A506 are not uniform across all field environments and seasons. Undoubtedly, variable efficacy is related to the capacity of A506 and *E. amylovora* to respond to varied field environments, the timing of BlightBan A506 application, and disease pressure. These and other sources of variation are likely to influence many biological control systems. **3) The effective**

use of biological control requires the grower to modify practices.
Unlike antibiotic sprays, BlightBan A506 must be applied early in the
bloom period to be effective, because the biocontrol agent must have time
to establish populations on stigmatic surfaces before *E. amylovora*
becomes established. This prophylactic approach to fire blight
management differs from the timing of traditional antibiotic sprays, which
are applied after weather conditions are conducive to infection when fire
blight poses a clear threat. Growers recognize the sporadic nature of fire
blight epidemics, and they balk at the prospect of applying products to
control fire blight unless they are fairly certain that the disease poses a
severe and immediate threat to their orchards.

Enhancement of Biological Control

Much effort has been devoted in recent years to the enhancement of
biological control as obstacles to its adoption (such as those described
above for BlightBan A506) are becoming recognized. Approaches for the
enhancement of biological control include the genetic manipulation of
biological control agents (21), strategies to integrate biological control
into disease management programs, use of mixtures of biological control
agents that can provide more consistent disease suppression, and
fermentation and formulation strategies to enhance survival or activities
of biological control agents.

INTEGRATING BIOLOGICAL CONTROL WITH OTHER DISEASE MANAGEMENT PRACTICES

Biological control can be an important complement to other
components of a disease management program, thereby decreasing the
dose or numbers of pesticide applications, more effectively suppressing
disease, or simply providing growers with a greater spectrum of inputs.
The enhanced efficacy achieved by combined applications of BlightBan
A506 and antibiotics, mentioned above, provides an example of this
approach (17,26). Other examples include: coupling biocontrol agents
with sub-lethal doses of fumigants (12), using seed-priming processes
(7), or utilizing cultural practices, such as soil solarization (38), for
suppression of soilborne diseases.

MIXTURES OF BIOLOGICAL CONTROL AGENTS

For the past several decades, investigations into biological control of plant disease have focused on the use of single microbial inoculants to suppress disease. However, the application of a single antagonist is not likely to be the best approach to provide sustainable disease management, especially under the full spectrum of environmental conditions conducive to disease. Biocontrol mixtures, comprised of compatible microbial antagonists that complement the activities of their co-inoculants, offer a promising approach to enhance the consistent efficacy of biological control. For example, BlightBan A506 has been combined with antagonistic strains of *Pantoea agglomerans* (formerly *Erwinia herbicola*), which suppress *E. amylovora* through antibiotic production (16,19,45). Applications of a mixture of A506 and *P. agglomerans* typically result in the establishment of at least one of the bacterial antagonists on a very high proportion of apple or pear blossoms (40). Because each blossom is a potential entry point for the fire blight pathogen, the efficacy of biological control should increase as the proportion of blossoms colonized by the biological control agents increases. There are hints in the literature that mixtures of A506 and *P. agglomerans* strains are not more effective than single antagonistic strains in suppressing fire blight, despite the greater proportion of blossoms colonized (V. O. Stockwell, unpublished). Reasons for the apparent discrepancy between the establishment and efficacy of mixtures are being evaluated.

The potential of using mixtures of antagonists has generated much excitement in the biological control research community, and such mixtures can be more effective (10,35) or more reliable than any individual microbial component therein. For example, the biocontrol agent *Pythium nunn* was more effective than *Trichoderma harzianum* T-95 in suppressing Pythium damping-off of cucumber in a Colorado soil amended with organic matter, but the converse was observed in the non-amended soil. A mixture of the two biocontrol agents was as effective as the better individual inoculant in both amended and non-amended soils (34). Therefore, the mixture provided more consistent disease sup-pression in soils varying in organic matter content. However, the success of strain mixtures depends upon complex interactions among microbial antagonists, the pathogen(s), and other components of the microflora, and mixtures are not always more effective than individual antagonists. Certain antagonists are incompatible with one another, and therefore

246

compromise the biocontrol activity of a mixture. For example, *Trichoderma hamatum* suppresses rot of pea seed caused by *Pythium* spp., but its biocontrol activity is compromised by fluorescent pseudomonads in soils containing relatively low levels of iron (15). The antagonism of *T. hamatum* by fluorescent pseudomonads, attributed to pyoverdine-mediated iron competition (15), is likely to compromise the success of a mixture comprised of microbial antagonists in these taxa.

FERMENTATION AND FORMULATION

Formulation technology provides opportunities to enhance numerous characteristics of biological control agents, including shelf life, efficacy, growth and survival in the environment, and compatibility with agricultural practices and machinery. For example, lyophilized cells of *P. fluorescens* A506 and *P. agglomerans* C9-1 established more consistent populations on pear blossoms in field trials than cells harvested directly from culture (41). Therefore, formulation technology can enhance the reliability of biological control by improving the consistent establishment of microbial antagonists on plant surfaces. Two excellent reviews describe recent innovations in formulating microbial antagonists to enhance efficacy of biocontrol (13,27).

Trends in Biological Control

Increasingly, researchers appreciate the inherent difficulties encountered in developing biological control agents that meet agriculture's expectations for highly effective and reliable disease-management products. During the past decade, multiple sources of variability in biological control have been elucidated. These include intra-specific and inter-specific variation in the ecology of the pathogen (29); variation in establishment of biological control agents in different field environments; variation in amenability of plant hosts to biological control, even at the genotype level (39); and variation in expression of traits required for biocontrol activity in response to plant and microbial signals (20,30,36) and environmental parameters (25). In response to the complexities of the field environment and the variations encountered therein, some biological control researchers have directed their efforts to more controlled agricultural environments (such as greenhouse environments or post-harvest disease problems) or to the development of disease management systems that provide an environment conducive to biological control. Others have departed from the focus that has dominated the field for the past 2 decades by examining biocontrol systems that do not operate by directly suppressing pathogen populations on plant surfaces, but utilize other mechanisms.

SPECIFIC APPLICATIONS

In field environments, the population sizes and activities of biological control agents are influenced by environmental factors that fluctuate over time and from site to site. In contrast to agricultural fields, environmental conditions in greenhouses are more uniform and can be controlled to favor the establishment and activities of microbial antagonists. Some hydroponic and soilless planting media, such as rockwool, support very small microbial populations when planting occurs, so microbial antagonists can become established without intense competition from indigenous microorganisms. Furthermore, inoculum of biocontrol agents can be introduced into irrigation or fertilization systems at a frequency that optimizes efficacy. Crops grown in hydroponics or in soilless culture generally have a high economic value and can support costs associated with introduction of biocontrol agents. For these reasons, these systems appear to be ideally suited for biological control to succeed (33). Indeed, a number of microbial antagonists, including *P. fluorescens* (33) and *G. virens* (Soilgard®) (28), suppress diseases of plants grown in these systems at levels considered to be acceptable economically.

Biological control shows great promise for suppression of post-harvest diseases due to their economic importance, the relatively controlled environments in storage facilities, and the discrete sites and timing of pathogen infection (46). The first biological control agents for post-harvest diseases have now been registered for commercial use in the United States, and efforts are underway to identify biocontrol agents for an increasing spectrum of post-harvest diseases (42).

CULTURAL PRACTICES THAT PROMOTE THE POPULATION SIZES AND ACTIVITIES OF BIOLOGICAL CONTROL AGENTS

One approach for enhancing efficacy is to place effective antagonists in environments that promote their timely expression of phenotypes critical to biological control. Such phenotypes are not expressed uniformly by microbial antagonists in all environments into which they are introduced (20,30,36). Although there has been great effort devoted to testing microbial antagonists inoculated into soil or onto seed planted in field soil, such inoculants have met with mixed success in their capacity to suppress disease, especially diseases caused by pathogens that infect plants for extended periods throughout one or many growing seasons.

Certain composts or peats used in potting mixtures suppress soilborne diseases. The disease-suppressive characteristics of composts and peats result from two fundamental components: the presence of effective microbial antagonists, and the energy to support the metabolic activities of these antagonists (4,5). Disease suppression in planting mixtures containing compost or suppressive peat has been attributed to the combined activities of the many components of microbial communities inhabiting these substrates, and the suppressiveness of some potting mixtures can be enhanced by adding specific microbial antagonists. Similarly, the activities of effective biocontrol agents can be enhanced by placing them into substrates such as composts that support microbial activity (4). During the past decade, tremendous gains have been made in our knowledge of the microbial ecology of disease suppressive systems, such as composts, and in our understanding of the population biology and biocontrol activities of individual microbial antagonists. Using concepts and techniques from microbial ecology, scientists have identified characteristics of planting mixtures that appear to promote the population sizes and biocontrol activities of antagonists. These studies are particularly valuable to the field of biocontrol because they illustrate the power of ecological approaches and concepts for building our collective understanding of *in situ* activities of biocontrol agents. There are now enormous opportunities to couple our knowledge of these areas in order to more effectively and consistently manage plant disease.

INCREASED ATTENTION TO BIOLOGICAL CONTROL EFFECTED THROUGH DIVERSE MECHANISMS

Certain biological control agents suppress diseases initiated from infection sites from which they are spatially separated. These agents clearly do not operate through direct effects on the target pathogen but by inducing resistance responses in plants. Studies have focused on induced systemic resistance (ISR) caused by rhizosphere bacterial biocontrol agents (44), but a number of non-pathogenic fungi (3) and disease-suppressive composts (48) also cause ISR. Like the more well-known systemic-acquired resistance (SAR) induced by salicylic acid or a necrosis-inducing pathogen, ISR is manifested as reduced symptoms on a portion of a plant distal to placement of the inducing agent. Further similarities between ISR and SAR are obscure, however, and evidence suggests that salicylic acid accumulation in the plant is not required for ISR caused by some rhizosphere bacteria (44). Regardless of the physiological mechanisms involved, ISR is an exciting area of biological control research. It has the capacity to suppress diseases caused by pathogens that infect the plant at locations distal to the application site of biological control agents, is active against a variety of diseases and pests, and has been effective in field settings.

Conclusions

In 1970, when the first symposium on integrated pest management took place (37), plant pathologists studying the ecology of soil fungi that cause plant disease had already developed and communicated a sophisticated body of knowledge documenting the importance of microbial interactions in plant disease epidemics. The writings of these plant pathologists emphasize their recognition of the complexity of the soil environment and paucity of knowledge of the soil microflora (1). When *A. radiobacter* K84 was discovered in the 1970s, and its remarkable success as a biological control agent was recognized, there was a visible shift in research emphasis from studies of ecological relationships between indigenous pathogenic and saprophytic soil microorganisms to those focused on introductions of single microbial strains with specific beneficial properties. The availability today of at least 37 different commercial products for the biological control of plant disease is a reflection of the emphasis that plant pathologists have placed on this aspect of biological control research. Many of the commercial biological control agents have been incorporated into disease management programs, but many have also met with obstacles to their widespread adoption by agricultural producers. Within the context of characteristics shared by many biological control agents, this chapter has discussed the obstacles as well as the factors contributing to adoption of one of the commercial biological control agents, BlightBan A506.. Widely used strategies to optimize biological control include integration of biocontrol with other management practices, improved fermentation and formulation processes, and applications of mixtures of compatible strains of biocontrol agents. Despite efforts to enhance these strategies, biological control in field environments continues to be plagued by variations in efficacy, posed in part by the environmental fluctuations encountered in agricultural fields and by the complexity of microbial communities resident in the natural environment, which form the context in which interactions between introduced biocontrol agents and target pathogens occur. Increasingly, attention is being directed to diseases that occur in controlled systems (e.g., post-harvest diseases and root diseases of plants grown in soilless substrates in greenhouses), where economic losses to disease are great, the environment is stable (relative to an agricultural field), and commercial levels of disease management have been attained through biological control. Efforts are also being directed toward the development of systems that promote both the activities and population sizes of biocontrol agents, demonstrating that adding effective antagonists, such as disease-suppressive composts, to these systems can promote plant health. During the past 10 years, the field of biological control has broadened from an emphasis on biocontrol agents that suppress pathogens on plant surfaces (such as *A. radiobacter* K84) to those that suppress diseases through other mechanisms. A notable example of this expanded view is the current work evaluating ISR caused by biological control agents.

Too often, critics consider commercially available microbial inoculants to be the sole outcome of biological control research. This view fails to recognize the full spectrum of fundamental and practical information that has been generated by these studies. Plant pathologists now typically consider the influences of the indigenous microflora on plant disease, and they recognize that plant disease epidemics occur within the context of resident microbial communities. Biological control is now acknowledged as a major contributing factor to disease suppression that results from many cultural practices, such as cover cropping, green manuring, and certain crop rotation strategies, that were mainstays of traditional agriculture and continue to be important components of disease management programs. Still, we recognize that our understanding of microbial communities is limited. Recent advances revealing the remarkable diversity of microorganisms on our planet and providing new tools for characterizing the components and functional activities of microbial communities present in agricultural systems have been met with great enthusiasm. Our understanding of the microorganisms that inhabit our world is currently in a period of

251

expansive growth, and the consequent knowledge promises to accelerate our progress in developing the concepts and practice of biological control as we move into the coming century.

Literature Cited

1. Baker, K. F., and Snyder, W. C. 1965. Ecology of Soil-borne Plant Pathogens, Prelude to Biological Control. University of California Press, Berkeley, California.
2. Baker, R. R., and Dunn, P. E. 1990. New Directions in Biological Control: Alternatives for Suppressing Agricultural Pests and Diseases. Alan R. Liss, Inc., New York, New York.
3. Benhamou, N., Rey, P., Chérif, M., Hockenhull, J., and Tirilly, Y. 1996. Treatment with the mycoparasite *Pythium oligandrum* triggers induction of defense-related reactions in tomato roots when challenged with *Fusarium oxysporum* f. sp. *radicis-lycopersici*. Phytopathology 87:108-122.
4. Boehm, M. J., Madden, L. V., and Hoitink, H. A. J. 1993. Effect of organic matter decomposition level on bacterial species diversity and composition in relationship to Pythium damping-off severity. Appl. Environ. Microbiol. 59:4171-4179.
5. Boehm, M. J., Wu, T., Stone, A. G., Kraakman, B., Iannotti, D. A., Wilson, G. E., Madden, L. V., and Hoitink, H. A. J. 1997. Cross-polarization magic-angle spinning ^{13}C nuclear magnetic resonance spectroscopic characterization of soil organic matter relative to culturable bacterial species composition and sustained biological control of Pythium root rot. Appl. Environ. Microbiol. 63:162-168.
6. Boland, G. J., and Kuykendall, L. D. 1998. Plant-Microbe Interactions and Biological Control. Marcel Dekker, New York, New York.
7. Callan, N. W., Mathre, D. E., Miller, J. B., and Vavrina, C. S. 1997. Biological seed treatments: Factors involved in efficacy. Hortscience 32:179-183.
8. Deacon, J. W. 1991. Significance of ecology in the development of biocontrol agents against soilborne plant pathogens. Biocont. Sci. Technol. 1:5-20.
9. Deacon, J. W., and Berry, L. A. 1993. Biocontrol of soil-borne plant pathogens: concepts and their application. Pestic. Sci. 37:417-426.
10. Duffy, B. K., Simon, A., and Weller, D. M. 1996. Combination of

Trichoderma koningii with fluorescent pseudomonads for control of take-all on wheat. Phytopathology 86:188-194.

11. Farrand, S. K. 1990. *Agrobacterium radiobacter* strain K84: A model biocontrol system. Pages 679-691 in: New Directions in Biological Control: Alternatives for Suppressing Agricultural Pests and Diseases. R. R. Baker and P. E. Dunn, eds. Alan R. Liss, Inc., New York, New York.

12. Fravel, D. R. 1996. Interaction of biocontrol fungi with sublethal rates of metham sodium for control of *Verticillium dahliae*. Crop Prot. 15:115-119.

13. Fravel, D. R., Connick, W. J., and Lewis, J. A. 1998. Formulation of microorganisms to control plant diseases. H. D. Burges, ed. Formulation of Microbial Biopesticides, Beneficial Microorganisms, and Nematodes. Chapman and Hall.

14. Hall, F. R., and Menn, J. J. 1999. Biopesticides. Humana Press Inc., Totowa, New Jersey, USA.

15. Hubbard, J. P., Harman, G. E., and Hadar, Y. 1983. Effect of soilborne *Pseudomonas* spp. of the biological control agent *Trichoderma hamatum* on pea seeds. Phytopathology 73:655-659.

16. Ishimaru, C. A., Klos, E. J., and Brubaker, R. R. 1988. Multiple antibiotic production by *Erwinia herbicola*. Phytopathology 78:746-750.

17. Johnson, K. B., and Stockwell, V. O. 1998. Management of fire blight: A case study in microbial ecology. Annu. Rev. Phytopathol. 36:227-248.

18. Johnson, K. B., Stockwell, V. O., McLaughlin, R. J., Sugar, D., Loper, J. E., and Roberts, R. G. 1993. Effect of bacterial antagonists on establishment of honey bee-dispersed *Erwinia amylovora* in pear blossoms and on fire blight control. Phytopathology 83:995-1002.

19. Kearns, L. P., and Mahanty, H. K. 1998. Antibiotic production by *Erwinia herbicola* Eh1087: Its role in inhibition of *Erwinia amylovora* and partial characterization of antibiotic biosynthesis genes. Appl. Environ. Microbiol. 64:1837-1844.

20. Kraus, J., and Loper, J. E. 1995. Characterization of a genomic region required for production of the antibiotic pyoluteorin by the biological control agent *Pseudomonas fluorescens* Pf-5. Appl. Environ. Microbiol. 61:849-854.

21. Ligon, J. M., Lam, S. T., Gaffney, T. D., Hill, D. S., Hammer, P. E., and Torkewitz, N. 1996. Biocontrol: Genetic modifications for enhanced antifungal activity. in: Biology of Plant-Microbe

253

Interactions. G. Stacey, B. Mullin, and P. M. Gresshoff, eds.International Society of Molecular Plant-Microbe Interactions, St. Paul, Minnesota.

22. Lindow, S. E. 1984. Integrated control and role of antibiosis in biological control of fire blight and frost injury. Pages 83-115 in: Biological Control on the Phylloplane. C. Windels and S. E. Lindow, eds. APS Press, St. Paul, Minnesota.

23. Lindow, S. E. 1995. Control of epiphytic ice nucleation-active bacteria for management of plant frost injury. Pages 239-256 in: Biological Ice Nucleation and its Applications. R. E. Lee, G. J. Warren, and L. V. Gusta, eds. APS Press, St. Paul, Minnesota.

24. Lindow, S. E., Desurmonth, C., Elkins, R., McGourty, G., Clark, E., and Brandl, M. T. 1998. Occurrence of indole-3-acetic acid-producing bacteria on pear trees and their association with fruit russet. Phytopathology 88:1149-1157.

25. Loper, J. E., and Henkels, M. D. 1997. Availability of iron to *Pseudomonas fluorescens* in rhizosphere and bulk soil evaluated with an ice nucleation reporter gene. Appl. Environ. Microbiol. 63:99-105.

26. Loper, J. E., and Lindow, S. E. 1997. Reporter gene systems useful in evaluating *in situ* gene expression by soil- and plant-associated bacteria. Pages 482-492 in: Manual of Environmental Microbiology. C. J. Hurst, G. R. Knudsen, M. J. McInerney, L. D. Stetzenbach, and M. V. Walter, eds. ASM Press, Washington.

27. Lumsden, R. D., Lewis, J. A., and Fravel, D. R. 1995. Formulation and delivery of biocontrol agents for use against soilborne plant pathogens. Pages 166-182 in: Biorational Pest Control Agents, Formulation and Delivery. F. R. Hall and J. W. Barry, eds. American Chemical Society, Washington, D.C.

28. Lumsden, R. D., Walter, J. F., and Baker, C. P. 1996. Development of *Gliocladium virens* for damping-off disease control. Can. J. Plant Pathol. 18:463-468.

29. Martin, F. N., and Loper, J. E. 1999. Soilborne plant diseases caused by *Pythium* spp.: Ecology, epidemiology, and prospects for biological control. Crit. Rev. Plant Sci. 18 (in press).

30. Maurhofer, M., Keel, C., Haas, D., and Défago, G. 1994. Pyoluteorin production by *Pseudomonas fluorescens* strain CHA0 is involved in the suppression of Pythium damping-off of cress but not of cucumber. Eur. J. Plant Pathol. 100:221-232.

31. Moore, L. W., and Warren, G. 1979. *Agrobacterium radiobacter*

strain K84 and biological control of crown gall. Annu. Rev. Phytopathol. 71:163-179.

32. Nuclo, R. L., Johnson, K. B., Stockwell, V. O., and Sugar, D. 1997. Secondary colonization of pear blossoms by two bacterial antagonists of the fire blight pathogen. Plant Dis. 82:661-668.

33. Paulitz, T. C. 1997. Biological control of root pathogens in soilless and hydroponic systems. Hortscience 32:193-196.

34. Paulitz, T. C., Ahmad, J. S., and Baker, R. 1990. Integration of *Pythium nunn* and *Trichoderma harzianum* isolate T-95 for the biological control of Pythium damping-off of cucumber. Plant Soil 121:243-250.

35. Pierson, E. A., and Weller, D. M. 1994. Use of mixtures of fluorescent pseudomonads to suppress take-all and improve the growth of wheat. Phytopathology 84:940-947.

36. Pierson, E. A., Wood, D. W., Cannon, J. A., Blachere, F. M., and Pierson, L. S. I. 1998. Interpopulation signaling via *N*-acyl-homoserine lactones among bacteria in the wheat rhizosphere. Mol. Plant-Microbe Interact. 11:1078-1084.

37. Rabb, R. L., and Guthrie, F. E. 1972. Concepts of Pest Management. North Carolina State University, Raleigh, North Carolina, USA.

38. Ristaino, J., Perry, K. P., and Lumsden, R. D. 1996. Soil solarization and *Gliocladium virens* reduce the incidence of southern blight in bell pepper in the field. Biocont. Sci. Technol. 6:583-593.

39. Smith, K. P., Handelsman, J., and Goodman, R. M. 1997. Modeling dose-response relationships in biological control: Partitioning host responses to the pathogen and biocontrol agent. Phytopathology 87:720-729.

40. Stockwell, V. O., Johnson, K. B., and Loper, J. E. 1996. Compatibility of bacterial antagonists of *Erwinia amylovora* with antibiotics used for fire blight control. Phytopathology 86:834-840.

41. Stockwell, V. O., Johnson, K. B., and Loper, J. E. 1998. Establishment of bacterial antagonists of *Erwinia amylovora* on pear and apple blossoms as influenced by inoculum preparation. Phytopathology 88:506-513.

42. Sugar, D., and Spotts, R. A. 1999. Control of post-harvest decay in pear by four laboratory-grown yeasts and two registered biocontrol products. Plant Dis. 83:155-158.

43. Tjamos, E. C., Papavizas, G. C., and Cook, R. J. 1992. Biological Control of Plant Diseases: Progress and Challenges for the Future. Plenum Press, New York, New York.

255

44. van Loon, L. C., Bakker, P. A. H. M., and Pieterse, C. M. J. 1998. Systemic resistance induced by rhizosphere bacteria. Annu. Rev. Phytopathol. 36:453-483.
45. Vanneste, J. L., Yu, J., and Beer, S. V. 1992. Role of antibiotic production by *Erwinia herbicola* Eh252 in biological control of *Erwinia amylovora*. J. Bacteriol. 174:2785-2796.
46. Wilson, C. L., and Wisniewski, M. E. 1994. Biological Control of Postharvest Diseases: Theory and Practice. CRC Press, Boca Raton, Florida.
47. Wilson, M., and Lindow, S. E. 1993. Interactions between the biological control agent *Pseudomonas fluorescens* strain A506 and *Erwinia amylovora* in pear blossoms. Phytopathology 83:117-123.
48. Zhang, W., Dick, W. A., and Hoitink, H. A. J. 1996. Compost-induced systemic acquired resistance in cucumber to Pythium root rot and anthracnose. Phytopathology 86:1066-1070.

Technologies for Strain Improvement for Biological Control of Plant Pathogens

James M. Ligon[1], Stephen T. Lam[1], Thomas D. Gaffney[1], D. Steven Hill[1], Phillip E. Hammer[1], Nancy Torkewitz[1], Tom Young[2], Dirk Hofmann[3], and Hans-Joachim Kempf[3]

[1]Novartis Agribusiness Biotechnology Research, Inc., Research Triangle Park, NC 27709 USA

[2]Novartis Crop Protection, Inc., Vero Beach, FL USA

[3]Novartis AG, CH-4002, Basel, Switzerland

The use of antagonistic microbes to prevent infection of plants by plant pathogens, commonly known as biocontrol, has been well documented. Several studies that have investigated the biological mechanisms underlying this phenomenon indicate that biocontrol activity is due largely to the production of antifungal compounds by the biocontrol antagonist. These include the production of: antibiotics such as phenazine-1-carboxylate (17), 2,4-diacetylphloroglucinol (11), and pyrrolnitrin ([3-chloro-4-(2'-nitro-3'-chlorophenyl)-pyrrole], abbreviated "Prn") (8); hydrolytic enzymes such as chitinase (15); hydrogen cyanide (18); and siderophores (2). We have recently characterized a *Pseudomonas fluorescens* strain, BL915, that is an effective biocontrol agent for the control of *Rhizoctonia solani*-induced seedling disease (7). This strain produces several antifungal compounds, including Prn, 2-hexyl-5-propyl resorcinol (Res), chitinase, and cyanide. Furthermore, the production of these compounds is coordinately regulated by a bacterial two-component regulatory system consisting of a receptor-kinase and response regulator (3). Prn was first described by Arima et al. (1) and is a

257

highly antifungal metabolite produced primarily by pseudomonads. Prn has a safe toxicological profile and is used as a clinical antifungal agent for the treatment of skin mycoses (16), and a phenylpyrrole derivative of Prn has been developed by Novartis Crop Protection as an agricultural fungicide (4). Res, another metabolite known to be produced by pseudomonads, has antifungal and anti-Gram-positive bacterial activity (10).

We have cloned and characterized the genes from strain BL915 that encode the sensor-kinase (*gacS*), the response regulator (*gacA*), chitinase, and enzymes involved in the synthesis of Prn (6). We have used these genes and our understanding of their role in biocontrol to genetically enhance the overall biocontrol activity of *P. fluorescens* strain BL915.

Production of Antifungal Activities is Coordinately Regulated

Mutagenesis of the wild-type *Pseudomonas fluorescens* strain BL915 with *N*-methyl-*N*'-nitro-*N*-nitrosoguanidine (NTG) and screening of the mutants in vitro for fungal antagonism resulted in the isolation of numerous mutants of strain BL915 that lacked the ability to inhibit *R. solani* on agar medium. Unlike the parent strain, these mutants did not produce Prn, chitinase, HCN, or Res. While strain BL915 provides excellent biocontrol of *R. solani* infections on seedlings, the antifungal mutants demonstrated no such activity (7). A genomic DNA library of strain BL915 was constructed in a broad host-range cosmid vector, and this was introduced into each of 20 antifungal-minus mutants of the strain. Functional complementation of the antifungal trait was accomplished for each of the mutants, and analysis of the resulting cosmid clones revealed that, based on restriction endonuclease profiles, they were distributed about equally into two distinct groups. In one group of clones, the gene region responsible for complementation of the antifungal trait was localized to an 11 kilobase (kb) EcoRI fragment. The DNA sequence of this fragment was determined and analyzed (3), revealing the presence of an open reading frame (ORF) that is highly homologous to known response regulator genes, including *gacA* from *P. fluorescens* (14). Precise deletions within the coding sequence of the *gacA*-homolog in the chromosome of strain BL915 were constructed, and these were shown not to produce antifungal metabolites or to provide biocontrol activity on plants. Similarly, the complementing locus from the other complementation group was isolated and sequenced and revealed the presence of an ORF with strong homology to the sensor kinase gene *lemA* (since renamed *gacS*) from *P. syringae,* which is involved in the regulation of pathogenic functions (9) in this phytopathogenic bacterium. Directed mutation of the *gacS* homolog in strain BL915 also resulted in loss of antifungal and biocontrol activities. These results indicate that the production of antifungal compounds by strain BL915 and the biocontrol activity that is a result of their production are coordinately regulated by a two-component regulatory system consisting of a sensor kinase encoded by a *gacS* homolog and a response regulator encoded by a *gacA* homolog.

Cloning and Characterization of the Prn Biosynthetic Genes

In an effort to isolate genes whose expression is regulated by the protein products of the *gacS* and *gacA* genes in a related *P. fluorescens* strain, strain BL914, we identified a Tn5/*lacZ* transposon mutant that was specifically affected in the synthesis of Prn (13). A cosmid clone from a genomic library containing DNA from strain BL915 was found that complemented the Prn⁻ phenotype of this mutant. Analysis of the cloned DNA in this cosmid by site-directed Tn5 mutagenesis demonstrated that a 6.2 kb region centrally located in the cloned DNA of this cosmid was involved in Prn biosynthesis (5). The nucleotide sequence of this region was determined, and analysis of it revealed the presence of four colinear ORFs organized on a single transcriptional unit (Fig. 1). Mutations in each of the ORFs were created in vitro by deletion of segments of DNA that are internal to the coding sequences, and then by subsequent ligation of a kanamycin resistance gene into the site of the deletion. Each deletion mutation was introduced into the chromosome of strain BL915 by homologous replacement using the kanamycin resistance marker for selection. A mutation in any one of the identified ORFs abolished Prn synthesis in BL915. In addition, the coding sequences of each of the four ORFs, including putative ribosome-binding sites, were amplified by PCR and fused to the *tac* promoter (from plasmid pKK223-3MCS, Pharmacia) in the proper juxtaposition to cause expression of the

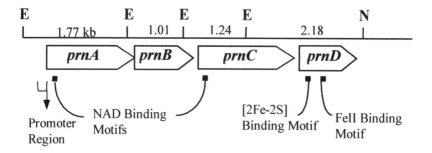

**Figure. 1. Genetic organization of the *prnABCD* gene cluster from
P. fluorescens strain BL915. A restriction map showing the
position of *EcoRI* (E) and *NotI* (N) sites and the distance between
them appears above open boxes that depict the coding region of
each gene. Special features identified in the nucleotide sequence of
the individual genes are shown below.**

ORFs from this promoter. Each *tac* promoter/ORF fusion was
introduced into the corresponding ORF deletion mutant, and in every
case, the mutant Prn⁻ phenotype was complemented. Based on this
evidence, the four ORFs were determined to encode enzymes directly
involved in the biosynthesis of Prn and were assigned the genetic
names *prnA, prnB, prnC,* and *prnD* for ORFs 1-4, respectively. The
prn gene deletion mutants and the *tac* promoter/*prn* gene fusions were
used to clearly elucidate the biochemical pathway for the synthesis of
Prn and the role of each of the enzymes encoded by the four *prn* genes
(12).

The *prnABCD* gene cluster was cloned and fused to the P*tac*
promoter in the proper orientation to effect transcription of the
operon. This construction was introduced on a broad host-range
plasmid into several different Gram-negative bacteria that were
incapable of Prn synthesis, including *E. coli, Enterobacter,* and other
Pseudomonas strains. In each case, bacteria containing the
tac/prnABCD fragment were shown to produce Prn (5,6). These
results indicate that the four *prn* genes identified represent the entire
biosynthetic pathway for Prn synthesis.

Genetic Modification of *P. fluorescens* for Increased Metabolite Production

In the past few years, we have attempted to utilize the recently acquired knowledge about the regulation, genetics, and biochemistry of the synthesis of antifungal metabolites by *P. fluorescens* strain BL915 to generate genetically modified versions of the strain with increased metabolite production for the purpose of improving the biocontrol activity of the strain. Some modifications have been directed at the two-component system that regulates metabolite synthesis, while others were specifically targeted at improving Prn production. Descriptions of the genetically modified derivatives of *P. fluorescens* strain BL915 appear in Table 1.

One example of genetic modifications affecting the regulatory system is the introduction of a plasmid containing the 11 kb EcoRI *gacA* gene fragment into strain BL915. In addition to the single copy of the *gacA* gene present in the chromosome, this modified strain, designated BL915(p-*gacA*), contains additional plasmid-borne copies of the *gacA* gene. BL915(p-*gacA*) produced about 2.5-fold more Prn than the parent strain (Fig. 2).

Subsequently, other modifications of the *gacA* gene were constructed and tested. The native *gacA* gene has an unusual TTG translation initiation codon that is known to result in reduced translational efficiency of genes compared to the more common ATG initiation codon. Using polymerase chain reaction (PCR) technology, we changed the first base in the coding sequence of the chromosomal *gacA* gene in *P. fluorescens* strain BL915 from a thymidine to an adenine, thereby changing the translation initiation codon of the *gacA* gene from the less efficient TTG codon to the more efficient ATG codon. The resulting strain, BL915(ATG/*gacA*), produced slightly more Prn than the parent strain (50 vs 40 mg/L).

In another modification of the *gacA* gene, we sought to increase the level of transcription of the gene by replacing the native promoter with the strong P_{tac} promoter from *E. coli*. The resulting modified strain, BL915(P_{tac}/*gacA*), produced about twice as much Prn as the parent strain (Fig. 2).

Table 1. Description of *P. fluorescens* strains.

Strain	Description
BL915	Wild-type *P. fluorescens* strain
BL915(p-*gacA*)	BL915; plasmid-borne copies of the *gacA* gene
BL915(ATG/*gacA*)	BL915; TTG initiation codon of *gacA* changed to ATG
BL915(P$_{tac}$/*gacA*)	BL915; chromosomal *gacA* gene expressed from the P$_{tac}$ promoter
BL915(p-P$_{tac}$/*prnABCD*)	BL915; plasmid-borne copies of the *prnABCD* cluster under control of the P$_{tac}$ promoter
BL915(P$_{tac}$/*gacA*, p-P$_{tac}$/*prnABCD*)	BL915; chromosomal *gacA* gene and plasmid-borne copies of the *prnABCD* cluster under control of the P$_{tac}$ promoter

Another class of genetically modified *P. fluorescens* strain was constructed by modification of the *prnABCD* gene cluster derived from this strain. In an approach similar to that used to construct the modified strain BL915(P$_{tac}$/*gacA*) described above, we replaced the native GacA-regulated promoter of the *prnABCD* gene cluster with the more active P$_{tac}$ promoter. A DNA fragment containing the modified P$_{tac}$/*prnABCD* genes was cloned into a broad host-range plasmid, and this was introduced into strain BL915. The resulting modified strain, BL915(P$_{tac}$/*prnABCD*), contains multiple plasmid-borne copies of the P$_{tac}$/*prnABCD* genes. It was shown to produce approximately four times as much Prn as the parent strain (Fig. 2). The modified version of the chromosomal *gacA* gene, P$_{tac}$/*gacA*, and the plasmid-borne P$_{tac}$/*prnABCD* gene cluster were combined into a single modified strain to create strain BL915(P$_{tac}$/*gacA*, p-P$_{tac}$/*prnABCD*). This strain was shown to produce almost 10-fold more Prn than the parent strain BL915 (Fig. 2).

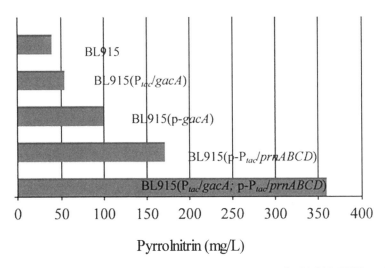

Pyrrolnitrin (mg/L)

Figure 2. Pyrrolnitrin production of the wild-type strain BL915 (WT) and genetically modified derivatives during growth in liquid medium. The genotypes used to identify the BL915 derivatives are described in the text.

Correlation of Prn Production and Biocontrol Activity

The genetically modified strains described above were tested for biocontrol activity in two pathosystems, including cucumber and impatiens, with *Rhizoctonia solani* as the pathogen. Since strain BL915 is very effective when used at the rate of 2×10^8 cells/g soil, and we were interested in determining whether the modified BL915 strains have increased biocontrol activity, we tested all modified strains at a rate one-tenth that of the normal rate (2×10^7 cells/g soil). The results, shown in Table 2, demonstrate that, in most cases, the modified strains exhibited increased biocontrol activity relative to parent strain BL915. Strain BL915(P_{tac}-*gacA*) showed slightly better activity compared with the parent strain BL915 in both pathosystems at the low-use rate (Table 2). Strain BL915(p-*gacA*) had activity similar to the parent strain on cucumber, but was significantly improved in the impatiens pathosystem (Table 2). Strain BL915(p-P_{tac}/*prnABCD*) at the low-application rate had activity in both pathosystems that was similar to the parent strain at the high-use rate

263

Table 2. Biocontrol activity of modified BL915 strains in two pathosystems against _R. solani_. Cucumber seeds or impatiens transplants were planted in infested soil, except in the case of the healthy control plants, which were planted in uninfested soil. All strains were applied as a drench at 2×10^7 cells/g soil, except for the high rate of BL915, which was applied at a 10-fold higher rate. Since the strains were tested in separate trials, results for the controls are shown with each test group. In the case of the healthy controls, the number of plants planted for each treatment is shown in parentheses.

Treatment	Stand (14 Days After Planting)	
	Cucumber	Impatiens
Healthy Control	113 (120)	120 (120)
Pathogen Control	86	92
BL915-High Rate	97	111
BL915	80	91
BL915(p-_gacA_)	75	103
BL915(p-P_{tac}/_prnABCD_)	101	105
Healthy Control	54 (60)	119 (120)
Pathogen Control	31	78
BL915-High Rate	54	106
BL915	41	73
BL915(P_{tac}/_gacA_)	46	80
BL915(P_{tac}/_gacA_, pP_{tac}/_prnABCD_)	55	115

and had significantly improved activity relative to the parent strain at the comparable low rate (Table 2). However, the plant stand from this strain was less than the stand of the healthy controls that were uninfected (Table 2). Strain BL915(P_{tac}/_gacA_, p-P_{tac}/_prnABCD_), the strain that produced the highest amount of Prn, gave excellent control, equal to that of the stand of the healthy controls in both pathosystems (Table 2). In a separate experiment, strain BL915 and different genetically modified derivatives were cultured in liquid medium for 48

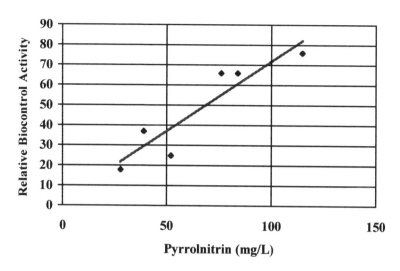

Figure. 3. Correlation between pyrrolnitrin production and bio-control activity by *P. fluorescens* strain BL915 and genetically modified derivatives.

hours, and the amount of Prn produced in culture was determined for each. In addition, the cultures were used as seed treatments in a cotton/*R. solani* biocontrol assay. The amount of Prn produced by each strain and the biocontrol activity of the corresponding strain were determined and compared. The results, shown in Figure 3, demonstrate a strong correlation between increased Prn production and increased biocontrol activity. These results demonstrate that the genetic modifications of strain BL915 described herein have resulted in increased production of Prn and in a concomitant increase in biocontrol activity.

In the summer of 1997, the genetically modified *Pseudomonas* strains that had shown the best biocontrol activity in greenhouse trials were tested under field conditions. Cotton trials with *R. solani* and *P. ultimum* as pathogens were conducted in the spring at the Novartis field test sites in Fresno, CA, and Vero Beach, FL. Cotton seeds were planted in a furrow and were inoculated with either *R. solani* or *P. ultimum* by spreading a fine, dried, fungal-infested wheat bran powder over the seeds. Prior to closure of the furrow, the seeds were

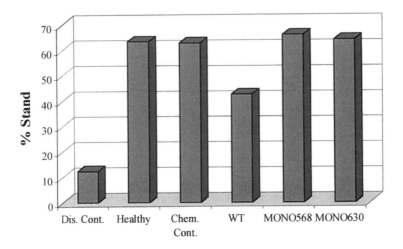

Figure 4. Biocontrol of *R. solani* infections of cotton in field trials conducted in Vero Beach, FL. The activity of each treatment, including strain BL915 (WT) and two modified strains (MONO568 and MONO630) is shown as percent stand 14 days after planting.

treated with a lyophilized preparation of one biocontrol bacterial strain or an appropriate chemical standard, or were left untreated to provide a disease check. Each treatment consisted of four rows of 60 plants each. Fourteen days after planting, the number of emerged seedlings was counted. Figure 4 depicts the results of the *R. solani* trial in Vero Beach, FL. In this trial, the parent strain BL915 provided protection of about 50% relative to the chemical control (pentachloronitrobenzene [PCNB]) and the disease check. However, both of the two modified strains that produce higher amounts of Prn than strain BL915 provided disease control that was signficantly better than strain BL915 and not significantly different from the chemical and healthy controls (Fig. 4). Similar results were obtained in the *P. ultimum* trials conducted in Fresno, CA (data not shown). One problem that has been typical of natural biocontrol strains is that, while they provide good, consistent control in controlled greenhouse environments, they are unreliable and less effective in field conditions. These results suggest that genetically modified strains that over-produce important antifungal metabolites may provide reliable disease control equal to that of standard chemical fungicides.

Literature Cited

1. Arima, K., Imanaka, H., Kousaka, M., Fukuda, A., and Tamura, C. 1964. Pyrrolnitrin, a new antibiotic substance, produced by *Pseudomonas*. Agr. & Biol. Chem. 28:575-576.
2. Becker, J. O., and Cook, R. J. 1988. Role of siderophores in suppression of *Pythium* species and production of increased growth response of wheat by fluorescent pseudomonads. Phytopathology 78:778-782.
3. Gaffney, T., Lam, S., Ligon, J., Gates, K., Frazelle, A., Di Maio, J., Hill, S., Goodwin, S., Torkewitz, N., Allshouse, A., Kempf, H. J., and Becker, J. 1994. Global regulation of expression of antifungal factors by a *Pseudomonas fluorescens* biological control strain. Mol. Plant-Microbe Interact. 7:455-463.
4. Gehmann, K., Neyfeler, R., Leadbeater, A., Nevill, D., and Sozzi, D. 1990. CGA173506: a new phenylpyrrole fungicide for broad-spectrum disease control. Proc. Brighton Crop Prot. Conf., Pest Dis. 2:399-406.
5. Hammer, P. E., Hill, D. S., Lam, S. T., Pée, -K. H.,-van., and Ligon, J. M. 1997. Four genes from *Pseudomonas fluorescens* that encode the biosynthesis of pyrrolnitrin. Appl. Environ. Microbiol. 63:2147-2154.
6. Hill, S., Lam, S., Hammer, P., and Ligon, J. 1995. Cloning, characterization, and heterologous expression of genes from *Pseudomonas fluorescens* involved in the synthesis of pyrrolnitrin. Phytopathology 85:1187.
7. Hill, D. S., Stein, J., Torkewitz, N., Morse, A., Howell, C., Pachlatko, J., Becker, J., and Ligon, J. 1994. Cloning of genes involved in the synthesis of pyrrolnitrin from *Pseudomonas fluorescens* and role of pyrrolnitrin synthesis in biological control of plant disease. Appl. Environ. Microbiol. 60:78-85.
8. Howell, C., and Stipanovic, R. 1979. Control of *Rhizoctonia solani* on cotton seedlings with *Pseudomonas fluorescens* and with an antibiotic produced by the bacterium. Phytopathology 77:480-482.
9. Hrabak, E. M., and Willis, D. K. 1992. The *lemA* gene required for pathogenicity of *Pseudomonas syringae* pv. *syringae* on bean

is a member of a family of two-component regulators. J. Bacteriology 174:3011-3020.

10. Kanda, N., Ishizaki, N., Inoue, N., Oshima, M., Handa, A., and Kitahara, T. 1975. DB-2073, a new alkylresorcinol antibiotic. I. Taxonomy, isolation, and characterization. J. Antibiot. 28:935-942.

11. Keel, C., Wirthner, P., Oberhansli, T., Voisard, C., Burger, U., Haas, D., and Défago, G. 1990. Pseudomonads as antagonists of plant pathogens in the rhizosphere: role of the antibiotic 2,4-diacetylphloroglucinol in the suppression of black rot of tobacco. Symbiosis 9:327-341.

12. Kirner, S., Hammer, P. E., Hill, D. S., Altmann, A., Fischer, I., Weislo, L. J., Lanahan, M., Pée, -K. H.,-van.,and Ligon, J. M. 1998. Functions encoded by pyrrolnitrin biosynthetic genes from *Pseudomonas fluorescens*. J. Bacteriology 180:1939-1943.

13. Lam, S., Frazelle, R., Torkewitz, N., and Gaffney, T. 1995. A genetic approach to identify pyrrolnitrin biosynthesis and other globally regulated genes in *Pseudomonas fluorescens*. Phytopathology 85:1163.

14. Laville, J., Voisard, C., Keel, C., Maurhofer, M., Défago, G., and Haas, D. 1992. Global control in *Pseudomonas fluorescens* mediating antibiotic synthesis and suppression of black root rot of tobacco. Proc. Natl. Acad. Sci. USA 89:1562-1566.

15. Shapira, R., Ordentlich, A., Chet, I., and Oppenheim, A. B. 1989. Control of plant diseases by chitinase expressed from cloned DNA in *Escherichia coli*. Phytopathology 79:1246-1249.

16. Tawara, S., Matsumoto, S., Hirose, T., Matsumoto, Y., Nakamoto, S., Mitsuno, M., and Kamimura, T. 1989. In vitro antifungal synergism between pyrrolnitrin and clotrimazole. Japanese J. Med. Mycol. 30:202-210.

17. Thomashow, L., and Weller, D. 1988. Role of phenazine antibiotic from *Pseudomonas fluorescens* in biological control of *Gaeumannomyces graminis* var. *tritici*. J. Bacteriology 170:3499-3508.

18. Voisard, C., Keel, C., Haas, D., and Défago, G. 1989. Cyanide production by *Pseudomonas fluorescens* helps suppress black root rot of tobacco under gnotobiotic conditions. EMBO J. 8:351-358.

Current Status of Biological Control of Weeds

Raghavan Charudattan,
Plant Pathology Department, University of Florida,
Gainesville, FL 32611-0680 USA

Biological control of weeds by using microbial plant pathogens, insects and mites, nematodes, and fish has had a long and successful record in the United States and several other countries. Although biological control plays only a minor role as a practical weed control strategy on a national and global scale, the value of many biological weed control programs on local and regional scales has been considerable in terms of the economic benefits and human welfare derived. In this regard, we should neither overlook nor understate the past successes and future potential of the different groups of agents: insects and mites as classical biological control agents (34,41,53), grass carp (*Ctenopharyngodon idella*) as a nonselective herbivore to manage submerged aquatic vegetation (37,40), plant-parasitic nematodes as classical or augmentative biocontrol agents (47), and microbial plant pathogens as classical and inundative biocontrol agents (discussed in this chapter).

Generally, classical (inoculative) and augmentation strategies have not been used as components of integrated weed management systems. This is due to the slowness of the biocontrol process relative to the short duration of the cropping season and the frequent disruptions to the activities of biocontrol agents caused by crop-management practices. High on the list of newer approaches available for integrated weed management is the use of plant pathogens as bioherbicides (i.e., as augmentative or inundative agents). Numerous reviews and treatises published in the last 2 decades have dealt with the principles and empirical successes in the use of plant pathogens as weed-control agents (8,18,63,72, and others), and these will not be repeated here. Instead, this chapter will review the current status of the use of plant pathogens as biological control agents for weeds (both as bioherbicide and classical agents), some examples of integrated control of weeds by using plant pathogen(s) as a component, and important constraints and needs with respect to development, implementation, and adoption of this technology.

Bioherbicides

A bioherbicide is a plant pathogen used as a weed-control agent through inundative and repeated applications of its inoculum or by augmentation of natural, seasonal disease levels through small releases of inoculum. In the United States and many other countries, the use of plant pathogens in prescriptive weed control is regarded as a "pesticidal use," and therefore these pathogens must be registered or approved as biopesticides (17).

As a novel weed-control technology, bioherbicides were introduced to commercial agriculture nearly 2 decades ago. Bioherbicides are still in use, albeit on a very small but noteworthy scale. Between 1980 and 1998, three bioherbicides were registered in the United States and one each was registered in Canada, Japan, and South Africa. Another fungal pathogen has been developed for unregistered use in the Netherlands. Contrary to some erroneous claims in literature (41), five of these seven bioherbicides are still commercially available for use, while two are unavailable due to technical difficulties in production or marketplace considerations, as explained below.

The first bioherbicide registered was DeVine® (Fig. 1a). It is composed of an indigenous isolate of *Phytophthora palmivora* and is registered for the control of *Morrenia odorata* (stranglervine) in citrus in Florida (Fig. 1a). Stranglervine, imported into Florida as a potential ornamental plant, has become a difficult-to-control weed in about 49,000 ha of Florida's citrus groves. The isolate of *P. palmivora* developed as the bioherbicide was isolated from dying stranglervine plants in Florida and determined to be a suitable bioherbicide agent on the basis of efficacy and safety evaluations (54,55). It was commercially developed and registered in 1980 by Abbott Laboratories, Chicago, IL. DeVine consists of chlamydospore concentrate and is applied as a postemergent, directed spray. Typically it yields 90% to 100% control of the vine with just one application, and the control lasts for at least 2 years following treatment (38). DeVine is produced and sold on a made-to-order basis and is used by citrus growers in Florida whenever the weed problem demands attention.

270

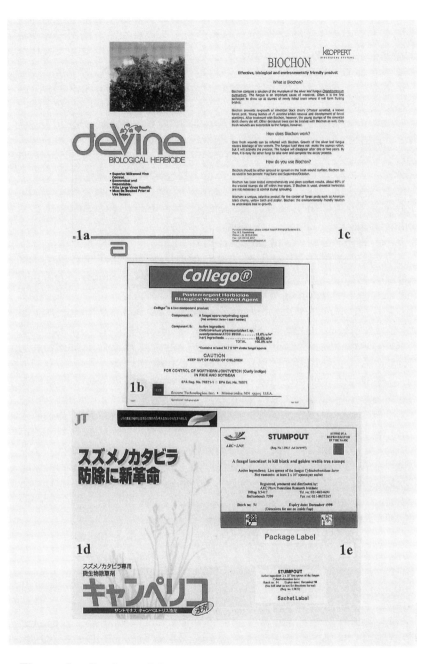

Figure 1. Registered labels or product brochures of presently available, commercial bioherbicides.

271

Collego® (Fig. 1b), based on the fungus *Colletotrichum gloeosporioides* f.sp. *aeschynomene,* was registered in 1981 by The Upjohn Company, Kalamazoo, MI. It was subsequently sold to other companies and is now produced and marketed by Encore Technologies, Minnetonka, MN. This native fungus was discovered and developed at the University of Arkansas by Templeton, TeBeest, and colleagues (66) to control northern jointvetch (*Aeschynomene virginica),* a leguminous weed in rice and soybean crops in Arkansas, Mississippi, and Louisiana. In addition to its competition with crops, the weed produces black, hard seeds that are difficult to separate from harvested rice and soybean. The presence of the contaminant seed lowers the quality and price of the harvested rice. In nearly 2 decades of combined experimental and commercial use, Collego has consistently yielded 90% to 100% weed control with no apparent adverse environmental impacts. This bioherbicide has been effectively integrated into rice and soybean pest management programs that involve applications of chemical fungicides, herbicides, and insecticides and cultural practices (61). Since 1997 when Encore Technologies began marketing Collego, the market size for this bioherbicide has increased steadily to 24,000 ha in 1999. The cost of Collego is currently $ 25.00 per ha (David Johnson, Encore Technologies, personal communication, 33).

BioMal® was the third bioherbicide registered in North America. The bioherbicidal preparation contained *Colletotrichum gloeosporioides* f.sp. *malvae,* a pathogen of round-leaved mallow (*Malva pusilla*). BioMal is registered for use in several crops in Canada, but due to technical difficulties in production and economic considerations, it has not been commercially sold since its registration by PhilomBios, Saskatchewan, Canada. However, attempts are underway to further develop and market this bioherbicide, possibly by Encore Technologies (David Johnson, personal communication).

Dr. BioSedge®, a bioherbicide based on the rust fungus *Puccinia canaliculata,* was registered in 1993 for the control of yellow nutsedge (*Cyperus esculentus*) in the United States. Extensive research by Phatak and coworkers (48,49) had established the potential of this native rust fungus as an augmentative agent to control yellow nutsedge. Dissemination of a small amount of uredospore inoculum (about 5 mg per hectare) in the spring and early summer, when the rust disease usually develops naturally in the field, can lead to severe rust epidemics (4,48). However, as the fungus is an obligate parasite, it cannot be mass-produced in vitro, and therefore it is presently not available for commercial use. Furthermore, Bruckart et al. (9) have determined that the rust strain registered as Dr. BioSedge was not effective against several regional biotypes of yellow nutsedge. In addition to the problem of host-specificity of strains, this rust pathogen was considered unsuitable for introduction into the Netherlands as a classical biocontrol agent because it attacked a native, nontarget species, *Cyperus fuscus* (59). Presently, Dr. BioSedge is unavailable for commercial use.

A wound-invading, wood-decaying basidiomycete, *Chondrostereum purpureum*, has been approved for treatment to prevent resprouting of black cherry (*Prunus serotina*) and to control this weedy, broad-leaved tree in conifer forests in the Netherlands. The fungus is sold by Koppert B.V. as BioChon®, a natural decay-promoter and a stump-treatment product (Fig. 1c). This fungus is presently under development for registration in Canada and the United States to prevent regrowth of unwanted hardwood trees in conifer forests (32,71). *Chondrostereum purpureum* is a weak parasite of wounded or weakened trees and is widespread in temperate regions of the world. It has a fairly wide host range among broad-leaved trees and causes an economically significant disease, the silverleaf disease, on several rosaceous fruit trees. However, in spite of the lack of a high degree of host specificity and the potential for damage to some fruit crops, it was determined, based on epidemiological models of aerial dispersal of inoculum and disease spread, that this pathogen could be used safely without added risk to cultivated plants (22).

The potential for risk from the biocontrol use of this fungus was assessed by de Jong et al. (23) also by estimation of the natural occurrence of *C. purpureum* in the forests of southern Vancouver Island, British Columbia, Canada, in relation to a proposed bioherbicidal use of this fungus. Surveys were done in randomly located, 1000-m^2 plots to determine the natural incidence of basidiocarps. The potential for increase in the occurrence of basidiocarps, which might occur as a result of the biocontrol use of the fungus, was also estimated. The surveys were made in forests, urban or agricultural areas, and other locations where the fungus would be expected to occur, such as woodpiles, silvicultural thinnings, and killed trees. Based on observations and projections, it was determined that the increased number of basidiocarps that might result from the biocontrol use of *C. purpureum* is of the same order of magnitude as the naturally occurring levels or even lower. In addition, it was determined that there was distinct geographical separation between predominantly forestry areas where the biocontrol use will take place and predominantly inhabited areas where fruit and ornamental trees are cultivated. Accordingly, it was concluded that this fungus is unlikely to pose a significant threat to fruit orchards and commercial forests.

An isolate of *Xanthomonas campestris* pv. *poae,* a wilt-inducing bacterium, isolated in Japan from annual bluegrass (*Poa annua*) has been developed and registered in Japan as the bioherbicide CAMPERICO® to control annual bluegrass in golf courses (Figure 1d). Fourteen isolates of this pathogen were isolated from annual bluegrass from throughout Japan and compared for pathogenicity and safety toward desirable turf grasses, such as creeping bentgrass (*Agrostis palustris*) and Kentucky bluegrass (*Poa pratensis*). The most effective isolate, JT-P482, did not cause disease on desirable turf grasses, but provided greater than 70% control of annual bluegrass when sprayed onto pre-cut annual bluegrass. Control in this case was determined on the basis of weight loss due to wilting and death of inoculated plants compared to uninoculated controls. The optimum inoculum concentration was about 10^8 colony forming units (CFU) sprayed in an application volume of 100 to 400 ml/m^2. The infection rates were increased by applying larger volumes of inoculum. At 25°C/20°C (day/night), annual bluegrass wilted severely in 7 to 10 days, but lower temperatures caused a loss of efficacy, showing that temperature is an important factor for effective control. Field tests in zoysia (*Zoysia tenuifolia*) greens in the fall with an inoculum rate of 10^9 CFU/ml at 400 ml/m^2 yielded in excess of 90% disease severity the following spring. Thus, this bacterium could be used to control annual bluegrass without harming the desirable grasses, including the closely related species Kentucky bluegrass (30).

Morris (43) has developed and registered a stump-treatment product based on the wood-infecting basidiomycete, *Cylindrobasidium laeve,* to control resprouting of cut trees. This product can be used to prevent regrowth of weedy, as well as harvested, timber trees. A laboratory-based inoculum production system to manufacture an oil-based, paste-formulation of the fungus has been set up, and the fungal product has been registered for use in South Africa under the name Stumpout® (Figure 1e).

Presently, the following pathogens are undergoing precommercial evaluation in the United States and other countries: *Alternaria destruens* for the control of dodder (*Cuscuta* spp.) in cranberries and carrots, *Ascochyta caulina* (common lambsquarters, *Chenopodium album*), *Colletotrichum truncatum* (hemp sesbania, *Sesbania exaltata*), *Dactylaria higginsii* (purple nutsedge, *Cyperus rotundus*), *Exserohilum monoceras* (barnyard grass, *Echinochloa crus-galli* and other species), *Fusarium* spp. for witchweed (*Striga hermonthica* and other species), *Pseudomonas syringae* pv. *tagetis* (various weeds in the Asteraceae and a few other families), *Phomopsis amaranthicola* (several pigweed and amaranth species, *Amaranthus* spp.), *Ralstonia solanacearum* (tropical soda apple, *Solanum viarum*), an unnamed fungus for dandelion (*Taraxacum officinale*) in turf, and others.

Classical Biocontrol Agents

One of the most successful examples of biological control of weeds resulted from the introduction of the rust fungus, *Puccinia chondrillina*, from the Mediterranean region into Australia to control rush skeletonweed (*Chondrilla juncea*). Skeletonweed, of Mediterranean origin, was introduced into Australia, where it became a serious weed in cereal crops and rangelands. The fungus was introduced, along with three insects, as a classical biocontrol agent. Following inoculative releases, the rust fungus disseminated rapidly and widely, created high levels of disease epidemics, and in this process infected, stressed, and killed the most common and susceptible biotype of the weed. As a result, the weed density in cereal crops decreased to less than 10 plants per m^2 from the level of ca. 200 plants per m^2 that existed before the rust introduction. Satisfactory levels of control were also obtained in pastures, suggesting an overall projected savings of $25.96 million per annum (21). It has been estimated that this highly successful biocontrol project has resulted in a cost-to-benefit ratio of 1:100 in Australia (21,56).

The rust fungus attacks one of three forms of the weed, the predominant type. Initially, as the population density of this susceptible type was reduced due to biocontrol, two other types that are more resistant to the rust became more widespread (21). Therefore, additional rust strains virulent on these more resistant forms were introduced from the Mediterranean region into Australia, and these strains are exerting a degree of control of the resistant forms (28). This case illustrates a potential problem in using biological control, namely, a shift in the weed population toward more resistant weed biotypes. However, it also illustrates the possibility of countering the predominance of resistant weeds by the introduction of pathogen strains that are effective against the resistant weed biotypes.

Puccinia chondrillina was also introduced into the United States to control skeletonweed in the western United States. However, unlike in Australia, the rust fungus has been only partially successful. Hence, in practice, the rust fungus is utilized along two insect biocontrol agents, *Cystiphora schmidti* (Diptera: Cecidomyiidae, a gall-forming midge) and *Eriophyes chondrillae* (previously referred to as *Aceria chondrillae*; Acarina: Eriophyidae, a gall-forming mite), and chemical herbicides in an integrated weed management program (39). As in Australia, the rust fungus has been the most effective biocontrol agent in the western United States (62).

Another example of a successful classical biocontrol program involves the use of a smut fungus, *Entyloma compositarum,* imported from Jamaica to control Hamakua pamakani (*Ageratina riparia*; Compositae) in Hawaiian forests and rangelands (69). This weed was introduced into Hawaii from Mexico in 1925 as an ornamental plant. By the 1970s, it had spread to an estimated 62,500 ha in the island of Hawaii and 10,000 ha in Oahu, and had invaded the island of Maui. The biocontrol fungus was introduced into Hawaii in 1974 after it was determined to be a highly host-specific and safe biocontrol agent for this weed. About 2 - 3 months after the pathogen was released in the field, devastating epidemics occurred in dense stands of *A. riparia* in cool, high-rainfall sites in Oahu, Hawaii, and Maui. The weed populations were reduced 80% in a 9-month period. Similar reductions in weed populations were recorded 3 - 4 years after the pathogen was released at all sites with adequate moisture. At sites with low temperatures and low rainfall, there was a greater than 50% reduction in the weed population in 8 years after the pathogen's release. It is estimated that more than 50,000 ha of pasture land have been rehabilitated to their full potential by this pathogen. No evidence of host resistance or of the presence of mutant strains of the pathogen has been encountered since its release (68).

Another rust fungus, *Puccinia carduorum*, has been imported from Turkey and released into the northeastern United States to control musk thistle, *Carduus thoermeri*. The fungus was confirmed to be pathogenic only to members of the genus *Carduus*; therefore, it was concluded that species other than *C. thoermeri* would not be threatened by the introduction of *P. carduorum* into North America. Accordingly, permission was granted to release the fungus in Virginia (10). Since its introduction, the fungus has spread, and the results have indicated that *P. carduorum* can indeed control musk thistle (3). Recently, the rust has been found in some western states, confirming its natural and widespread dissemination (William Bruckart, personal communication).

Morris (44) has established another highly successful classical biocontrol program in South Africa, where he has introduced a gall-forming rust fungus, *Uromycladium tepperianum,* to control an invasive tree species, *Acacia saligna.* The fungus causes extensive gall formation on branches and twigs, accompanied by a significant energy loss. Heavily infected trees are eventually killed. The fungus was introduced from Australia into the Western Cape Province between 1987 and 1989, and in about 8 years, the disease has become widespread in the province, and the tree density has decreased by at least 80% in rust-established sites. The number of seeds in soil seed-banks has also stabilized at most sites. Large numbers of trees have begun to die, and this process is continuing. Thus, *U. tepperianum* is providing very effective biocontrol.

Common groundsel (*Senecio vulgaris*), a native of Eurasia, is a problem weed in more than 40 horticultural crops in 65 countries, including the United States. It is resistant to triazine herbicides, and the use of highly specific herbicides tends to select for this annual weed. An autoecious rust fungus, *Puccinia lagenophorae*, which was accidentally introduced into France in the early 1960s, possibly from Australia, has spread rapidly on common groundsel throughout Central Europe, UK, and Ireland, and has become the most ubiquitous pathogen associated with common groundsel in these areas. Crop losses from competition from common groundsel may be reduced by nearly 50% due to rust infection. In studies in Europe, the fungus has shown promise as a biological control agent (46,73). These results, together with the limited host range and natural potential for rapid spread of *P. lagenophorae*, suggest that this pathogen could be an effective classical biocontrol agent of common groundsel in the United States (74).

Integrated Control of Weeds by Using Plant Pathogens and Other Control Agents

Typically, biological control agents are effective when used in an integrated approach with other biocontrol agents as well as chemical, mechanical, and cultural controls. Indeed, in many successful weed biocontrol systems, insect and mites, plant pathogens and secondary microbial invaders interact to enhance the overall efficacy of the biocontrol system. It is also common to find unintended interactions of biological agents with cultural, chemical, or mechanical weed control practices, or with associated biotic agents. Some of these interactions can provide beneficial weed-control effects (16,11,12).

There have been numerous experimental attempts to integrate weed-control pathogens into production systems (16,61). The registered bioherbicides DeVine and Collego are routinely used in conjunction with chemical pesticides and other practices to control the weeds, insect pests, and diseases that affect citrus and rice production, respectively (61). Difficulties arise when some chemical pesticides prevent the bioherbicides' activities. Several herbicides can inhibit spore germination and growth of weed-control pathogens when tank-mixed (9,16,25,52). In general, it is more difficult to integrate the use of bioherbicides with fungicides than with insecticides or herbicides. However, it has been possible to apply bioherbicides sequentially, through proper timing of applications, to avoid adverse effects of chemicals (61). Studies have demonstrated that it is possible to integrate two or more pathogens (5,13,27,42), pathogens and insect biocontrol agents (14,62), pathogens and weed-crop competition (51,35), and pathogens and mowing (24,30).

Müller-Schärer and Frantzen (45) have proposed a system-management approach to weed control based on augmentation of the biotrophic rust pathogen *Puccinia lagenophorae*, a naturalized pathogen of common groundsel in Europe. This approach may be well suited to situations where it is necessary to control a single weed species, control is not required immediately or completely, production of large amounts of inoculum is limited due to the biotrophic nature of the pathogen, and importation or utilization of other pathogens as classical biocontrol agents is not possible. The main emphasis in this method is given to the manipulation of the infection window, the genetic structure of the plant and pathogen population, the infection conditions to maximize disease spread and the impact on the weed, and to minimize the development of resistant weed populations. Integration of pathogens with low doses of chemical herbicides, necrotrophic pathogens, or biochemicals that can interfere with the weed's defense mechanisms is also envisioned. Here the pathogens are viewed as stress factors, not as weed-killers, and biological control is viewed as an integral part of a well-designed pest management strategy.

Constraints and Needs for the Adoption of Plant Pathogens as a Component of Integrated Weed Management

BIOHERBICIDES

By the early 1990s, about 250 bioherbicide agents with proven efficacies had been reported, but the majority of these agents have since failed to reach the marketplace (8,15,72). Among the reasons given for this deficiency is the lack of industrial participation in technology transfer. Generally, many agrichemical companies do not find biocontrol to be an economically attractive proposition. With a few exceptions, the current attempts to develop plant pathogens as weed control agents are supported by public institutions. In the United States and Canada, federal agencies such as USDA-ARS, Agriculture and Agri Food Canada, and Forestry Canada have made significant investments in this field. In addition, a few companies in Europe, Japan, and the United States are engaged in in-house efforts to develop bioherbicides.

The most common reason given for the limited commercial interest is that the market size for biocontrol agents is typically small and the market is often too regional, and consequently the financial returns from biocontrol agents are too small for big industries. The complexities in production and assurance of efficacy and shelf-life of inoculum can further stifle bioherbicide development (2). For instance, the inability to mass-produce inoculum needed for large-scale experiments or commercial use is a serious limitation that has led to the abandonment of several promising agents (15). Nonetheless, experience indicates that the economic returns from a bioherbicide such as Collego can support small companies and small-scale, publicly funded or private ventures (29). Such enterprises could succeed not only in industrial nations but also in developing countries by adopting a cottage-industry model. In this regard, the recent focus on alternative paradigms for production and implementation of biocontrol agents is a welcome step (1).

From an economic standpoint, it would be attractive to develop a bioherbicide that can control several closely related weed species even if these weeds are not commonly found in the same locations or crops. The potential to use a bioherbicide in diverse crops and against several weeds might stimulate commercial interest in this technology. Accordingly, in our research we are targeting closely related weeds such as pigweeds (*Amaranthus* spp.) and nutsedges (*Cyperus* spp.) with fungal pathogens that have broad host range at the generic level (i.e., restricted to several species of a genus) (36,57). Furthermore, to enhance the effectiveness and acceptability of bioherbicide agents, we are developing a novel approach that combines several host-specific pathogens into a single application. In this multiple-pathogen strategy, two or more pathogens are combined at optimum inoculum levels and sprayed onto the weeds in postemergent or preemergent applications. The feasibility of this approach has been demonstrated in greenhouse and field trials, where it was shown that seven weedy grasses could be controlled (killed) with three fungal pathogens, *Drechslera gigantea, Exserohilum longirostratum,* and *E. rostratum* applied in an emulsion (13).

Research on formulation technology is proceeding in several laboratories, prompted by the need to develop specially designed inoculum products that will assure high levels of field activity, handling properties, shelf-life, etc. (6,26). Of the various materials shown to be useful for formulating bioherbicides, recent attention has focused on emulsions, vegetable oils, hydrophilic gels composed of various natural and synthetic polymers, flours, pregelatinized starch, and surfactants (20, 26,60). Results indicate that improvements in the level of control, consistency of performance, and host range are possible through the use of specific formulations.

Some progress has been made to apply bioherbicides using alternatives to foliar spray. For example, it is possible to apply foliar pathogens through preemergent soil incorporation (50,70) and thus overcome the need for an extended dew period necessary for foliar infection. Jackson et al. (31) have been able to produce microsclerotia of a foliar pathogen, *Colletotrichum truncatum,* for soil application. Microsclerotia placed on the soil surface or incorporated into soil were effective in causing disease on emerging seedlings of hemp sesbania, *Sesbania exaltata.* Soil application can also help to overcome the need for directed application of inoculum to the foliage and increase the application window. However, the level of weed control obtained through soil inoculation of a foliar pathogen may not result in high levels of disease and weed control (50).

A newer tool, the Wet Blade™ mower system, has been shown to be highly suitable for application of *Ralstonia (=Pseudomonas) solanacearum* for the control of tropical soda apple (*Solanum viarum*), an exotic invasive weed in pastures in Florida and several other southern states (24).

Improvement in efficacy of plant pathogens used for weed control is possible by recombinant DNA methods, but practical use of genetically engineered pathogens may be difficult due to regulatory restrictions. Nonetheless, attempts are being made to genetically select pathogens with desired traits (67,64,7). Sands and coworkers (58) have produced nutritionally conditioned mutants of *Sclerotinia sclerotiorum* that are capable of killing a target host such as Canada thistle (*Cirsium arvense*) without adversely affecting nontarget plants in the vicinity. To facilitate selection of genetic recombinants, transformation systems have been developed by using the β-tubulin gene for use in selecting for benomyl resistance (65) and the *bar* gene (7) for resistance to bialaphos, a herbicidal compound produced by *Streptomyces hygroscopicus*. An attempt was made in our laboratory to modify the host range and improve the virulence of *Xanthomonas campestris* pv. *campestris* by using genes encoding bialaphos production (19). The transformed bacterium showed an ability to alter the type of hypersensitive response it elicited in certain nonhost plants while retaining pathogenicity to its normal hosts. More work is needed to characterize genes that may be useful for improving the efficacy of bioherbicidal pathogens (e.g., pathogenicity, virulence, host range, and production of enzymes, toxins, and hormonal compounds).

CLASSICAL BIOCONTROL

The major constraint facing classical biocontrol of weed by means of plant pathogens is institutional and regulatory in nature. Presently, there are an insufficient number of containment facilities in the continental United States to expand research on microbial weed control agents. More quarantine facilities specifically devoted to research on microbial pathogens and situated on a regional basis are needed. More rapid approval of pathogens for importation into quarantine facilities and for interstate movement of approved pathogens for cooperative research is also needed. More scientists should be assigned to work on pathogens for classical biocontrol of weeds. It should become the policy, not the exception, to include a plant pathologist in every classical biological weed control project sponsored by state or federal agencies. Finally, adequate resources should be directed to classical biocontrol. Given the fact that invasive weeds are a major threat to North American ecosystems, we cannot afford to overlook the immense potential of classical biocontrol as an important tactic.

Conclusions

Plant pathogens, when used as classical biocontrol agents or as bioherbicides, can provide an effective, safe, and viable method of weed control. Of seven bioherbicides developed and registered since 1980, five are still available for commercial use. In spite of the general dearth of commercial interest in bioherbicides, research in this field has been sustained by the public's demand for nonchemical weed control alternatives. It is likely that two or three new bioherbicides will enter the market in various countries during the next 5 years. This optimistic outlook is based on the encouraging progress of the ongoing research, continued funding by agencies, and public interest in biocontrols in the United States and abroad. Considering the small number of potential biocontrol pathogens explored so far, about 250 by some recent estimates, the return on the investment made in bioherbicide research has been quite good.

Literature Cited

1. Anonymous. 1999. Alternative Paradigms for Commercializing Biological Control Workshop. Web page at **http://www-rci.rutgers.edu/~insects/contrlp.htm.**
2. Auld, B. A., and Morin, L. 1995. Constraints in the development of bioherbicides. Weed Technol. 9:638-652.
3. Baudoin, A. B. A. M., Abad, R. G., Kok, L. T., and Bruckart, W. L. 1993. Field evaluation of *Puccinia carduorum* for biological control of musk thistle. Biol. Control 3:53-60.
4. Bewick, T. A., Colvin, D. L., Degner, R. L., Charudattan, R., Stall, W. M., and Shelby, M. E. 1993. Profitability of chemical and biological control of yellow nutsedge (*Cyperus esculentus*) in soybean, peanut, and bush bean. Proc. SWSS 46:234.
5. Boyette, C. D., Templeton, G. E., and Smith, R. J., Jr. 1979. Control of winged waterprimrose (*Jussiaea decurrens*) and northern jointvetch (*Aeschynomene virginica*) with fungal pathogens. Weed Sci. 27:497-501.
6. Boyette, C. D., Quimby, P. C., Jr., Caesar, A. J., Birdsall, J. L., Connick, W. J.. Jr., Daigle, D. J., Jackson, M. A., Egley, G. H., and Abbas, H. K. 1994. Adjuvants, formulations, and spraying systems for improvement of mycoherbicides. Weed Technol. 10:637-644.

7. Brooker, N. L., Mischke, C. F., Patterson, C. D., Mischke, S., Bruckart, W. L., and Lydon, J. 1996. Pathogenicity of bar-transformed *Colletotrichum gloeosporioides* f.sp. *aeschynomene.* Biol. Control 7:159-166.

8. Bruckart, W. L., and Hasan, S. 1991. Options with plant pathogens intended for classical control of range and pasture weeds. Pages 69-79 in: Microbial Control of Weeds. D. O. TeBeest, ed. Chapman and Hall, New York.

9. Bruckart, W. L., Johnson, D. R., and Frank, J. R. 1988. Bentazon reduces rust-induced disease in yellow nutsedge, *Cyperus esculentus.* Weed Technol. 2:299-303.

10. Bruckart, W. L., Politis, D. J., Defago, G., Rosenthal, S. S., and Supkoff, D. M. 1996. Susceptibility of *Carduus, Cirsium,* and *Cynara* species artificially inoculated with *Puccinia carduorum* from musk thistle. Biol. Control 6: 215-221.

11. Caesar, A. J. 1996. Identification, pathogenicity and comparative virulence of *Fusarium* spp. associated with stand declines of leafy spurge (*Euphorbia esula*) in the Northern Plains. Plant Dis. 80:1395-1398.

12. Caesar, A. J. 1998. Insect-pathogen synergisms for biological control of rangeland weeds. 7th Int. Congr. Plant Pathology, Edinburgh, Scotland. Abstract 2.10.5.

13. Chandramohan, S., and Charudattan, R. 1996. Multiple-pathogen strategy for bioherbicidal control of several weeds. WSSA Abstracts 36:49.

14. Charudattan, R. 1986. Integrated control of water hyacinth (*Eichhornia crassipes*) with a pathogen, insects, and herbicides. Weed Sci. 34 (Suppl. 1):26-30.

15. Charudattan, R. 1991. The mycoherbicide approach with plant pathogens. Pages 24-57 in: Microbial Control of Weeds, D. O. TeBeest, ed. Chapman and Hall, New York.

16. Charudattan, R. 1993. The role of pesticides in altering biocontrol efficacy. Pages 421-432 in: Pesticide Interactions in Crop Production. J. Altman, ed. CRC Press, Boca Raton, FL.

17. Charudattan, R., and Browning, W. H. 1992. Regulations and Guidelines: Critical Issues in Biological Control. Proceedings of a USDA/CSRS National Workshop, June 10-12, 1991, Vienna, VA, Institute of Food and Agricultural Sciences, University of Florida, Gainesville, FL 32611.

18. Charudattan, R., and Walker, H. L. 1982. Biological Control of

Weeds with Plant Pathogens. John Wiley, New York.

19. Charudattan, R., Prange, V. J., and DeValerio, J. T. 1996. Exploration of the use of the "bialaphos genes" for improving bioherbicide efficacy. Weed Technol. 10:625-636.

20. Connick, W. J., Jr., Daigle, D. J., Pepperman, A. B., Hebbar, K. P., Lumsden, R. D., Anderson, T. W., and Sands, D. C. 1998. Preparation of stable, granular formulations containing *Fusarium oxysporum* pathogenic to narcotic plants. Biol. Control 13:79-84.

21. Cullen, J. M. 1985. Bringing the cost benefit analysis of biological control of Chondrilla juncea up to date. Pages 145-152 in: Proceedings of the VI International Symposium on Biological Control of Weeds. E. S. Delfosse, ed. Agriculture Canada, Ottawa.

22. de Jong, M. D., Scheepens, P. C., and Zadoks, J. C. 1990. Risk analysis for biological control: A Dutch case study in biocontrol of Prunus serotina by the fungus *Chondrostereum purpureum*. Plant Dis. 74:189-194.

23. de Jong, M. D., Sela, E., Shamoun, S. F., and Wall, R. E. 1996. Natural occurrence of *Chondrostereum purpureum* in relation to its use as a biological control agent in Canadian forests. Biol. Control 6:347-352.

24. DeValerio, J. T., and Charudattan, R. 1999. Field testing of *Ralstonia solanacearum* [Smith] Yabuuchi et al. as a biocontrol agent for tropical soda apple (*Solanum viarum* Dunal). WSSA Abstracts 39:70.

25. Grant, N. T., Prusinkiewicz, E., Makowski, R. M. D., Holstrom-Ruddick, B., and Mortensen, K. 1990. Effect of selected pesticides on survival of *Colletotrichum gloeosporioides* f.sp. *malvae*, a bioherbicide for round-leaved mallow (*Malva pusilla*). Weed Technol. 4:701-715.

26. Green, S., Stewart-Wade, S. M., Boland, G. J., Teshler, M. P., and Liu, S. H. 1998. Formulating microorganisms for biological control of weeds. Pages 249-281 in: Plant-Microbe Interactions and Biological Control. G. J. Boland and L. D. Kuykendall, eds. Marcel Dekker, New York.

27. Hallett, S. G., Paul, N. D., and Ayres, P. G. 1995. A dual pathogen strategy for the biological control of groundsel Senecio vulgaris. Biological Control of Weed, VII International Symposium, 2-7 February 1992, Canterbury, New Zealand. E. S. Delfosse, and R. R. Scott, eds. DSIR/CSIRO, Melbourne, p.82 (Abstract).

28. Hasan, S. 1985. Search in Greece and Turkey for *Puccinia*

chondrillina strains suitable to Australian forms of skeleton weed. Pages 625-632 in: Proceedings of the VI International Symposium on Biological Control of Weeds. E.S. Delfosse, ed. Agriculture Canada, Ottawa.

29. Heiny, D. K., and Templeton, G. E. 1993. Economic com-parisons of mycoherbicides to conventional herbicides. Pages 395-408 in: Pesticide Interactions in Crop Production: Beneficial and Deleterious Effects. J. Altman, ed. CRC Press, Boca Raton, FL.

30. Imaizumi, S., Nishino, T., Miyabe, K., Fujimori, T., and Yamada, M. 1997. Biological control of annual bluegrass (*Poa annua* L.) with a Japanese isolate of *Xanthomonas campestris* pv. *poae* (JT-P482). Biol. Control 8:7-14.

31. Jackson, M. A., Shasha, B. S., and Schisler, D. A. 1996. Formulation of *Colletotrichum truncatum* microsclerotia for improved biocontrol of the weed hemp sesbania (*Sesbania exaltata*). Biol. Control 7: 107-113.

32. Jobidon, R. 1998. Comparative efficacy of biological and chemical control of vegetative reproduction in *Betula papyrifera* and *Prunus pennsylvanica*. Biol. Control. 11:22-28.

33. Johnson, D. R. 1999. Bioherbicide research - Status reports, Encore Technologies. IBG News 7(2):6.

34. Julien, M. H., and Griffiths, M. W. 1998. Biological Control of Weeds: A World Catalogue of Agents and Their Target Weeds. 4th Edn., CABI Publishing, CAB International, Wallingford, UK.

35. Kadir, J. B. 1997. Development of a Bioherbicide for the Control of Purple Nutsedge. Ph.D. Dissertation, University of Florida, Gainesville.

36. Kadir, J., and Charudattan, R. 1996. *Dactylaria higginsii* (Luttrell) M. B. Ellis: A potential bioherbicide for nutsedges (*Cyperus* spp.). WSSA Abstracts 36:49.

37. Kay, S. H., and Rice, J. A. 1991. Using Grass Carp for Aquatic Weed Management. North Carolina Coop. Ext. Serv., N.C. State Univ., Raleigh.

38. Kenney, D. S. 1986. DeVine -- the way it was developed -- an industrialist's view. Weed Sci. 34 (Suppl. 1):15-16.

39. Lee, G. A. 1986. Progress on classical biological and integrated control of rush skeletonweed (*Chondrilla juncea*) in the western U.S. Weed Sci. 34 (Suppl. 1):2-6.

40. Leslie, A. J., Cassani, J. R., and Wattendorf, R. J. 1996. An introduction to grass carp: Biology and history in the United States.

Pages 1-13 in: Managing Aquatic Vegetation with Grass Carp: A Guide for Water Resource Managers. Am. Fisheries Soc., Introduced Fish Section, Bethesda, MD.

41. McFadyen, R. E. C. 1998. Biological control of weeds. Annu. Rev. Entomol. 43:369-393.

42. Morin, L., Auld, B. A., and Brown, J. F. 1993. Synergy between *Puccinia xanthii* and *Colletotrichum orbiculare* on *Xanthium occidentale*. Biol. Control 3:296-310.

43. Morris, M. J. 1996. The development of mycoherbicides for an invasive shrub, *Hakea sericea*, and a tree, *Acacia mearnsii*, in South Africa. Page 547 in: Proceedings of the IX International Symposium on Biological Control of Weeds. V. C. Moran and J. H. Hoffmann, eds. University of Cape Town, South Africa.

44. Morris, M. J. 1997. Impact of the gall-forming rust fungus *Uromycladium tepperianum* on the invasive tree *Acacia saligna* in South Africa. Biol. Control 10:75-82.

45. Müller-Schärer, H., and Frantzen, J. 1996. An emerging system management approach for biological weed control in crops: *Senecio vulgaris* as a research model. Weed Res. 36:483-491.

46. Paul, N. D., Ayres, P. G., and Hallett, S. G. 1993. Myco-herbicides and other biocontrol agents for *Senecio* spp. Pesticide Sci. 37:323-329.

47. Parker, P. E. 1991. Nematodes as biological control agents of weeds. Pages 58-68 in: Microbial Control of Weeds. D. O. TeBeest, ed. Chapman and Hall, New York.

48. Phatak, S. C., Sumner, D. R., Wells, H. D., Bell, D. K., and Glaze, N. C. 1983. Biological control of yellow nutsedge with the indigenous rust fungus *Puccinia canaliculata*. Science 219:1446-1447.

49. Phatak, S. C., Callaway, M. B., and Vavrina, C. S. 1987. Biological control and its integration in weed management systems for purple and yellow nutsedge (*Cyperus rotundus* and *C. esculentus*). Weed Technol. 1:84-91.

50. Pitelli, R. A, Charudattan, R., and DeValerio, J. T. 1994. Control of sicklepod (*Cassia obtusifolia* L.) by preemergent soil-incorporation of Alternaria cassiae spores and flumetsulam. WSSA Abstracts 34:49.

51. Pitelli, R. A., Charudattan, R., and DeValerio, J. T. 1997. Effect of *Alternaria cassiae, Pseudocercospora nigricans*, and soybean (*Glycine max*) planting density on the biological control of sicklepod (*Senna obtusifolia*). Weed Technol. 12:37-40.

52. Prasad, R. 1994. Influences of several pesticides and adjuvants on *Chondrostereum purpureum* - a bioherbicide agent for control of forest weeds. Weed Technol. 8:445-449.
53. Rees, N. E., Quimby, P. C., Jr., Piper, G. L., Coombs, E. M., Turner, C. E., Spencer, N. E., and Knutson, L. V. 1996. Biological Control of Weeds in the West. Western Weed Science Society, U.S. Department of Agriculture, Agricultural Research Service, Montana Department of Agriculture, and Montana State University, Bozeman, MT.
54. Ridings, W. H. 1986. Biological control of stranglervine in citrus--a researcher's view. Weed Sci. 34 (Suppl. 1):31-32.
55. Ridings, W. H., Mitchell, D. J., Schoulties, C. L., and El-Gholl, N. E. 1976. Biological control of milkweed vine in Florida citrus groves with a pathotype of *Phytophthora citrophthora*. Pages 224-240 in: Proceedings of the IV International Symposium on Biological Control of Weeds. T.E. Freeman. ed. Univ. Florida, Gainesville, FL.
56. Room, P. M. 1980. Biological control of weeds - modest investments can give large returns. Page 291 in: Proceedings of the Australian Agronomy Conference: Pathways to Productivity. I.M. Wood, ed. Australian Society for Agronomy, Australia.
57. Rosskopf, E. N., Charudattan, R., DeValerio, J. T., and Stall, W. M. 1996. Control of pigweeds and amaranths (*Amaranthus* spp.) with a fungus: Three years of field experimentation. WSSA Abstracts 36:48.
58. Sands, D. C., Ford, E. J., and Miller, R. V. 1990. Genetic manipulation of broad host-range fungi for biological control of weeds. Weed Technol. 4:471-474.
59. Scheepens, P. C., and Hoogerbrugge, A. 1991. Host specific-ity of *Puccinia canaliculata*, a potential biocontrol agent for *Cyperus esculentus*. Netherlands J. Plant Pathol. 97:245-250.
60. Shabana, Y. M., Charudattan, R., DeValerio, J. T., and Elwakil, M. A. 1997. An evaluation of hydrophilic polymers for formulating the bioherbicide agents *Alternaria cassiae* and *A. eichhorniae*. Weed Technol. 11:212-220.
61. Smith, R. J., Jr. 1991. Integration of biological control agents with chemical pesticides. Pages 189-208 in: Microbial Control of Weeds. D. O. TeBeest, ed., Chapman and Hall, New York.
62. Supkoff, D. M., Joley, D. B., and Marois, J. J. 1988. Effect of introduced biological control organisms on the density of *Chondrilla juncea* in California. J. Appl. Ecol. 25:1089-1095.

63. TeBeest, D. O. 1991. Microbial Control of Weeds. Chapman and Hall, New York.

64. TeBeest, D. O. 1993. Biological control of weeds: Potential for genetically modified strains. Pages 147-163 in: Advanced Engineered Pesticides. L. Kim, ed. Marcel Dekker, New York.

65. TeBeest, D. O. and Dickman, M. B. 1989. Transformation of *Colletotrichum gloeosporioides* f.sp. *aeschynomene.* Phytopathology 79:1173.

66. TeBeest, D. O., and Templeton, G. E. 1985. Mycoherbicides: progress in the biological control of weeds. Plant Dis. 69:6-10.

67. TeBeest, D. O., Yang, X. B., and Cisar, C. R. 1992. The status of biological control of weeds with fungal pathogens. Annu Rev. Phytopathol. 30:637-657.

68. Trujillo, E. E. 1985. Biological control of Hamakua pa-makani with *Cercosporella* sp. in Hawaii. Pages 66-671 in: Proceedings of the VI International Symposium on Biological Control of Weeds. E. S. Delfosse, ed. Agriculture Canada, Ottawa.

69. Trujillo, E. E., Aragaki, M., and Shoemaker, R. A. 1988. Infection, disease development, and axenic culture of *Entyloma compositarum*, the cause of Hamakua pamakani blight in Hawaii. Plant Dis. 72:355-357.

70. Vogelsang, S., Watson, A. K., and DiTommasso, A. 1998. Effect of soil incorporation and dose on control of field bindweed (*Convolvulus arvensis*) with the preemergence bioherbicide *Phomopsis convolvulus*). Weed Sci. 46:690-697.

71. Wall, R. E. 1994. Biological control of red alder using stem treatments with the fungus *Chondrostereum purpureum*. Can. J. For. Res. 24:1527-1530.

72. Watson, A. K. 1991. The classical approach with plant pathogens. Pages 3-23 in: Microbial Control of Weeds. D. O. TeBeest, ed. Chapman and Hall, New York.

73. Wyss, G. S. 1997. Quantitative Resistance in the Weed-Pathosystem *Senecio vulgaris* L.-*Puccinia lagenophorae* Cooke. A dissertation, Swiss Federal Institute of Technology, Zürich.

74. Wyss, G. S. and Müller-Schärer, H. 1998. *Puccinia lagenophorae* as a classical biocontrol agent for common groundsel (*Senecio vulgaris*) in the United States? Phytopathology 88(Suppl.):S99.

Challenges to the Commercialization of Biological Control Technologies for IPM

J. C. Meneley

AgBio Development, Inc., Westminster, CO 80030 USA

Ten years ago, the author prepared a review of commercial biological control technologies. The emphasis was on microbial products and was entitled "Biopesticides: A Shining Star in the Future of the Agricultural Industry." The presentation was partly based on enthusiastic projections by market consulting firms, who estimated that the biopesticide market in 20 years would reach 40% to 50% of the total pesticides sold. Today, that is equivalent to $12 billion to $15 billion in a total pesticide market of about $31 billion. Conservatively, the author estimated that biopesticides would reach $1 billion to $2 billion in sales, or 3% to 5% of the market.

How have biopesticides fared during this time frame? This review will attempt to analyze the current market for biopesticides and the challenges that must be overcome for a successful product to be developed.

Definitions

The National Academy of Sciences defined biocontrol as "the use of natural or modified organisms, genes, or gene products to reduce the effects of undesirable organisms (pests) and to favor desirable organisms such as crops, trees, animals and beneficial insects and microorganisms" (1).

The U.S. Environmental Protection Agency defined a biological control agent in 40 CFR 152.3 as "any living organism applied to or introduced into the environment that is intended to function as a pesticide against another organism declared to be a pest by the Administrator." A pesticide is "any substance or mixture of substances intended for preventing, destroying, repelling, or mitigating any pest." A pest is defined as any organism deleterious to humans or the environment. This generally includes weeds, insects, plant pathogens, and unwanted vertebrate animals. Pests that directly affect humans, animals, or the products they consume are regulated by the United States Food and Drug Administration.

Within the definition of biocontrol products are biochemical and microbial pesticides as described in 40 CFR 158.65. This group is "distinguished by their unique modes of action, low use volume, target species specificity or natural occurrence." Biochemical pesticides are products such as insect pheromones, juvenile hormones, and plant-derived biochemicals (which include toxins, insect growth regulators, and enzymes). Low-risk products, such as insecticidal and herbicidal soaps, light oils, and bicarbonate products, also come under this classification. A microbial toxin may or may not be classified as a biochemical pesticide depending on its toxicology spectrum.

Bacteria, fungi, viruses, and protozoa, genetically engineered or not, constitute the microbial pesticide group. The microbes in this group may or may not be living entities or have the capability of reproducing. Genetically engineered microbes are subject to additional registration requirements. In the broadest sense, crops that have been engineered to produce a protein toxin originally isolated from the bacterium *Bacillus* could be considered biopesticides. Biocontrol products excluded from EPA regulations are entomophagous nematodes, insect predators and parasites, and macroscopic parasites.

Biological controls are a component of IPM, which has been described as a system that uses a broad range of pest controls, such as cultural, physical, chemical, and biologicals, to minimize effects on the environment and maintain agroecosystems. A recommended Web site for IPM and its components is **http://www.IPMnet.org.**

In any discussion of biological controls, it is worth comparing biopesticides to conventional pesticides and evaluating how they are viewed in the marketplace.

Comparison of Biopesticides and Synthetic Chemicals

DISADVANTAGES OF BIOPESTICIDES

Compared to traditional synthetic chemicals, biopesticides generally have the following disadvantages:

1   Time needed to control pest
    ▪   Biopesticides are almost always slower to kill or disable a pest because of the modes of action.
2   Biological efficacy
    ▪   Often, biopesticides are less efficacious in controlling a pest population, especially under high pest pressure. For example, control provided by microbial fungicides may be reduced in situations when pathogen populations are high or when environmental conditions are not favorable for growth of the plant, such as low light or excessive moisture. Conversely, high insect populations and/or high humidity may be required for microbial insecticides to be most effective. Competent growers strive to provide optimum conditions for production of their crops, but this is not always possible.
    ▪   Consumers have come to expect blemish-free fruits, vegetables, and ornamentals. A perfect-looking product may be hard to produce with biological controls.
3   Spectrum of activity
    ▪   Most biopesticides control a narrow range of pests. If broad-spectrum control is needed or the grower does not know the identity of the pest requiring control, then a chemical pesticide or combination of chemicals with a wide host range may be selected.
4   Storage and shelf life
    ▪   Living microbial pesticides may require moderate or cold temperature storage. Temperatures encountered in warehouses or delivery trucks in summer will shorten shelf life. Under recommended storage conditions, shelf life may still be less than 1 year. Liquid products may be difficult to stabilize or keep free of unwanted microbial growth.
5   Application
    ▪   Mixing or application of the biopesticide may have special requirements. With microbial pesticides, large particles may plug pump screens or nozzles in application equipment.
6   Compatibility
    ▪   Living microbial pesticides may not be compatible with chemical pesticides or fertilizers in a tank mix or with chemicals applied to the crop. For example, it may be necessary to wait a certain number of days between application of a chemical fungicide and

a fungal biopesticide.

7   Grower education
- Teaching the grower how to store, mix, and apply the product may take extra effort.
- Teaching the grower when it is best to use the product and how to evaluate biocontrol effects may be needed.

8   Product cost
- Treating a specified area with a biopesticide may be more costly than the chemical alternative.
- Cost/benefit analysis will usually favor the chemical pesticide.

ADVANTAGES OF BIOPESTICIDES

Compared to traditional synthetic chemicals, biopesticides are generally assigned the following benefits:

1) Safety
- Safer to transport, handle, apply
- Much less concern for workers entering field or greenhouse after application.
- Little concern for short- or long-term toxic or chronic residues when crop is either consumed or handled by humans, making them suitable for the growing popularity of "natural" and "organic" foods and fibers.

2) Environmentally benign
- Biopesticides usually have a narrow host range and generally show biocontrol activity on a limited group of pests. Non-target insects and beneficial microbes are seldom affected.
- Little concern about exposure of wildlife, soils, bodies of water, or disruption of other ecosystems.

3) Government regulations less cumbersome
- Requirements for EPA and state registrations are much less costly to meet.
- Transportation, storage, and application regulations are much less strict.
- Re-entry intervals and time between application and harvest are much shorter.

4) Pest resistance
- The chance of resistance developing to biopesticides is minimal. Only in a few situations has resistance to *Bacillus thuringiensis*

292

(Bt)(a protein toxin) occurred where concentrated applications were made over several years or in enclosed environments. However, resistance to Bt-engineered crops is a real threat, and systems designed to manage resistance are being implemented.

5) Biotechnology
   - Biopesticides have the potential to be improved through genetic engineering. This extends not only to biological efficacy but also to new methods of identifying, producing, storing, and applying biopesticides to the crop.

6) Integrated Pest Management
   - IPM has been mandated to reduce the use of chemical pesticides. The goal is to convert 75% of agriculture to IPM. Movements such as "sustainable agriculture" have been created to protect human health, wildlife, soil, water quality, and ecosystems by reducing traditional chemical inputs and replacing them with natural alternatives. Biocontrols will be an integral part in this conversion process as more uses are developed that can reliably replace chemical controls.

The Process of Discovery, Development, and Commercialization

DISCOVERY

The process to discover a chemical pesticide is almost cookbook. Market opportunities are well understood in the agrichemical industry, and the goal is to invent a "better mousetrap." A better product is one that is more cost effective and safer to humans and the environment. The effectiveness of a company research program, simply stated, becomes a matter of how many chemicals a company can screen or how clever they are in identifying active chemical candidates in their laboratories before proceeding to the development stage.

Biopesticide discovery is still rather immature in that most companies in the business do not have the financial resources to establish large and intensive discovery programs. The exceptions, of course, are the few major agrichemical companies that have active programs in discovery and development of biopesticides. Typically, active biologicals selected for development have come from government-supported researchers who had programs to search for novel bioactives or more active species/strains of known biologicals. Many of the biologicals that are now commercial products had previously been identified, and most were well described in the literature, such as *Bacillus, Beauveria, Metarhizium, Trichoderma, Gliocladium, Phlebia, Colletotrichum, Streptomyces, Agrobacterium, Pseudomonas, Steinernema*, baculoviruses, beneficial insects, oils, and soaps.

DEVELOPMENT

Once an active chemical or biological has been identified, then the long and arduous development stage takes place. Depending on the nature of the active, early testing is started in growth chambers, greenhouses, or small field plots. This stage could be considered part of the screening process, since many actives do not progress further. For chemical pesticides, testing is accomplished at laboratories or research farms owned or controlled by the agrichemical companies. For biologicals, much of this early testing has been performed by the research institute that identified the active biochemical or microbial. The biopesticide company may become involved, but more likely, this stage of R&D is still controlled by the discovery researcher.

After a chemical passes the early trial stage and has progressed to more intensive field and greenhouse evaluation, toxicology testing may begin. At a minimum, a company will want to know acute and dermal toxicology profiles in mice before investing more development time and finances. For a biopesticide, company participation is usually required at this point. A company should closely examine the potential market for the biopesticide as a top priority. It is the experience of the author that all too often a research institute may have spent much time in identifying, evaluating, and pushing for development of a biological without having taken a close look at what the market wanted. The company interested in the biological might quickly come to the conclusion that the market opportunity is too small, complicated, or competitive for commercialization. In defense of the researchers, conducting a market evaluation was never part of their mandate, and one can often get caught up in the excitement of the discovery and demonstration of biological activity without much investigation of what the grower wants. Research institutes need to involve an agricultural economist or other market evaluator before offering to license the technology. This would benefit all parties involved.

If the market evaluation is successful, then a complete examination of biological efficacy is required. A review of known toxicology and projected registration requirements is needed, as well as careful consideration of how to produce, stabilize, formulate, and package the product. Once the company concludes that the biological can be economically produced, distributed, and registered, and that it has a market, then the process can continue toward commercialization.

All avenues of development should be proceeding in full at the next stage. Larger field trials, process development, production scale-up, and toxicology tests begin. Before any toxicology testing is started, EPA recommends that pre-registration discussions be held with the agency to make sure the potential registrant clearly understands data requirements and how to proceed. EPA will assign a product manager to the project. Occasionally, some data requirements may be waived if toxicology reports have been published or ingredients generally recognized as safe (GRAS) are involved.

Toxicology testing to meet Tier 1 data requirements for biopesticide registration with EPA will cost $250,000 to $500,000. Tier 1 data requirements are generally the extent of toxicology and environmental data needed for registration. California EPA has its own registration division and must be treated separately. Concurrent submission of data to U.S. EPA and CalEPA was recently approved. Employing the services of a registration consulting firm can easily add another $50,000 to $100,000 to costs. In contrast, the cost to register a chemical pesticide is estimated at more than $8 million.

As the development process comes to completion, the company should have created an extensive database on the biopesticide's biological efficacy, toxicology, commercial production method, formulation, compatibility with other products, storage stability, and application methods.

COMMERCIALIZATION

The commercialization process should actually begin in the development stage. A solid knowledge of the market is essential. Pricing at wholesale and retail level should be established. Sales people and others who will provide technical recommendations on the product should become competent to answer questions that will arise. The approved EPA and final commercial label will need to be submitted and accepted by registration agencies within each state. Supportive technical literature, advertisements, press releases, distributor and grower meetings, and any other method that can be used to announce the availability and purpose of the product should be employed.

It is during this period that any remaining problems in formulation, packaging, application, and marketing are hopefully solved. Ideally, no surprises will arise after the product is released for sale.

Markets and Products

CURRENT MARKETS

At the risk of presenting inaccurate figures on the size of the biopesticide market, the author has compiled the following values based on discussions with various manufacturers and consultants. The purpose is to indicate relative importance and growth of biopesticides in the market (Tables 1 and 2).

Biopesticides have demonstrated growth during the last 10 years, but they still account for a minuscule share of the total pesticide market. In 1988, biopesticides were less than 0.2% of total pesticide sales, and today they are somewhere around 0.3-0.4% (Table 2).

The bioinsecticides in both 1988 and 1998 were predominantly *B. thuringiensis* products. Biofungicides and bioherbicides still account for a small part of the biopesticide market, although they have grown in sales because of new products that have come to market since 1988.

The biopesticide market is not so clear-cut as it was 10 to 15 years ago. New insect growth regulators (developed by various manufacturers) and microbial fermentation products (such as spinosad from Dow AgroSciences and abamectin from Novartis) all come under EPA's

Table 1. Estimated global end-user sales (billions of U.S. $) of herbicides, insecticides, and fungicides in 1988 and 1998.

	1988		1998	
Pesticide Group	U.S. $	Percent	U.S. $	Percent
Herbicides	9	43	15	48
Insecticides	6	29	9	29
Fungicides	4	19	5	16
Other	2	9	2	7
Total	21	100	31	100

Table 2. Estimated global end-user sales (millions of U.S. $) of bioherbicides, bioinsecticides, and biofungicides in 1988 and 1998.

Biopesticide Group	1988		1998	
	U.S. $	Percent	U.S. $	Percent
Bioherbicides	<1	<2	2-4	2-3
Bioinsecticides[a]	50	96	85-120	90-92
Biofungicides	<1	<2	5-10	5
Total	<52		92-134	

a/ Does not include: insect growth regulators, spinosad, abamectin, pheromones.

definition of a biochemical pesticide. Add transgenic crops containing *Bacillus* genes, and the biopesticide market today could be in the range of $750 million to $1 billion.

BIOPESTICIDE PRODUCTS

In 1988, there were 16 microbial pesticide active ingredients commercially registered under various trade names in the United States. The list included four varieties of *B. thuringiensis* insecticides, three other *Bacillus* species, two other bacteria, four fungi, two viruses, and one protozoan. In 1998, there were the same number of *Bacillus* actives, four other bacteria, seven fungi, four viruses, one protozoan, one yeast, and one actinomycete, for a total of 25. Some of the active registrations in 1988 were later discontinued.

In 1988, biochemical pesticides numbered only a handful. Since then, several insect growth regulators, plant extracts, oils, and soaps have become available.

An extensive list of worldwide microbial fungicides is maintained by D. Fravel, U.S. Department of Agriculture, ARS, at the following Web site: **http://www.barc.usda.gov/psi/bpdl/bioprod.htm**. Most of the products are for control of root diseases. R. Charudattan, University of Florida, has presented a list of bioherbicides, and M. Hoy, University of Florida, has provided a review of insect predators/parasites in this publication. A list of 142 commercial suppliers of 130 beneficial insects

Table 3. Commercial bioinsecticides and nematicides in the United States.

Active Ingredient	Trade Names	Type
Abamectin	Agrimec, Avid	toxin
Azadirachtin	Azatin, Ecozin, Neemex	IGR, repellent
B. thuringiensis endotoxins	numerous	bacterium
B. poppilliae	Milky Spore Powder	bacterium
B. sphaericus	VectoLex	bacterium
Baculoviruses	Spod-X, Gemstar.	polyhedrosis virus
	Gypcheck,	polyhedrosis virus
	Cyd-X	granulosis virus
Beauvaria bassiana	Botanigard, Mycotrol, Naturalis-O	fungus
Capsaicin	Hot Pepper Wax	plant extract
Cinnamaldehyde	Cinnamite, Cinnacure, Hefty/Sergeant Repellent	toxin, repellent
Cyromozine	Citation, Trigard, Larvadex	IGR
Diatomaceous earth	various formulations	abrasive
Diflubenzuron	Dimilin	IGR
Fenoxycarb	Precision, Comply	IGR
Horticultural oils	Sunspray UF, Stylet Oil	light oils
Hydroprene	GenTrol, Mator	IGR
Kinoprene	Enstar II	IGR
Logenidium giganteum	Laginex	fungus
Metarhizium anisopliae	Bio-Blast Termiticide	fungus
Methoprene	Altosid	IGR
Myrothecium verrucaria	DiTera	nematicide
Nosema locustae	Nolo-Bait, Semaspore	protozoan
Orange oil	Power Plant, Scent Off	repellent

298

Table 3 continued:

Active Ingredient	Trade Names	Type
Pyrethrum	various formulations	toxin
Pyriproxyfen	Distance, Pyrigro, Knack	IGR
Sabadilla	Red Devil, Natural Guard	alkaloid
Soap	M-Pede	insecticidal soap
Spinosad	Conserve, SpinTor, Success	toxin
Sulfur/lime sulfur	various formulations	toxin

in the United States, Mexico, and Canada is located at **http://www. cdpr.ca.gov/docs/ipminov/bensuppl.htm**. A review of biocontrol nematodes is found at **http://www.nysaes.cornell.edu/ent/biocontrol/ pathogens/nematodes.html**. For a broad listing of biocontrol Web sites, see: **http://www.nysaes.cornell.edu/ent/biocontrol/websites.html**.

A compilation of EPA-registered bioinsecticides is presented in Table 3. Included are IGRs (insect growth regulators) and biochemical products produced by fermentation, but not pheromones. CalEPA lists 37 registered IGR actives. Many of these products have growth regulator activity but are primarily recognized as toxins. They are not listed in Table 3. At EPA, pyrethrum and sulfur products are registered as traditional chemical insecticides but are considered low risk. (The author has attempted to present all actives and some product names, but undoubtedly has missed some.)

A list of EPA-registered biofungicides is shown in Table 4. Even though the chemicals potassium bicarbonate and light petroleum oils are not of plant origin, they are included because of low risk and environmental safety.

The Challenge

PRODUCT PERFORMANCE

In simple terms, the purpose of a business is to make and keep customers. This can be accomplished with either a service or a product. A customer will buy a product if it delivers a certain value above the total cost of the product, i.e., will the product make or save money or time? An important intangible in the valuation is "peace of mind." Will the customer have confidence that the product will solve the problem?

The product must deliver value to the customer. The grower or processor of a crop, the professional pest control applicator, and even the homeowner tending a garden or household plants will ultimately choose products that solve pest problems or make production easier. Except for organic certified crops, a biocontrol product must compete against chemical pesticides that generally have very positive cost/benefit ratios and have been used for many years, with the result that growers feel confident in using them. Biocontrols must equal or come close to the chemical benchmark. The challenge in conventional agricultural markets is to discover, develop, and produce a biocontrol that will perform as well as the chemical pesticide in biological efficacy, reliability ("peace of mind"), ease of use, and storage.

SNAKE OILS

Those in the business of biopesticides and natural products constantly see dubious claims made on all types of products. There seems to be little awareness by regulators at EPA and state agencies of the illegal pesticide claims made by many of these products in their marketing literature. This disregard can even extend into verbal recommendations or product lists of biocontrols made by USDA, university, or extension personnel. These products have a clear advantage in the marketplace, because the cost and time associated with meeting and maintaining registration requirements is very burdensome to small companies. In addition, many of these products do have biological activity and provide a certain degree of pest control. Most, however, have little or no activity and reflect poorly on legitimate efforts in biocontrol.

Table 4. Commercial biofungicides in the United States

Active Ingredient	Trade Names
Bacteria	
Agrobacterium radiobacter	Galltrol-A, Norbac 84C
Bacillus cercus	Pix Plus
Bacillus subtilis	Epic, Kodiak, System 3
Burkholderia cepacia	Intercept, Deny
Pseudomonas fluorescens	BlightBan A50G
Pseudomonas syringae	Bio-save 100, Bio-save 110
Streptomyces griseoviridis	Mycostop
Fungi	
Ampelomyces quisqualis	AQ10
Candida oleophila	Aspire
Gliocladium catenulatum	Primastop
Gliocladium virens	SoilGard
Trichoderma harzianum	RootShield T-22
Biochemical and Chemical	
Light petroleum oil	Sun Spray Ultra Fine
Neem oil extract	Triact
Potassium bicarbonate	Armicarb, Remedy

LICENSING, DEVELOPMENT, REGISTRATION COSTS

The cliches "money makes the world go 'round" and "money talks" are still truisms. Seldom can a small company afford to do a CRADA (cooperative research and development agreement) with the USDA or afford to license a promising technology from a university research foundation. The companies with the money acquire the best projects, even though they may not be the most qualified or motivated to develop them.

Once a potential biopesticide has been identified either in-house or from a research institute, funds will be required to develop it into a product. Some funds may be available from the USDA's SBIR program, but other sources of funds will have to be acquired. This is a major challenge.

As described previously, the cost of registering a biopesticide can be $500,000 or more. For a small company with limited capital, it is a tremendous burden to raise the funds. Except for some assistance from IR-4 in submitting registration packages to EPA, there is no public source of funding to generate toxicology and environmental data needed for registration.

Renewal of EPA and state registrations is another burden that the company endures annually. The actual cash costs are around $12,000. A few states charge less than $50, while many others are more than five times that amount. Some states actually demand that groundwater contamination fees be paid! The argument that a biopesticide should not be thrown into the chemical pesticide heap is ignored.

SPREAD THIN

Within a large company, personnel are employed based on education and experience in a specialization. People are hired to do a certain job or task. In the small company, employees may have been trained in a certain specialty but because of a shortage of personnel, they are compelled out of necessity to perform a variety of jobs. Expertise and experience are often lacking. Some of the longer-term areas, such as marketing and planning, can often get pushed aside in order to address the problem of the week.

The Opportunity

TRADITIONAL MARKETS, OATMEAL, REDUCED FAT, AND GREEN TEA

Within the arena of the traditional, chemically oriented grower, there is increasing interest in biological products that promise to solve pest problems. This is a consequence of the broad public movement toward better nutrition, fitness, and health. The product tag of "organic," "natural," "green," "pesticide-free," "environmentally friendly," and "recycled" are common in today's world. No longer are the "hippie" organic food stores located only in the older, seedy sections of town. They have moved to the suburbs and have become comfortable places in which to shop. This has not gone unnoticed by major chain grocery stores, which now offer a variety of "organic" foods. Television and radio ads promoting garlic, ginseng, and other health supplements are common. Newspaper and magazine articles describe results of promising medical studies on natural products. Even the medical "expert" on the nightly news may discuss a recent discovery on the health benefits of a plant extract.

The cultural trend toward "natural" products and away from "synthetics" undoubtedly influences the conventional grower. Many are now open to trying biological alternatives to chemical pesticides. If the results, ease of use, and costs are comparable, the switch to the biological will probably occur.

IPM, Sustainable Agriculture, FQPA

The federal government is moving ahead to reduce the use of many conventional pesticides. The persistent and environmentally damaging chlorinated hydrocarbons of the past are gone. The use of soil fumigants has steadily declined. Methyl bromide will be phased out. The FQPA (Food Quality Protection Act) will force down the majority of pesticide tolerances to much lower levels. IPM, sustainable agriculture, and other environmental movements are working to reduce reliance on chemical pesticides and substitute biocontrol technologies.

If today's conventional growers have not already considered biocontrol technologies, they will want or be forced to consider lower-risk and biological alternatives in the future.

ORGANIC MARKETS

Sales of organic foods are growing 15% to 20% annually, according to industry surveys. This is creating an additional demand for innovative methods or improvements in traditional organic methods for controlling crop-damaging pests. Biocontrol technologies will play a major part in meeting this need. Pesticides that are used in growing or processing certified organic foods must be approved by certification agencies, and synthetic chemicals are not approved for such use. The federal government entered into regulation of organic foods when the Organic

Food Production Act was passed by Congress in 1990.

As a consequence, the USDA created the National Organic Standards Board to work with growers, consumers, retailers, certification agencies, environmental experts, and scientists to set national standards for products and procedures used in organic culture. Organic trade associations are growing larger and gaining more political power. It appears that the growth of organic products will continue, which will increase the need for biocontrols.

THE FUTURE

In summary, the purpose of a product is to fill a market need. With the public movement toward safer and more environmentally friendly products, the demand side of the equation is established and growing. The challenge is to develop and deliver effective biocontrol products to meet that need.

Literature Cited

1. Anonymous. 1987. Report of the Research Briefing Panel on Biological Control in Managed Ecosystems, National Academy Press: Washington, DC.

Section 5

Pesticide Technology

Pesticides have long been a dominant component of crop protection because of their efficacy, ease of use, and low cost. Human health and environmental problems associated with pesticide use, pest control crises associated with pesticide resistance, and pesticide-induced outbreaks of secondary pests led directly to the development and growth of IPM. During the past 20 years, there have been dramatic changes in the types of insecticides, fungicides, and herbicides that are available and in the ways that they are used in crop protection. Regulatory changes that place a premium on the human health and environmental safety aspects of pesticide use have resulted in the loss of registration or the imposition of severe use restrictions for many older fungicides, nematicides, herbicides, and insecticides. These changes have provided a powerful incentive for agrichemical companies to develop novel pesticides that are highly active against the target organisms but have minimal mammalian toxicity, environmental impact, and potential for resistance development. Pesticides with these attributes are highly suited for use in IPM programs. In this section, we provide an overview of recent advances in pesticide technology as they relate to IPM.

In the first chapter of this section, Paul Jepson discusses the lack of progress that has been made toward improving the efficiency of pesticide transfer from the spray tank to the target organism. He identifies the need to develop both theory and practical applications that will lead to improvements in the efficiency of pesticide use and associated reductions in environmental impacts. In addition, he calls attention to the importance of developing a landscape perspective if we are to fully understand and mitigate unintended environmental effects associated with pesticide use. Franklin Hall then provides an overview of emerging pesticide application technologies and provides a discussion of the changes in agriculture and crop protection that are influencing the development of new application technologies. In the next chapter, Beth Carroll describes the innovations that Novartis

Crop Protection has implemented in the development of pesticides and crop protection packages. Eric Tedford and Richard Brown then discuss advances in fungicide technologies and challenges to integrating fungicide technologies into IPM.

There have been dramatic advances in weed control technology (including postemergence herbicides and transgenic, herbicide-tolerant crops) and in insecticide chemistry that provide a range of new weed and insect management options. Jeffrey Gunsolus et al. discuss the importance of accounting for biological time constraints (time of weed emergence, rate of weed growth, and time of weed removal as they affect crop yield) and risk management in selecting weed management strategies. J. R. Bradley, Jr. then reviews the array of novel, selective insecticides that are in advanced stages of development or have been registered recently and have the potential to facilitate the transition to more biologically intensive IPM systems. He further discusses the reasons why insecticides will remain critical components of IPM programs in the future, and identifies obstacles that exist to the integration of new insecticide technologies into IPM. In the final chapter of this section, Timothy Dennehy discusses the critical importance of managing pesticide inputs and of resistance management if we are to preserve the advances that are being made in IPM. He describes impediments that limit the ability to manage pesticide use in intensive agricultural production systems and discusses impediments and misconceptions that have the potential to limit the future success of resistance management.

Pesticides and IPM: Concepts and Reality

Paul C. Jepson
Department of Entomology and Department of Environmental and
Molecular Toxicology, Oregon State University,
Corvallis, OR 97331 USA

Pesticides continue to evolve under a shifting range of selection pressures. Pressures for the development of active ingredients with novel modes of action arise from resistance and cross-resistance in target organisms. Newly developed pesticides must also conform to strictly enforced limits to the chemical and physico-chemical characteristics that minimize risks of uptake and accumulation through food chains and the risks that they will be excessively mobile in the environment. In addition, these compounds must not pose unacceptable risks to human consumers or wildlife or to ecological functioning in terrestrial and aquatic systems. Their intrinsic toxicity, spectrum of activity, chemical fate, and behavior and bioavailability in a variety of environmental compartments are subject to intense scrutiny throughout developmental and regulatory processes (40). The accumulated impact of these selection pressures, by the 1980s, was to reduce the rate of new product introductions to virtually zero (30). More recently, however, a range of new insecticides, fungicides, and herbicides has been introduced (e.g., 8,16) and pest, disease, and weed management specialists are examining the performance of a more diverse array of synthetic pesticides than has ever been possible in the past.

With more than 50 years of experience with synthetic insecticides, pest management specialists should be in a position to evaluate the role and potential of these new materials within economically and ecologically sustainable crop protection programs. This should particularly be the case when we consider that at least some of the damaging side effects of the modern generation of synthetic materials were recognized in the earliest stages of research in the 1940s (e.g., beneficial insect mortality in carrots sprayed with DDT in 1945 [47]; mortality of predators and parasites of insect and mite pests in apple in 1944, with consequent resurgence [37]). Despite very considerable advances in the application technology, environmental toxicology, environmental chemistry, and basic ecology and its applications in pest management theory, we fall short of what our scientific predecessors might have justifiably expected. Although potent tools for the evaluation of pesticides within IPM systems have been developed in these various disciplines, our capacity to integrate and synthesize this knowledge for the effective exploitation of pesticides is still limited. This chapter explores some of the challenges that this limitation places upon us, and draws attention to some possible pathways for improvement.

The Efficiency of Pesticide Use

MEASUREMENTS OF EFFICIENCY

Efficiency can be expressed in terms of the proportion of original active ingredient entering the treated system that is delivered to the target organism. One limitation to effective deployment of pesticides is that we judge the effectiveness of pesticide use in terms of impact upon the target organism and not in terms of the efficiency of transfer and delivery to the organism for a given level of effect. "High levels of pest suppression with pesticides" and "efficient pesticide use" are by no means synonymous terms.

Published estimates of the efficiency of pesticide spray application (defined as above) vary from 0.03% for a foliar spray of dimethoate against aphids infesting field beans to 6% for aerial spraying of gamma-HCH against locust swarms (12). Even when grass weeds are sprayed with paraquat in the laboratory, a maximum of 30% of the applied dose was taken up by the sprayed plants (12). Given the variable distributions of pests, diseases, and weeds, the complex architectures of crop canopies that pesticides must permeate, and the inherent inefficiency in any of the placement and delivery technology that is available at present, it would be unreasonable to expect efficiency to be high. The fact that we do not attempt to quantify efficiency on a routine basis, however, leaves us without any basis for improvement. A doubling of efficiency could, theoretically, permit a halving of the applied dose to yield the same effect.

Measurements of the efficiency of spray application, expressed in terms of the ability of spray applicators to apply the correct dose, are also rare. Farm measurements made in Oregon, USA (2,3), demonstrated that, of 44 sprayers checked (for growers of 14,000 ha, spraying 3 to 5 times a year), sprayer operation or application practices could be improved in 38 cases. For example, one grower over-applied by 14.7% over 1000 ha, equating to a $40,973 loss. Overall, 6.6% of the hectares received >110% of the recommended dose rate. In examining sprayers, 28 had worn nozzles, 12 dripping check valves, 8 excess pressure, 9 uneven boom heights, 5 uneven nozzle spacing, and 3 mismatched nozzles.

Surveys of spray application timing relative to pest, disease, or weed epidemiology are even more rare, although in some cases they have revealed that growers may apply insecticides in no clear relation to pest infestation levels, and suffer reduced profits or economic losses as a result (42,45,46).

THEORETICAL EXPLORATION OF DOSE DELIVERY AND UPTAKE

The processes of dose delivery to the target are enormously variable. They range from genetic expression of toxic transgene products in plant foliage to the atomization of a liquid and the deposition of spray drops on foliage. Research into the phenomena that determine efficiency of utilization requires the various steps in the process of delivery to be examined quantitatively. The longest tradition of research in the various modes of delivery addresses liquid atomization and spraying systems, and this has been selected as an example.

Although techniques for measuring physical aspects of transfer to the target have been developed, including atomization and events leading up to impaction and deposit formation, they are not systematically employed on a wide scale to investigate utilization efficiency (49). The three areas of pest control where significant research has attempted to link drop transport and behavior to toxic impact concern insect pests (locusts, tsetse flies, and mosquitoes) that may be controlled by means of direct exposure to spray drops, rather than exposure to a surface deposit or residue. The need for investigation of the deposition process was recognized in the earliest stages of research, with modern synthetic insecticides as control agents for these pests (24,27). Theoretical models have been developed to determine the importance of primary application variables (i.e., volume and mass application rates, drop size distribution, emission height, and wind speed) on the dose received by settled locusts exposed to ultra-low volume spraying (22). Environmental monitoring of the spray cloud has also been used to determine the availability of pesticide to resting tsetse flies (23). This research has been supported by the development of physically based models of the deposition of pesticides onto resting and flying insects by the processes of impaction and sedimentation (4).

Direct exposure to spraying is a primary route of uptake for herbicides, and may be a significant route of exposure for pathogens and many target and non-target invertebrate species in agricultural crops. Despite this, research into spray deposition and delivery to these targets is not routinely undertaken in the development of IPM systems. This may partly be explained by the potential complexity of considering multiple routes of exposure to the active ingredient, and of including interactions between the active ingredient and substrates such as the leaf surface (43).

The work of Salt and Ford (e.g., 33,34) is unique in its detailed exploration of the kinetics of insecticide action from the point following drop impaction through to uptake and toxicological impact within the target organism. Salt and Ford (33) explore the factors affecting behavior and efficacy of ULV deposits applied to plant surfaces for protection against lepidopteran pests using models. They relate pesticide distribution, pickup, and accumulation to knockdown and mortality by generating a random walk by an individual insect, which rests and feeds in a predictable way on a pesticide-treated surface. The model simulates the processes of encounter, uptake, penetration, and elimination, and uses stochastic simulation techniques to select values for each parameter from their statistical distributions in order to reflect observed variability of individual characters and responses. This permits the user to investigate, theoretically, the interplay between deposit characteristics, pest behavior, and detailed aspects of the poisoning process. More recently, Salt and Ford (34) have employed deterministic models to identify combinations of deposit size, density, and concentration, which result in effective pest control using the minimum amount of insecticide. These unique investigations provide a theoretical basis for the quantitative analysis of utilization efficiency, but they have received very limited attention to date.

DIRECT MEASUREMENTS OF DISTRIBUTION AND EXPOSURE

Even without the rigorous application of theoretical models of droplet transport, deposit formation, and uptake, simple and non-costly techniques can be used to determine pesticide distributions for the estimation of short-term effects on target and non-target taxa. Fluorescent tracer techniques have quantified spray drift into field boundaries to determine exposure levels for non-target Lepidoptera (9,28). These have supported mitigation tactics, including the use of unsprayed field-boundary zones for synthetic pyrethroids. They have also been used to determine the exposure of Coccinellidae (44) and Braconidae (29) in wheat crop canopies to assist in the estimation of selective dose rates for aphicides. Finally, they have determined differential deposition rates and distributions achieved with novel spray application systems in cereals, and demonstrated that utilization efficiency could be improved with enhanced exposure of pests in the canopy, even at reduced pesticide application rates (18).

In conclusion, we do not lack an appreciation of the physical, chemical, and biological processes that contribute to the utilization efficiency of pesticides. We do, however, lack a rigorous approach to the measurement of efficiency, and we seem to share a collective amnesia concerning the critical importance of the stages in pesticide dose delivery that lead up to the uptake of toxin and the toxic effect within the target organism. This is despite evidence that utilization efficiency is poor, and that it can be improved by attention to the details of dose transfer. Very few surveys of pesticide application are undertaken in practice, and where they are, they reveal considerable scope for improvement, both in terms of application equipment and calibration and in the timing of spray application. In an era when mode of action is under intense scrutiny and numerous pesticides are slated for removal, we might be wise to pay more attention to mode of use, and actively seek to develop both theory and practical

311

applications that can contribute directly to greater efficiency and reduced environmental impact.

Scaling Up Pesticide Management to the Agroecosystem Level

A number of phenomena associated with pesticide use, pesticide impacts on target and non-target organisms, and pesticide interactions with a variety of environmental compartments have influences beyond the initial area of treatment. Pesticide residues may accumulate away from the treated area as a result of leaching, vapor transfer, or even the transport of aerosols as drift. Organisms with tolerant genotypes may be favored and disperse from the treated area, and less tolerant organisms may enter the treated area and succumb to toxic residues. The dynamic biological and chemical interactions between pesticides and the surrounding landscape have been a subject of considerable debate and controversy, and they present immense practical, philosophical, and methodological difficulties that must be addressed.

Of greatest concern are those processes that are scale dependent in nature, which cannot be investigated experimentally on conventional experimental scales. These include ecological processes that evolve on a landscape scale, particularly the population processes of many pest and beneficial species in farmland. Recent research and theoretical modeling have investigated the longer-term, larger-scale impacts of pesticides and have reinforced the need to develop a landscape perspective of the costs and benefits of pesticide use.

ON-FARM MONITORING AND TIME-SERIES ANALYSIS

Potts and Vickerman (31) reported a pioneering program in agricultural ecology that has inspired many of the most recent developments in our understanding of the ecological impacts of pesticides on a landscape scale. For more than 30 years, the invertebrate fauna of the cereal ecosystem in W. Sussex, UK, has been monitored on a 62 sq km scale. In their initial report, Potts and Vickerman demonstrated negative relationships between cereal aphid pest population densities and the diversity of arthropods, particularly polyphagous predatory taxa, including Carabidae, Staphylinidae and Linyphiidae. These findings, and those that emerged from intense

312

research in the cereal ecosystem during the next 20 years in Europe, established a clear emphasis upon the preservation of invertebrate biodiversity within the agroecosystem, rather than simply preservation of specific natural enemy taxa. They also helped place an emphasis upon the need to undertake research at the farm system level, and established the cereal system as something of a model for the investigation of pesticide side effects.

One direct way in which the W. Sussex monitoring data can be exploited in the evaluation of pesticide impacts is by the use of time-series modeling (1). In 1989, a 7 sq km area within this monitoring site was first sprayed with dimethoate. Based upon monitoring data from the 20 years prior to this first use, annual sawfly (Hymenoptera: Sympyta) densities were found to be related, with a 1-year time lag, to the proportion of cereal fields that were undersown, and to summer rainfall and temperature, with a strong autoregressive component. Sawflies constitute an important game-bird chick food species. In 1989, the sawfly density in the sprayed area was less than one tenth of that predicted by time-series analysis.

LARGE-SCALE EXPERIMENTAL SYSTEMS

A number of research projects investigating the ecology and economics of arable farming systems were initiated throughout the 1980s and 1990s (17). Pesticide impacts were investigated in a number of these, and comparisons between conventional and reduced pesticide inputs were made in at least 13 studies in Germany, France, the Netherlands, Switzerland, and the UK. In all cases, beneficial non-target invertebrate densities were higher in integrated or reduced input areas. In the larger-scale studies (e.g., The Boxworth study [13]), with 5.6-15.7 ha treatment-area sizes, some beneficial species were rendered locally extinct for the full 5-year treatment phase of the project, and recovery was slow. The level of impact on a given species was shown to be a function of life history attributes that affected pesticide exposure and capacity to re-invade (7,41). Predatory capacity was inhibited in the highest pesticide regimes (7), and there was evidence that this contributed to higher pest densities in some years. In one of the studies (the SCARAB project) reported by Holland et al. (17), continued monitoring has demonstrated long-term depletions (>4 years) of Collembola following single organophosphate pesticide applications.

These investigations have revealed the subtler impact of second and third generation active ingredients, foreseen by Potts and Vickerman (31). They have reinforced the need to measure ecological impacts on scales that reflect agricultural practice in the real world and that are tuned to the scale of dispersive movement of non-target taxa. They have also triggered considerable interest in the mechanisms that underlie long-term depletions and even local extinction of certain species. Both scale of treatment and the mode and rate of dispersal of arthropods influence rates of population recovery following pesticide use (10,11,20,39). In addition, the proximity of local refugia from which recovery can occur and landscape features conducive to movement and colonization are both important factors underlying extinction risk. Modelling (14,35) may provide an appropriate tool for testing our understanding of invertebrate population processes at the agroecosystem level, but it does not substitute for the need to undertake a far greater number of manipulative experiments and monitoring programs that examine the spatial dynamics of non-target organisms in sprayed farming systems.

DETECTING ECOLOGICAL IMPACTS ON AVIAN AND AQUATIC WILDLIFE

The requirement for large-scale, field-based monitoring and experimentation is not restricted to terrestrial non-target invertebrates. Avian toxicologists have long recognized that field-based studies provide unique and indispensable data for interpretation of pesticide effects on birds (6). Many of the most important effects of pesticides on birds were originally detected and documented as a result of field-based observations. These effects include eggshell thinning, endocrine disruption, food supply reduction or alteration, variation in sensitivity between species, and the synergistic effects of exposure to multiple compounds.

Similar arguments may be applied to the investigation of pesticide effects on aquatic organisms. The recent decision in the USA to reduce the requirements for ecological data from multi-species test systems in the regulatory process has been criticized because it will reduce the probability of detecting biologically significant effects

314

(32,38). These effects include indirect trophic-level impacts, compensatory shifts within a trophic level, responses that are associated with seasonal trends in populations, chemical transformations effected by organisms in the exposed system, and impacts that result from long persistence of either the parent product or toxic breakdown products.

In conclusion, there is now abundant evidence that pesticide impacts can evolve at the agroecosystem scale, and that this requires the development of appropriately scaled monitoring or experimental systems. Toxicological data are of fundamental value in the initial evaluation of pesticides, but in their real world applications, pesticides may also elicit ecological effects that reverberate through the system, long after the chemical residues have become undetectable. The inclusion of natural enemy toxicological test data in the regulatory package for pesticides in Europe has significant potential benefit for IPM through the renewed emphasis upon physiological selectivity (5). Laboratory-based, single-species tests are, however, limited in their predictive power for impacts in the field (36), and further development of the theory and methodology associated with ecological effects is required (25).

Improving Decisions About Pesticides in IPM

It is questionable whether we apply our knowledge about pesticides effectively at any stage following the regulatory permission to use a product in a specified way. Given the need to combine knowledge of chemical properties, fate, and behavior and environmental attributes such as soil type, toxicology and ecology, the challenge is considerable. In each of these disciplinary areas, however, there are considerable databases of information and predictive models that can be used to interpret how a pesticide will behave in a given set of field conditions. What was computationally challenging a decade ago is now commonplace, and the advent of Internet-accessible databases and modeling tools should enhance the prospect for much greater exploitation of available information about pesticides.

Models and decision-making systems for determining the likelihood that pesticides will contaminate ground or surface waters are now in common use (e.g., 19,21). These combine a knowledge of the tendencies of specific soil types to allow pesticides to be transported through the soil profile to groundwater with basic pesticide properties that affect chemical persistence and mobility. Factors such as soil permeability, soil depth to bedrock or water table, soil slope, and the balance between infiltration and runoff contribute to leaching potential. These are equated with sorption potential, a function of organic matter and clay content, and hydraulic loading, which is determined by rainfall and irrigation patterns, to define the properties of a given soil. Chemical half-life estimates and sorption coefficients are then used to determine a groundwater vulnerability rating that may assist growers or advisors to make informed decisions about the appropriateness of a given pesticide on a specified soil type.

Environmental Impact Quotients

Attempts have also been made to summarize clusters of pesticide properties in databases that may be used to compare the environmental and ecological risks posed by lists of candidate compounds for specific uses. These databases are less tuned to local conditions, compared with the models outlined above; they do, however, combine many more factors in the ranking, including risks to wildlife, and enhance the capacity of the end-user to make better informed decisions about particular uses. The most significant development in this field was the proposal to derive Environmental Impact Quotients (EIQ) for compounds from established databases of pesticide properties (26). The quotient combined farm worker effects (built from acute and chronic toxicity and plant surface half-life) with consumer effects (built from systemicity, soil and leaf surface persistence, and potential for groundwater contamination) and ecological effects (built from aquatic and terrestrial ecotoxicology). The originality and potential of this approach has been understated because of concerns that the methodology was not built upon a sound base of toxicological theory. There would, however, seem to be considerable scope for toxicologists to develop approaches such as

316

this, from first principles, in order to make data available in a form that can be used to assist decisions in the field. Guided by EIQs, growers and their advisors could then plan pest management tactics that attempted to minimize non-target impacts, and track their progress with a rigorous and quantitative methodology that could also be explained to the consumer.

TOOLS TO PREDICT ECOLOGICAL RECOVERY IN SOIL AND AQUATIC SYSTEMS

Recent attempts have been made to bridge the gap that exists between risk assessment approaches derived from basic toxicology and environmental chemistry, and the potential ecological impact of pesticides. Van Straalen and van Rijn (48) computed species sensitivity distributions for soil fauna, based upon laboratory test data for toxic effects in soil, and calculated the concentration that would be below the no-effect concentration (NOEC) for reproductive inhibition for 95% of the species that are theoretically present within the soil arthropod community. Using a simple model for pesticide fate based upon exponential decay, they then calculated, for a series of compounds, the time that each product would take to reach the "95% protection level." This time, referred to as the "ecotoxicological recovery time," can be used as a basis for calculating the minimum time for ecological recovery to take place after pesticide application. There is considerable scope for further development of this technique, built upon readily collected toxicological datasets. Such data are, however, surprisingly difficult to find in the published literature, and they are not routinely collected for suites of compounds in such a way that they may be exploited within this new approach as a decision-making aid.

Finally, in aquatic ecotoxicology, approaches have recently been developed that combine databases of invertebrate distributions in pond systems with models of chemical fate and population processes (15). Although these integrated modeling efforts are still in their infancy, they may offer the prospect for linking the distributions of specific assemblages of beneficial organisms or wildlife to potential patterns of the use and effects of pesticides.

In all three of the areas that this brief review has addressed, the efficiency of pesticide use, the scaling up of pesticide management to the agroecosystem scale, and the enhancement of local decision-making by end-users, it has been possible to find evidence of good theoretical and technical knowledge. None of the areas are, however, as well developed as one might expect for such a sophisticated technology, particularly one that has been in large-scale use for more than 50 years. In the latter area (decision-making), this deficiency is most damaging in the short term. IPM systems rely upon an ability to quantify the relative contributions that different pest control tactics make to effective pest suppression. There is scope for rapid advances in the quality and reliability of decision-making aids, based upon current knowledge. The advent of so many novel modes of action in the most recently marketed pesticides, including the delivery of pesticides as transgenic gene products, in no way reduces the urgency or importance of continuing to build the scientific and technical basis for understanding the performance and behavior of these materials beyond the narrow requirements of regulatory toxicology and basic marketing and commercialization. Without these developments, pest managers will not be able to determine the most effective ways to exploit chemical means of pest, disease, and weed suppression.

Literature Cited

1. Aebischer, N. J. 1990. Assessing pesticide effects on non-target invertebrates using long-term monitoring and time series modeling. Functional Ecol. 4:369-373.
2. Aldrich-Markham, S. 1998. Pesticide sprayer tune up. Oregon State University Extension Service.
3. Aldrich-Markham, S., Pirelli, G. J., and Shenk, M. D. 1998. Pesticide sprayer tune-up program. Oregon State University Extension Service.
4. Bache, D. 1994. The trapping of spray droplets by insects. Pesticide Sci. 41:351-357.
5. Barrett, K. L., Grandy, N., Harrison, E. G., Hassan, S. A., and Oomen, P. A. 1994. Guidance document on regulatory testing procedures for pesticides and non-target invertebrates. Society of Environmental Toxicology and Chemistry-Europe.

6. Blus, L. J., and Henny, C. J. 1997. Field studies on pesticides and birds: unexpected and unique relations. Ecol. Applic. 7:1125-1132.

7. Burn, A. J. 1992. Interactions between cereal pests and their predators and parasites. Pages 110-131 in: Pesticides, cereal farming and the environment. P. W. Greig-Smith, G. K. Frampton, and A. R. Hardy, eds.. London, HMSO.

8. Casida, J. E., and Quistad, G. B. 1998. Golden age of insecticide research: past, present and future. Annu. Rev. of Entomol. 43:1-16.

9. Cilgi, T., and Jepson, P. C. 1995. The risks posed by deltamethrin drift to hedgerow butterflies. Environ. Pollution 87:1-9.

10. Duffield, S. J, and Aebischer, N. J. 1994. The effect of spatial scale of treatment with dimethoate on invertebrate population recovery in winter wheat. Journal of Appl. Ecol. 31:263-281.

11. Duffield, S. J., Jepson, P. C., Wratten, S. D., and Sotherton, N. W. 1996. Spatial changes in invertebrate predation rate in winter wheat following treatment with dimethoate. Entomol. Exp. Appl. 78:9-17.

12. Graham-Bryce, I. J. 1977. Crop protection: a consideration of the effectiveness and disadvantages of current methods and of the scope for improvement. Philosophical Transactions of the Royal Society, London. B, 281:163-179.

13. Greig-Smith, P. W., Frampton, G. K., Hardy, A. R., eds. Pesticides, cereal farming and the environment. London, HMSO.

14. Halley, J. M, Thomas, C. F. G., and Jepson, P. C. 1996. A model for the spatial dynamics of linyphiid spiders in farmland. J. Appl. Ecol. 33:471-492.

15. Heneghan, P. A., Biggs, J., Jepson, P. C., Kedwards, T., Maund, S. J., Sherratt, T. N., Shillabeer, N., Stickland, T. R., and Williams, P. 1999. Pond-FX: ecotoxicology from pH to population recovery [online]. 1st edition. Oregon State University: Department of Entomology. 11 February 1999 [cited 04 August 1999]. Available from Internet: <http://www.ent.orst.edu/PondFX>.

16. Hewitt, H. G. 1998. Fungicides in crop protection. CAB International, Wallingford & New York.

17. Holland, J. M., Frampton, G. K., Cilgi, T., and Wratten, S. D. 1994. Arable acronyms analysed - a review of integrated arable

farming systems research in W. Europe. Ann. of Appl. Biol. 125:399-438.

18. Holland, J. M., Jepson, P. C., Jones, E. C., and Turner, C. 1997. A comparison of spinning disc atomisers and flat fan pressure nozzles in terms of pesticide deposition and biological efficacy within cereal crops. Crop Protection 16:179-185.

19. Huddleston, J. H. 1998. Oregon Water Quality Decision Aid. Oregon State University Extension Service.

20. Jepson, P. C., and Thacker, J. R. M. 1990. Analysis of the spatial component of pesticide side-effects on non-target invertebrate populations and its relevance to hazard analysis. Functional Ecol. 4:349-355.

21. Jenkins, J. J., Bagdon, J. K., Esser, A. J., Lyons, J., and Hugo, J. 1990. National Pesticide/Soils Database and user decision support system for risk assessment of ground and surface water contamination. University of Massachusetts Cooperative Extension Service.

22. Johnstone, D. R. 1991. Variations in insecticide dose received by settled locusts: a computer model for ultra-low-volume spraying. Crop Protection 10:183-194.

23. Johnstone, D. R., Cooper, J. F., and Dobson, H. M. 1987. The availability of fall-out of an insecticidal aerosol dispersed from aircraft during operations for control of tsetse fly in Zimbabwe. Atmospheric Environ. 21:2311-2321.

24. Kennedy, J. S., Ainsworth, M., and Toms, B. A. 1948. Laboratory studies on the spraying of locusts at rest and in flight. Anti-Locust Bulletin No 2. Anti-Locust Research Center, London.

25. Kimball, K. D., and Levin, S. A. 1985. Limitations of laboratory bioassays: the need for ecosystem-level testing. Bioscience 35:165-171.

26. Kovach, J., Petzoldt, C., Deghni, J., and Tette, J. 1992. A method to measure the environmental impact of pesticides. New York State Agricultural Experiment Station, Bulletin 139.

27. Latta, R., Anderson, L. D., Rogers, E. E., LaMer, V. K., Hochberg, S., Lauterback, H., and Johnson, I. 1947. The effect of particle size and velocity of movement of DDT aerosols in a wind tunnel on the mortality of mosquitoes. Journal of the Washington Acadamy of Sciences 37:397-407.

320

28. Longley, M., Cilgi, T., Jepson, P. C., and Sotherton, N. W. 1997. Measurements of pesticide spray drift deposition into field boundaries and hedgerows: 1. summer applications. Environ. Toxicol. and Chem. 16:165-172.

29. Longley, M., and Jepson, P. C. 1997. Cereal aphid and parasitoid survival in a logarithmically diluted deltamethrin spray transect in winter wheat: field-based risk assessment. Environ. Toxicol. and Chem. 8:1761-1767.

30. National Research Council. 1986. Pesticide Resistance: Strategies and Tactics for Management. National Academy Press, Wash. DC.

31. Potts, G. R., and Vickerman, G. P. 1974. Studies on the cereal ecosystem. Adv. Ecol. Res. 2:107-197.

32. Pratt, J. R., Melendez, A. E., Barreiro, R., and Bowers, N. J. 1997. Predicting the ecological effects of herbicides. Ecol. Appl. 7:1117-1124.

33. Salt, D. W., and Ford, M. G. 1984. The kinetics of insecticide action. Part III: The use of stochastic modeling to investigate the pick-up of insecticides from ULV-treated surfaces by larvae of *Spodoptera littoralis* Boisd. Pesticide Science 15:382-410.

34. Salt, D. W., and Ford, M. G. 1996. The kinetics of insecticide action. Part V: deterministic models to simulate the movement of pesticide from discrete deposits and to predict optimum deposit characteristics on leaf surfaces for control of sedentary crop pests. Pesticide Science 48:77-87.

35. Sherratt, T. N., and Jepson, P. C. 1993. A metapopulation approach to modelling the long-term impact of pesticides on invertebrates. J. Appl. Ecol. 30:696-705.

36. Stark, J. D., Jepson, P. C., and Mayer, D. F. 1995. Limitations to use of topical toxicity data for predictions of pesticide side-effects in the field. J. of Econ. Entomol. 88:1081-1088.

37. Steiner, L. F., Arnold, C. H., and Summerland, S. A. 1944. Laboratory and field tests of D.D.T. for control of the Codling moth. J. Econ. Entomol. 37:156.

38. Taub, F. 1997. Unique information contributed by multispecies systems: examples from the standardized aquatic mesocosm. Ecol. Appl. 7:1103-1110.

39. Thomas, C. F. G., Hol, E. H. A., and Everts, J. W. 1990. Modelling the diffusion component of dispersal during the

recovery of a population of linyphiid spiders from exposure to an insecticide. Functional Ecol. 4:357-368.

40. Touart, L. W., and Maciorowski, A. F. 1997. Information needs for pesticide registration in the United States. Ecol. Appl. 7:1086-1093.

41. Vickerman, G. P. 1992. The effects of different pesticide regimes on the invertebrate fauna of winter wheat. Pages 82-118 in: Pesticides, cereal farming and the environment. P. W. Greig-Smith, G. K. Frampton, and A. R. Hardy, eds. London, HMSO.

42. Watt, A. D., Vickerman, G. P., and Wratten, S. D. 1984. The effect of the grain aphid Sitobion avenae (F.) on winter wheat in England: an analysis of the economics of control practice and forecasting systems. Crop Protection 3:209-222.

43. Wiles, J. A., and Jepson, P .C. 1994. Substrate-mediated toxicity of deltamethrin residues to beneficial invertebrates: estimation of toxicity factors to aid risk assessment. Archives of Environ. Contam. and Toxicol. 27:384-391.

44. Wiles, J. A., and Jepson, P. C. 1995. Dosage reduction to improve the selectivity of deltamethrin between aphids and coccinellids in cereals. Entomol. Exp. et Appl. 76:83-96.

45. Wratten, S. D., and Mann, B. P. 1988. A survey of aphicide use on winter wheat in the summer of 1988. Pages 979-984 in: Proceedings of the 1988 British Crop Protection Conference-Pests and Diseases. BCPC Publications.

46. Wratten, S. D., Watt A. D., Carter, N., and Entwistle, J. C. 1990. Economic consequences of pesticide use for grain aphid control on winter wheat in 1984 in England. Crop Protection 9:73-77.

47. Wright, D. W., and Ashby, D. G. 1945. The control of the carrot fly (Psila rosae Fab.) (Diptera) with D.D.T. Bull. Entomol. Res. 36:253.

48. Van Straalen, N. M., and van Rijn, J. P. 1998. Ecotoxicological risk assessment of soil fauna recovery from pesticide application. Reviews of Environ. Contam. and Toxicol..

49. Young, B. W. 1986. The need for a greater understanding in the application of pesticides. Outlook on Agriculture 15:80-87.

Delivering New Crop Protection Agents Within An IPM Environment

Franklin R. Hall
Laboratory for Pest Control Application Technology
The Ohio State University, Wooster, OH 44691 USA

Delivering crop protection agents (cpa's) within an IPM environment requires that one select the appropriate material that has the least cost to the environment and beneficial organisms and that one optimally places the correct dose in the plant canopy. A basic assumption for this discussion is that the use of crop protection agents will continue to be at least one of the viable options for implementation of IPM and integrated crop management (ICM) programs. However, serious criticisms remain about the loss of pesticides into the environment (2,15). Most agricultural crop protection agents are delivered as liquid sprays. Unlike conventional polydisperse sprays, the newer "dial-in" sprays now require that the biologist provide the optimum criteria to maximize the biological effect. The recent changes in spray delivery technology and further changes that are likely in the future will have a significant impact on the amount and kind of information needed to realize the full potential benefits of advances in application technology. Understanding how we arrived at this point and the implications of these changes for the development of unique crop protection tactics is a key to predicting the needs for the future.

Factors Influencing the Evolution of Spray Technology

The long history of technological developments in pesticide application has been influenced significantly by changes in demo-

323

graphics, policies/regulations, new chemistry, terminology and expectations, global events, and developments in science. For example, world food needs based on current population estimates of 10 billion by 2050 will require either increased crop yields and/or increased use of sub-optimal agricultural land. Demographic studies already demonstrate that the urban/rural interface is experiencing an increase in problems involving dust, noise, odor, and pesticide drift. These factors are currently driving much research/regulatory activity on pesticide application technology (PAT) that is directed toward drift mitigation. Much of the recent drift management research on nozzle technology was accomplished because of public perceptions and public policy. The pesticide user has become the focal point for modifying pesticide delivery, improving application efficiency, and reducing risks of off-target movement.

The second area impacting PAT is that of pesticide policy and regulation, which have placed increasing restrictions on delivery processes, including restrictions on application sites and timing, as well as on the use of specific pesticides. The Food Quality and Protection Act (FQPA), which eliminated the Delaney paradox (12), clearly has not yet had its full impact on fruit and vegetable crops. However, implementation of the "risk cup" concept continues to evolve and probably will result in the elimination of significant choices in pesticide chemistries for many of these crops. Recent revisions in the U.S. farm support program have been predicted to impact planting decisions. As support dollars decline, crop diversity will likely increase and, hence, complicate decision-making at the farm level. In conjunction with continued weather perturbations and price competition, these factors all increase risk at the farm level. Farmers will be pressed to become more creative with respect to inputs, outputs, and overall farm economic health. This complicates IPM strategies, but it also offers unique opportunities for those who develop whole-farm crop protection strategies (ICM tactics). With the increased number of crops and consultants and with new technologies of precision agriculture (GPS/GIS), farming is entering a new arena of information technology that will support and complement improved PAT adjustments. The economics of delivering alternative crop protection agents at the farm level remain uncertain. Risk reduction and safer pesticide policies of EPA have increased the potential for industry to register safer products at a faster pace.

These alternatives, however, remain more vulnerable to environmental degradation (UV and rainfastness) and require much more information on delivery tactics.

The new chemistries of crop protection agents increase selectivity, which, while appropriate for ICM strategies, challenges the traditional paradigm for successful pesticides (i.e., effective against a wide spectrum of pests on a large number of crops, easy to use, safe). Pesticides are spectacularly effective and easy to use, and they reduce uncertainties in an already risky enterprise - - agriculture (8). The recent evolution of pesticide active ingredients over the last 10-20 years has allowed users to rely on broad-spectrum and prolonged residual activity to reduce the number of trips through the field to apply crop protection agents. The shorter residual activity of newer compounds necessitates that we know more about when applications are economically justified than was the case when previous crop protection agents allowed users to demand and to achieve a "biological desert" and the "piece of mind" that comes with it.

One intriguing characteristic of newer crop protection agents is a decline in the efficiency with which activity in the laboratory translates into activity in the field. As shown by Graham-Bryce (7) in Table 1 and our continuing research inquiries (4,9), as activity in the laboratory increases, we tend to see a greater proportional loss of

Table 1. A comparison of the loss of relative activity of three insecticides during the transfer from the laboratory to the field (adapted from 7).

Insecticide	Application rate (g/ha)	Relative[a] application rate	Relative toxicity[a,b]
DDT	1000	50	3700
Dimethoate	500	25	550
Deltamethrin	20	1	1

a/ values expressed relative to Deltamenthrin
b/ Based on LD_{50} of *Phaedon cochleariae*

activity in the transfer from lab to field conditions. There is a need to understand these losses as regulations, restrictions and resources continue to constrain crop protection tactics and the use of pesticides.

Terminologies and changes in societal expectations and their resulting impacts on successful crop protection tactics have subtly made their way into the language of crop protection and of society in general (Table 2). These changes place more emphasis on the acquisition of knowledge, thus increasing the complexity of decisions at the farm level. In order to attain higher levels of crop protection within an IPM environment through the use of newer technologies in PAT, organization of information into an easily understood form is increasingly critical. How are we to achieve this? Who will do it? Who will manage and integrate this complex, multi-level information? Who will design practical tactics for use in the field?

Global competition and resulting farm economics will play an increasing role in the development and acceptance of new strategies.

Table 2. Evolution of terminologies and tactics

Directional Changes	
From	**To**
Changes in Society	
Industrial society	Information society
Forced technologies	High tech / High touch
National economy	Global economy
Institutional help	Self help
Crop Protection Changes	
Independence	Interdependence
Short-term	Long-term
Control	Management
Single tactic	Multi-tactic
Insurance	As needed
Conventional	Alternative
Simple	Complex

The current low grain prices (lowest in 2 decades) require that farmers critically review pesticide inputs and farm management strategies. Vertical integration of the farm industry suggests that agricultural marketing will be dramatically different in 10 years. Consolidations in the food, transport, agro-marketing, farm equipment, and food retail industries continue to restructure business linkages. Mergers and coalitions between the seed industries and the increasingly larger agrichemical industries suggest control of value-added seed from seed to retail market is close at hand. These changes also suggest a larger corporate control of PAT and pesticide strategies, as well as of agriculture itself.

Emerging Spray Application Technologies

The objectives of the application process in crop protection are to deliver the agent in its most effective form to the target area/pest with the least off-target impact for the lowest cost (i.e., effective, economical, and environmentally friendly). Advances in application technology are vast and imply significant changes in both strategies and tactics. The significant emerging trends (Table 3) are a result of some of the aforementioned factors. In particular, issues associated with off-target drift of pesticides have driven many advances in nozzle and precision-farming technology. Drift mitigation through nozzle design and shielding represents one of the most recent innovations (16). Typical hydraulic nozzles produce a wide spectrum of droplet sizes. This results in a great deal of waste, because large drops do not arrive at certain target sites in a well-distributed pattern, and small drops are more prone to drift, although they have great bio-efficiency for many pests.

As shown by Adams (1), Hislop (10), Wolf and Downer (17), and others, there is no easy way to design the ideal spray. The ability to control spray quality independently from spray volume using twin nozzles and variable rate nozzles (6) offers new options for the user. These "dial-in" capabilities also pose a dilemma for biologists: we now have to define the optimum spray quality and deposit parameters for each pest scenario. They also offer significant opportunities for an integrated approach to the use of crop protection agents. The new

327

standards of nozzle classification for spray quality, and now drift (17), are increasing the number of decisions that a pesticide applicator must make, and they are leading to concern over the practical aspects of the need to change nozzles between (or within) fields as site requirements and pest populations change. Injector nozzles, air assistance, new sensors, drop-size control nozzles, and site-specific equipment all have potential to more accurately deliver crop protection agents, reduce risk of drift, and generally improve the efficiency and safety of the application process. However, all have significantly higher costs. Air induction nozzles can reduce drift, but issues relating to coverage for some active ingredients are still unresolved. Standardization and inspections of new and used equipment (European Union countries) increase savings and improve accuracy of toxin delivery, but the costs to government and/or farmer are yet to be resolved. Increased standards for water quality and pollution prevention in Germany and the Netherlands will likely increase attention to precision delivery systems, as will requirements of IPM.

Table 3. Emerging technologies in pesticide application.

Function	Technology
Drift reduction	Air-assist, shrouds, low pressure, adjuvants, electrostatic application
Control of spray volume and droplet size	Twin fluid technology, variable rate nozzles
Monitoring, site specificity, mapping	GPS / GIS
Pest detection	Lasers, monitors
On-the-go, site-specific applications	Variable-rate technology
Management assistance and decision aids	Decision models
Pest resistance and value-added traits	Biotechnology

Variable-rate technology (VRT) and site-specific technologies capable of on-the-go changes in delivery made possible by developments in GPS/GIS offer greater control over the placement of crop protection agents (14) (Fig. 1). GIS, for example, holds considerable promise in terms of identifying site-specific needs and precisely delivering precisely the optimal dose of a crop protection agent or other production input to the target area (13,14). The greater complexity offered by computer-controlled machinery and pesticide formulations that have a range of physico-chemical properties requires a broad range of expertise on the part of users of these technologies. Acquiring some of the needed information has been aided by additional technology (e.g., soil sensors, yield monitors, etc.). However, determining pest and weed numbers per site still requires a significant amount of labor and analysis. In addition, there are other unresolved issues in precision farming that involve data ownership and control and the retention by farmers of the power to make decisions (Hall, unpublished). Key questions relating to precision farming technologies are: Who will deliver the needed information in a timely manner? Who has the necessary information on farm health and pest management needs, as well as the experience to guide the user through the myriad of potential changes that these new technologies will bring about? The power of the technology is quite impressive, but we have yet to effectively harness it. Finally, implementing this technology on the farm will face the hard test of cost-effectiveness. Benefits must be clearly identified and well documented.

"Wow" technologies tend to remove the incentives to understand a system because the impact of new tools is striking. In addition, technologies are typically inherent fixation points for humans. In conjunction with the quickening pace of technological change and change in general, these attributes are reducing the incentives for innovative change on the farm. Technological change, coupled with policy and regulatory changes, population demographics, the vagaries of weather, and global competition, contribute to an increase in risk and the perception of risk for farmers.

GIS promises to reduce these risks, but it will take more than pretty maps to accomplish this (13). Decision models allow farmers

329

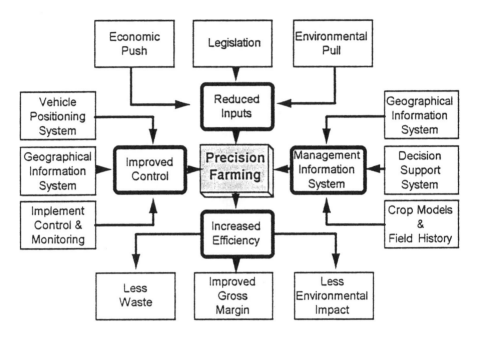

Figure 1. The elements of Precision Farming (PF) (by permission 14).

to adopt a "wait and see" approach before making a decision to treat for a pest problem. This requires good management of scarce farm resources to focus on applying crop protection agents on a very tight weather-moderated scale, sometimes at rates of travel faster than desired. This adds to the dilemma of how to manage critical events with available technologies and brings into question the merits of maintaining "clean" fields as opposed to "dirty" fields that may reflect an economically sustainable approach to pest management but have historically conveyed a poor image of a farmers management skills. Integration of farmer goals, technology, and IPM strategies will require different approaches to these more complex situations than we have thus far provided.

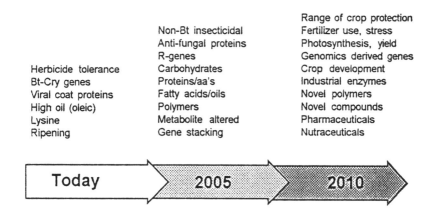

| Herbicide tolerance
Bt-Cry genes
Viral coat proteins
High oil (oleic)
Lysine
Ripening | Non-Bt insecticidal
Anti-fungal proteins
R-genes
Carbohydrates
Proteins/aa's
Fatty acids/oils
Polymers
Metabolite altered
Gene stacking | Range of crop protection
Fertilizer use, stress
Photosynthesis, yield
Genomics derived genes
Crop development
Industrial enzymes
Novel polymers
Novel compounds
Pharmaceuticals
Nutraceuticals |

Today ▷ 2005 ▷ 2010 ▷

Figure 2. Evolution of Biotechnology for agriculture (by permission 11).

Biotechnology provides new options for toxin delivery and crop protection. The production of genetically engineered crops having input traits is already well under way, and the production of engineered crops with output traits is close at hand (11) (Fig. 2). Technology experts suggest that in the future, combinations of value-added traits introduced into plants through biotechnology will increase in importance.

New Directions?

We have been through the phase of toxin substitution and, because of pesticide resistance, combinatorial approaches to pest control. We are now entering a phase of alternatives and potential redesign of crop protection and farming systems that are based on biotechnology. Cardina et al. (3) suggest that we are going through a series of phases (Fig. 3). Progress through these phases is accompanied by an increasing need for interdisciplinary approaches and is enabling us to ask more fundamental questions about population survival and movement, and to better understand plant/pest interactions than has been possible in the past. In addition, we can

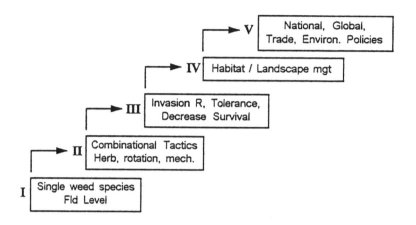

Figure 3. Advancing phases in weed IPM strategies (modified from 3).

now more readily address important, practical questions, such as: Does the disruption of weed patches increase or decrease weed aggressiveness in a field or landscape?

GIS and integrated site analyses represent valuable new tools with which to approach pest management on the broader spatial and temporal scales that characterize landscapes and regions (3,13,14).

Seeds are becoming more than commodity items. They are beginning to offer value-added qualities that clearly need protection for consistent and vigorous early plant health. Perhaps controlled release of crop protection agents and nitrogen administered through seed treatment will now provide economic advantages associated with optimal targeting efficiencies. The restructuring of the seed industry that is currently underway suggests that the trend toward packaging value-added qualities with seeds is already underway.

The mediation of plant health through resistance to stress is really in its infancy. Many advances will accrue as new knowledge of plant stress mechanisms and timed expression by the plant of protection proteins is better understood. The use of plant tolerance to stress in conjunction with pest population thresholds and sensor-based sampling (not labor-intensive counts) will add to the power of delivering, in a timely manner, crop protection agents to proper target sites at the level needed. Accomplishing this will require much greater

332

organization of information than we have thus far experienced. The requirements for additional information, at levels not yet achieved, remain as significant constraints to using this technology at a level commensurate with its power.

Spray coverage issues remain unresolved. Spray quality (and resulting deposits) can be varied by the spray delivery system or by the active addition of adjuvants (Fig. 4). Air induction nozzles are rapidly replacing conventional spray nozzles in the countries of the European Union in an effort to address concerns over spray drift. Air induction nozzles produce spray deposits that are larger and more sparse than those produced by conventional nozzles. Just how this altered deposit structure interacts with drift management adjuvants and with new crop protection agents that have different modes of action is not clear, but represents an important research opportunity.

There are some intriguing questions concerning toxin/pest interactions. For example, how do a pest's biology and activities

Effect of Adjuvant on Spray Quality

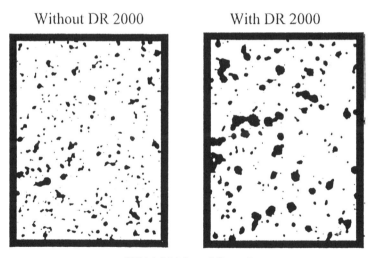

Without DR 2000   With DR 2000

TT110015 at 10 mph

Figure 4. Spray quality at the target surfaces.

Cabbage looper feeding on cabbage leaf disks

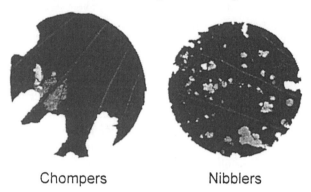

Chompers   Nibblers

Figure 5. Chompers and Nibblers; an example of cabbage looper (*Trichoplusia ni*) larvae feeding on cabbage leaves.

influence potential encounters with a toxin?   Utilizing computer modeling tools and new experimental designs, interactions involving feeding behaviors, dose, and deposit structure as related to crop protection agents can be exploited and better understood (Fig. 5).

The influence of feeding behavior on encounters with toxin deposits is illustrated by the cabbage looper (*Trichoplusia ni*).   Some cabbage looper larvae feed on foliage by nibbling at a number of sites, while others are "chompers" and concentrate their feeding at localized sites on leaves.   Nibblers have greater variability in their response to toxin presentation than do chompers.   Small numbers of large deposits have greater effectiveness against nibblers, whereas large numbers of small deposits are more effective on chompers.

Percent control is not always strongly correlated with crop protection.   Thus, elucidation of specific crop protection goals is critical in order to provide optimal toxin delivery processes (5). Technologies to alter the bioavailability of crop protection agents remain somewhat unexplored, and the efficiency of the dose-transfer process remains low relative to the power of the new active ingredients and our ability to manipulate spray deposits.   What will be our expectations for crop protection in the future?   Will we strive to maintain a "biological desert" in our crop fields, or will economics, restrictions, crop prices, and reduced reliance on crop protection agents for IPM become the primary determinants of our crop protection goals? Where are all of the pesticide experts, and who will identify and deliver to the farmers critical information on integrating new tools into their crop production and pest management systems?

Can we identify areas that will be critical to the increased adoption of IPM? Yes, but there are many needs that must be addressed and that will be affected by new technologies. These technologies, as affected by energy, water quality, and regulatory factors (mandated IPM?), will critically influence crop protection strategies in the next millenium. Critical areas include:

1. Information technology - the way we communicate; speed and transfer loads; new pulsing technology.

2. Pace and scope of changes - transfer of information.

3. Biotechnology - basic inquiries on photosynthesis, plant health, plant and pest genomes, gene switching, plant stress identification, and switches in plant tolerance expression; vertical integration of agriculture.

4. Sensors - take higher labor costs out of the fields; reduce costs of scouting.

5. Bioinformatics - changes in how we organize information, critical for GIS and development of holistic, ecosystem thrusts of crop protection.

6. Policy - risk reduction, fine tuning of FQPA; world trade policy; crop support; insurance for risky innovations.

7. Resources - availability of people with knowledge of field events; ability/time/resources to develop on-farm solutions; trends towards molecular.

New crop protection delivery tools and systems offer more control and improved targeting options for our new crop protection agents. The biologist must show how farmers can take advantage of these options to obtain real value. To change current reliance on agricultural pesticides, we must better understand farmer behavior, expectations, management skills, goals, decision-making processes, critical resource constraints, capital requirements, and aversion to risk. If commodity prices remain low and variable, then the appeal of new technologies will remain high, although costs and benefits of new technologies must be presented more clearly to potential users. Economic issues will remain dominant forces determining acceptance of new strategic practices for U.S. farmers.

The emerging IPM paradigms use new technology transfer methodologies, include participatory efforts by farmers, contain a basket of choices, and empower the user with increased competence. The diverse, complex strategies of the "on-the-go tool set" must be made (to appear) simple in structure. Without clear rules, economic benefits, and well-organized tactics, innovative, but "fuzzy," systems will not be readily accepted by farmers, who are still accustomed to having simple, but robust and successful, pesticide-driven tactics. The opportunities seem abundantly clear.

The prospects of biotechnology to eliminate or reduce external uses of pesticides have substantial appeal. With 2 more years of field experimentation, we will know how real that promise is, given the potential for pests to develop resistance and for other issues, as yet unforeseen, to emerge. Combined with new delivery technologies, these tactics may revolutionize toxin presentations, thereby challenging current crop protection expectations.

Opportunities now exist for all farmers to assess the health of their farms relative to resources, management styles, and goals, and to identify site-specific options for their farms. The goals of farmers, as well as the pace of change in crop production and IPM, are likely to vary among crops, sectors of the country, and socio-economic levels. Herein lies an opportunity for partnering between public sectors and private industries to achieve a participatory empowerment through the organization of information. We must develop critical economic analyses for crop protection strategy options in order to advance the state of crop protection in the next millennium. The tools for more precise delivery of external crop protection agents are advancing rapidly. Let us make sure that we use them wisely!! That will take all of us working together.

336

Literature Cited

1. Adams, A., Chapple, A., and Hall F. 1990. Agricultural sprays: lessons and implications of drop size spectra and biological effects. Pages 156-169 In: Pesticide formulations and applications systems. L. Bode, J. Hazen, and D. Chasin, eds. ASTM STP 1978. Amer. Soc. For Testing and Materials.
2. Benbrook, C. 1996. Pest management at the crossroads. Consumers Union, Yonkers, NY.
3. Cardina, J., Webster, T., Herms, C., and Regnier, E. 1999. Development of weed IPM: levels of integration for weed management. J. Crop Prod. 2:239-267.
4. Copping, L., Merritt, C., Grayson, T., Wakerley, S., and Reay, R., eds. 1989. Comparing laboratory and field pesticide performance. Assoc. of Applied Biologists, Wellsbourne, UK Aspects of Appl. Bio. #221.
5. Ebert, T., Taylor, R., and Hall, F. 1998. Deposit structure interacting with insect behavior modifies efficacy. IUPAC Poster 2C-013. London, UK, August, 1998.
6. Giles, D., and Comino. J. 1990. Droplet size and spray pattern characteristics of an electronic flow controller for spray nozzles. J. Agri. Eng. Res. 47:249-267.
7. Graham-Bryce, I. 1983. Pesticide research for improvement of human welfare. Pages 21-42 in: Proceedings of the 5th International Congress of Pesticide Chemistry. Vol. 1:21-42.
8. Hall, R. 1996. Challenges and prospects of integrated pest management. Pages 1-19 in: Novel approaches to integrated pest management, Revini, ed., CRC Press, Baco Raton, FL
9. Hewitt, H., Caseley, J., Copping, L., Grayson, T., and Tyson, D. 1994. Comparing glasshouse and field pesticide performance II. BCPC Mono. #59.
10. Hislop E. 1987. Can we define and achieve optimum pesticide deposits? Aspects Appl. Biol 14:153-172.
11. McLaren, J. 1999. The success of transgenic crops in the USA. Pesticide Outlook, Vol. 9:36-41.
12. National Research Council. 1987. Regulating pesticides in food: the Delaney paradox. National Academy Press. Washington, DC.
13. Nelson, L. M., Orum, T., Jaime-Garcia, R., and Nadeem, A. 1999. Applications of geographic information systems and

geostatistics in plant disease epidemiology and management. Plant Disease 83: 308-319.

14. Parkin, C., and Blackmore. B. 1995. A precision farming approach to the application of agrochemicals. in: Web site: **http://www.silsoe.cranfield.ac.uk/cpf/papers/BAAS/BAAS. htm**

15. Pimentel, D., and Levitan. L. 1986. Pesticides: amounts applied and amounts reaching pests. Bioscience 36:86-91.

16. Wolf, T., Grover, R., Wallace, K., Shewchuk, S., and Maybank, J. 1993. Effect of protective shields on drift and deposition characteristics of field sprayers. Canad. J. Plant Sci. 73:1261-1273.

17. Wolf, T. and Downer, R. 1998. Developing criteria for the ideal agricultural spray – a biologists perspective. in: Proc. ILASS, 11th Ann. Conf. On Liquid Atomization and Spraying Systems. Sacramento, CA, May, 1998.

Pesticide Development in Transition: Crop Protection and More

N. Beth Carroll
Novartis Crop Protection
Greensboro, NC 27419-8300 USA

The crop protection industry is continually challenged to provide the technologies and products to satisfy the increasing global population needs for food and fiber, despite decreasing availability of arable land for agricultural production. Effective crop management solutions will flow from the integration of our resources for managing crop pests, resulting in increased crop yields and improved quality. Novartis Crop Protection discovers, develops, manufactures, and markets innovative biological and chemical solutions to help growers produce high-quality, higher-yielding crops with minimal impact on the environment (1).

Novartis Crop Protection was formed from a 1997 combination of Ciba Crop Protection and Sandoz Agro, followed by the acquisition of Merck Crop Protection (2). The result of joining these three companies is the assemblage of an encompassing product portfolio for all phases of crop protection. Novartis Crop Protection and Novartis Seeds jointly endeavor to develop innovative solutions for agricultural production. This requires pooling our resources to design systems to protect crops against diseases, insects, and weeds, while also improving varieties, quality, and yield enhancement in foods, maintaining affordable prices without putting more land under cultivation, and adhering to high environmental standards (3).

Food Security

It is a widely held belief that there is enough food already being produced and that the world hunger problem is chiefly caused by a lack of fair distribution. The crop protection industry is fully aware that food production is highly complex and that an increase in food production does not necessarily correlate with reducing or ending hunger. However, industry-based farm inputs are important elements of maintaining food security (3).

Machinery, seeds, fertilizer, and crop protection product inputs are there to achieve the output, which is a stable, affordable food supply of high quality, suitable to be stored, transported, and processed. All along the way, mechanisms are in action that define the agricultural environment, including regulatory policies (lack of clarity and harmonization), free-trade agreements (NAFTA), farm subsidies (or the lack thereof), world economic issues, environmental considerations, socio-cultural pressures, and, of course, the media. Other factors are also driving significant strategic changes in agricultural production. These include: a growing and aging population; concern for wellness, food safety, the environment, and sustainability; development of functional foods; and bio-based energy (3).

Crop Solutions and IPM

Crop solutions as defined by Novartis are whole crop production packages. They must be customized to address the specific crop production systems and geographic areas (4). To address how crop solutions may add value to integrated pest management (IPM) systems, there first must be definition of sustainability and related concepts. Sustainable agriculture is defined as the use of practices and systems that maintain and enhance the sufficient and affordable supply of high-quality food and fiber, the economic viability of agriculture, the natural resource base of agriculture and its environment, and the ability of people and communities to provide for their social well-being (3).

Good Agricultural Practice (GAP), Integrated Crop Management (ICM), and IPM are the infrastructure for Sustainable Agriculture (Fig. 1). GAP is defined as efficient production of good-quality food, feed, and fiber while maintaining natural resources and optimizing crop inputs to minimize environmental impacts and ensure responsibility for the health and safety of farmers. ICM is a Whole Farm Strategy that manages crop profitability, uses IPM as a component, safeguards the farm's natural assets, minimizes waste

340

and pollution, and enhances energy efficiency. ICM systems are dynamic, adapting and making sensible use of the latest research, technology, advice, and experience (5).

Novartis Crop Protection defines IPM as a component of ICM (Fig. 1). Novartis supports the definition of IPM set by FAO's (Food and Agriculture Organization of the United Nations) International Code of Conduct on Distribution and Use of Pesticides (Article 2): *a pest management system that, in the context of the associated environment and the population dynamics of the pest species, utilizes all suitable techniques and methods in as compatible a manner as possible and maintains the pest populations at levels below those causing economically unacceptable damage or loss* (6,7).

Integrated Crop Management

ICM is sustainable agriculture put into practice. It considers economy and ecology as equal components. Novartis Crop Protection is focusing on several areas to bring together agricultural solutions for fully integrated and highly efficient crop production systems that result in safe, high-quality, affordable food and fiber in an environmentally sound manner that fully supports sustainability. Our crop solutions' scope encompasses 1) new product systems and

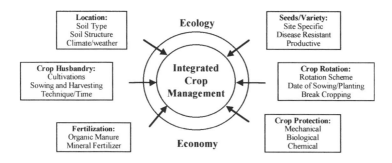

Figure 1. Crop and Pest Management practices are the components of Sustainable Agricultural Systems (4,5).

integration of new technologies, 2) services and information, 3) integrated systems research across disciplines, 4) stewardship, and 5) education.

New Product Systems and Integration of New Technologies

It is obvious that if the crop protection industry is to support sustainability, the criteria for development of new active ingredients must consist of characteristics that are consistent with IPM and ICM. Novartis evaluates new active ingredients on the basis of reduced risk factors, which include: IPM compatibility and flexibility, lower use rates, human and environmental safety, high biological efficacy, selective solutions for specific pests, and safety to beneficial or naturally occurring insects. When a new active ingredient meets these standards, it is developed for use in integrated systems. During the entire life cycle of our products, IPM is taken into account. As product knowledge increases, its use in sustainable agriculture is continually developed and improved (6).

INNOVATION IN BIOLOGICALLY BASED PRODUCTS

Novartis' agrichemical-based research seeks to develop biologically based products and to integrate those with chemical controls. An example of this is an approach taken to address problems with two-spotted spider mites in California strawberry production areas. Strawberries have little tolerance for damage from spider mites, and the two-spotted spider mite has historically been difficult to control. Therefore, some growers have taken an aggressive approach to controlling mites using numerous sprays, often resulting in the development of resistance in the pest. Combining a chemical component, such as Agri-Mek®, with a biological component, such as the predaceous mite *Phytoseiulus persimilis,* gives a solution to the miticide resistance problem and the mite population explosion cycle. Agri-Mek® is very efficacious on target pests but does not harm beneficial organisms such as predaceous mites once the sprays have dried. *Phytoseiulus persimilis* feeds strictly on spider mites, not on plant parts or on other beneficial organisms, and this predaceous mite species completes generations more quickly than spider mites. Novartis also provides

342

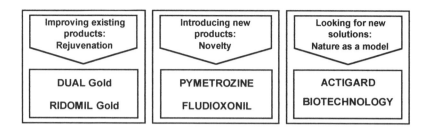

Improving existing products: Rejuvenation	Introducing new products: Novelty	Looking for new solutions: Nature as a model
DUAL Gold RIDOMIL Gold	PYMETROZINE FLUDIOXONIL	ACTIGARD BIOTECHNOLOGY

Figure 2.   Novartis seeks innovation in chemistry by several avenues to developing new crop protection solutions.

guidance on scouting and application timing with this pest management solution (8).

INNOVATION IN CHEMISTRY

Novartis approaches innovation by three complementary avenues to innovation. These are i) **Rejuvenation** to improve existing products, ii) **Novelty** to introduce new compounds, and iii) **Nature as a model** in seeking new solutions (Fig. 2)(9).

Rejuvenation. An example of Rejuvenation is the development of Ridomil Gold®, or mefenoxam, the active isomer of metalaxyl known as Ridomil®.   Some chemical crop protection products contain more than one isomer. Isomers are reversed images of the same molecule that cannot be manipulated to be identical. Often in an "active ingredient," there are numerous isomers of the same molecule with varying levels of activity against the target pest. Identification of the most active isomer and development of that isomer as the new "active ingredient" results in a number of benefits (Fig. 3)(9).

Development of Ridomil Gold® from Ridomil®resulted in a 50% reduction of pounds applied, as well as reduced package waste and disposal compared to metalaxyl. The U.S. Environmental Protection Agency (EPA) classifies mefenoxam as a new "active ingredient" (10).   Historically, selective production of individual isomers was often not feasible or was too expensive for commercialization. Today, with state-of-the-art technology and an incentive from EPA to designate a specific isomer as a new active ingredient, it is possible to identify and produce only one isomer at an affordable cost. EPA also classified mefenoxam as a reduced-risk product, and its development is consistent with EPA's objective to reduce pesticide load going into the environment by 50% in 2005. Since mefenoxam reduces residues by 50%, it allows for expanded use of this product to minor crops. With metalaxyl, higher residues occupy more of the acceptable daily intake (ADI). When 100% of the ADI is occupied, additional crops may not be added, and minor-use crops may be at risk if the ADI is exceeded.

Development of Dual Magnum® from Dual® is another example of rejuvenation of a product. Dual Magnum selectively contains the active isomer S-metolachlor. Development of this product resulted in reduction of use rates by approximately 40%, with significantly less environmental loads and smaller production units, which means less energy consumption, less trash, less transportation and storage costs, and fewer risks (Fig. 3)(11).

Novelty. Pymetrozine, to be sold under the trade name of Fulfill®, is an example of Novelty in active ingredient development. It is a highly specific insecticide, efficacious only against sucking pests such as aphids and whiteflies. It has low mammalian toxicity,

Use the Phenomenon 'Isomerism'

RIDOMIL Gold selectively contains the active isomer *R-metalaxyl* and can therefore be applied at use rates of only 50%.

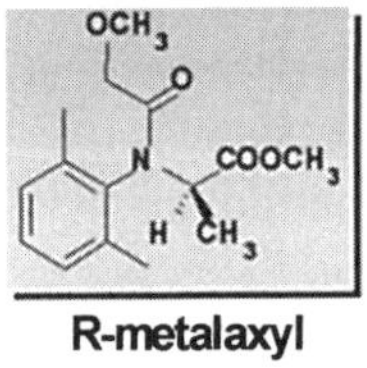

R-metalaxyl

Lowering the use rates by 40% (DUAL) or even 50% (RIDOMIL) certainly is a significant contribution to sustainable agriculture.

Figure 3. Novartis has developed isomers of two active ingredients into new products (9).

Pymetrozine - a Novel Insecticide for IPM

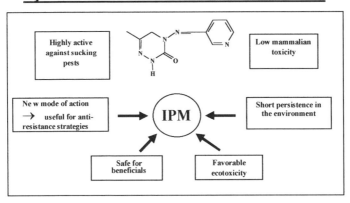

Figure 4. Pymetrozine, to be sold under the tradename of Fulfill®, is a highly specific aphicide that is classified as "reduced-risk" by EPA (9).

a favorable ecotoxicity profile, short environmental persistence, and is safe for beneficial organisms. Due to these numerous beneficial qualities, Fulfill® has already been classified as a reduced-risk compound by EPA. In addition, its new mode of action and specificity will be very useful in resistance management strategies in IPM systems (Fig. 4)(12). However, because of the more selective and subtle action of many new products, better user education may be necessary. This is particularly true with Fulfill®. Although it quickly stops aphids from feeding on plant tissue, it does not kill them on contact, and the aphids may live for several days after they first feed on the compound. If ample time for the aphids to die is not allowed before taking additional population counts, the scout or grower could make an erroneous decision to re-spray. Therefore, education of field scouts on proper evaluation of aphid populations after application is critical to the success of this IPM method.

Fludioxonil, known as Maxim®, is a seed treatment fungicide developed from pyrrolnitrin, a natural antibiotic (Fig. 5)(13). Fenpiclonil, a chemical variation of pyrrolnitrin, was the precursor to synthesis of fludioxonil. This fungicide, with its new mode of action, is setting a new standard for low-use rates at 9.9 to 19.8 grams of

Products Derived from Natural Compounds

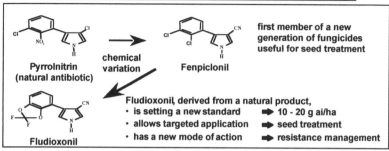

Pyrrolnitrin
(natural antibiotic)

chemical variation

Fenpiclonil

first member of a new generation of fungicides useful for seed treatment

Fludioxonil

Fludioxonil, derived from a natural product,
- is setting a new standard ➡ 10 - 20 g ai/ha
- allows targeted application ➡ seed treatment
- has a new mode of action ➡ resistance management

Figure 5. Fludioxonil, the active ingredient of Maxim® seed treatment fungicide, is derived from the natural product pyrrolnitrin (9).

'Systemic Acquired Resistance' (SAR) induced by Actigard: Nature as a model

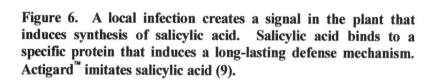

- **fundamentally new technology for the control of plant diseases**
- **activates natural defense mechanisms ('immunization')**
- **has no direct fungicidal activity**
- **broad spectrum (fungi, bacteria, viruses)**

Resistance

Signal

Induction

Figure 6. A local infection creates a signal in the plant that induces synthesis of salicylic acid. Salicylic acid binds to a specific protein that induces a long-lasting defense mechanism. Actigard™ imitates salicylic acid (9).

active ingredient per hectare (13). As the first member of a new generation of fungicides useful for seed treatments and as a derivative of a natural product, development of fludioxonil exemplifies both the **Novelty** and the **Nature as a Model** avenues of innovation.

Nature as a Model. Another example of **Nature as a Model** is Actigard™, based on plant activator chemistry also known as systemic acquired resistance (SAR). Use of SAR is fundamentally new technology for use in managing plant diseases. Actigard™ activates the natural defense mechanisms in the plant to immunize it against a broad spectrum of plant pathogens (fungi, bacteria, and viruses) (Fig. 6)(14). Since the plant's immune system must be activated before the pathogen infects the plant, it must be applied as a preventive measure. IPM systems that employ this technology must apply Actigard™ early, rather than waiting until economic threshold levels of the disease are reached. Again, this fundamental change in disease management technology may require education of growers and scouts.

SCREENING CONCEPTS

Novartis Crop Protection now employs an improved method for identifying chemicals for development as crop protection products. Previously, Novartis (Ciba) screened 10,000 to 12,000 compounds per year in vivo and developed one or two candidates per year (Fig. 7)(9). Recognition of the need to look not only at random chemistry and chemical design but also at natural products, enzymes and receptors, third party products, and combinatorial chemistry led to development of a new screening system (Fig. 7)(9). Currently, 100,000 or more compounds are screened yearly in a High Throughput Screening (HTS) method using both in vivo and in vitro technologies. High Throughput Screening allows pre-selection of promising chemistries with improved characteristics (novelty, activity, selectivity, etc.) to test in vivo (Fig. 7)(9). Although HTS allows Novartis to identify development candidates that are much more likely to satisfy the standards set for efficacy, specificity, and human and environmental safety, other limitations prevent an increase in the number of candidates selected each year. Pesticide

A Modern Screening Concept

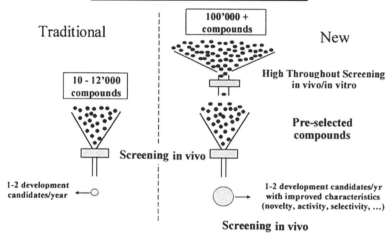

Figure 7. Our screening concept is based on identification of new targets through random chemistry, combinatorial chemistry, biosynthesis, chemical design, and natural products (9).

development still costs manufacturers from $35-$50 million per compound over 8-10 years to achieve registration in the USA.

INNOVATION IN SEEDS

Use of biotechnology to produce genetically modified plants is one of the most promising techniques in crop production. It not only achieves development much more quickly than traditional plant breeding; it also allows the "ideal" plant to be constructed. Novartis seed biotechnology has generally focused on input traits such as those that confer insect resistance (Maximizer®, YieldGard®, KnockOut®)(15) and herbicide tolerance (Acuron™ technology)(16). Future research will address other aspects of crop production, such as higher yields, drought and salt resistance, wider geographic adaptability, disease resistance, and increased nutritional value. There is also a shift occurring in agriculture focusing on "industrial crops" (optimized for quantity and quality), "designer crops"

There are "Crops" and

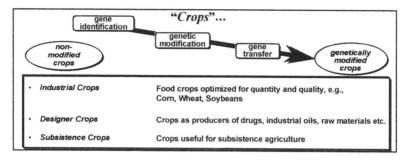

Figure 8. Biotechnology will allow development of ideal plants (9).

(crops as producers of drugs), industrial oils, raw materials, etc., and "subsistence crops," which are useful in subsistence agriculture (Fig. 8)(9). Changes in seed technology will have a significant effect on IPM systems and on future definitions of IPM.

Services and Information

Novartis is committed to the development of fully integrated systems to provide the support and services needed for customized crop solution programs to meet growers' needs. We have ongoing research in diagnostics, decision support systems and predictive models, global information systems (GIS), precision farming, and optimization of application technology.

One example of where "Services and Information" are employed is in our "Citrus Solutions," an integrated system that Novartis currently is testing with selected Florida citrus growers. "Citrus Solutions" is a systems approach to managing insects, plant pathogens, and weeds in citrus groves. The grower has the option to choose one of several levels of citrus pest management. There is a guaranteed one-price-per-hectare, with threshold assurance for mite control, root disease control, and weed control that is dependent

349

upon the level of management the grower has selected. One option specifies that supplemental treatments are provided at no additional charge when based on scouting and diagnostics. As an example of how services and information become a part of the solutions approach, "Citrus Solutions" utilizes geo-referenced soil sampling and grid maps, and scouting by trained and certified field scouts. Scouts are provided computer access in the field, and the information is delivered via the Internet through an Entonet Expert System providing "real-time" grove analyses and reports (Claude Flueckinger, personal communication).

Other Novartis services include closed-system packaging, residue analysis methodology, and application technologies. An example of an application method developed by Novartis is the Crop Adapted Spraying (CAS) system designed to adapt spray volumes and dose rates per hectare for growing plants. The CAS system provides equal deposition on leaves and fruits, predictable biological efficacy, and predictable residues throughout the growing season of the crop (9).

Integrated Systems Research

If IPM and ICM are to be realized at the desired levels in global farming systems, working across disciplines is required. Often it is difficult to identify university researchers who have the ability and the support of their administration to assist in development of whole programs. Private-sector research involving crop consultants who can address IPM holistically seems to be emerging. Integrated systems researchers must be flexible in their definition of IPM if they are to adapt threshold and scouting procedures to accommodate new technologies. With increased specificity of numerous new products and an emphasis on resistance management, development of integrated recommendations is critical to rapidly reaching the utility potential of new technologies, as well as providing for their longevity as tools for crop protection.

Environmental Stewardship

Our commitment to environmental stewardship is evident in the numerous research programs and cooperative efforts in which Novartis is active. Conservation tillage research to reduce soil erosion, surface runoff, and soil nutrient loss is ongoing, especially in the midwestern region of North America. Novartis funds Best Management Programs (BMPs) for education and implementation of cost-effective practices to reduce or eliminate runoff into watersheds (17). An example of these efforts is cooperation with the "Iowa Trees Forever" project, in which Novartis provided monetary and educational support for buffer plantings of trees between corn fields and water resources. The soil conservation improvements and runoff reductions realized through this project are significant.

Water monitoring programs are ongoing for a number of our products and will provide important information for implementation of the Food Quality Protection Act (FQPA). FQPA requires EPA to assess the aggregate risk of a product to include water and non-occupational exposures in addition to dietary exposure (18). There is very little information on exposure to pesticides via water or non-occupational routes, and these Novartis programs will assist in addressing the shortfall of knowledge that currently exists. Novartis is also a leader in packaging and formulation technology for reduction of waste, worker exposure, and point-source contamination.

Education

Continuing education is always important. Programs are in place for the ongoing training of Novartis technical staff, who, in turn, train university cooperators and growers. Training programs for farmers and scouts are critical to the success of our integrated solutions programs. Demonstrations are used as valuable training tools for farmers, university cooperators, and food industry decision-makers, as well as those who influence regulatory and policy decisions, and others involved in the chain of food and fiber production.

The most important cooperative educational effort the industry undertakes is directed toward policy-makers. Industry cooperates with EPA and USDA to continually bring sound agricultural science to the decision-makers in our government. This often involves re-education of decision-makers who have been provided information that is not based on sound science. For example, The National Research Council's Board on Agriculture (which is currently addressing "The Future of Pesticides in U.S. Agriculture") was given a presentation on apple scab during its deliberations. The

351

presenter indicated that apple scab is merely a cosmetic problem of apple fruit and that education of consumers so they knew the fruit was safe to purchase and consume was all that was needed to solve the problem of needing fungicide applications to control apple scab. In reality, apple scab is a devastating disease that, if left unchecked, will completely defoliate apple trees. To make this valid point with the committee required a re-education process by involved farmers, university scientists, and industry.

Conclusions

Novartis Crop Protection is a leader in developing safe, effective crop solutions for sustainable agriculture. Our ongoing effort is toward establishing confidence and trust in sustainable agri-chemical and seed technologies. Clarification of agricultural policy, direction, and regulatory harmonization, as well as a better understanding of the impact and benefits of biotechnology, are critical to our success.

While individual agribusiness companies face challenges from stiff competition by fewer but stronger competitors, there are also numerous external challenges to the success of farmers and our industry. Growing economic pressure on farmers is coming from decreasing farm subsidies and global trade agreements such as NAFTA and GATT. The public concerns about food safety and environmentally sound production practices are often exacerbated by the media's preoccupation with half-truths. Registration requirements are increasingly more complex. The methodology and costs for assessing aggregate and cumulative risk as required under FQPA have not yet been elucidated. Global harmonization of regulations is still an unrealized goal (9).

The conditions and methods of crop protection are continually changing based on discovery, adaptation, and development of new technologies. In the near future, chemicals will continue to be the basis for development of crop protection solutions, with genetically modified crops becoming increasingly important. Biological solutions will continue to be developed but will most likely only impact niche markets. Information technology and management will increasingly support integrated crop management systems (8). The challenge of the 21st century will be to achieve the required increased production while reducing adverse environmental effects.

This can only be done if agriculture is managed in a yield-intensive and sustainable fashion through use of emerging technologies (1).

Literature Cited

1. Carta Nova — The Charter of Novartis Crop Protection http://www.cp.novartis.com/b.frame.htm [Accessed March 1999]
2. History, A Merger of Historical Significance. http://www.cp.us.novartis.com/corpinfo/history.shtml [Accessed 2 March 1999]
3. Benefits of Our Products. http://www.cp.novartis.com/c4_cont.htm [Accessed 2 March 1999]
4. Crop Protection Glossary, Sustainable Agriculture. http://www.cp.novartis.com/sa_cont.htm [Accessed 2 March 1999]
5. Crop Protection Glossary, Integrated Crop Management. http://www.cp.us.novartis.com/cl_cont.htm [Accessed 2 March 1999]
6. Crop Protection Glossary, Integrated Pest Management. http://www.cp.us.novartis.com/ipm_cont.htm [Accessed 2 March 1999]
7. Food and Agriculture Organization of the United Nations; International Code of Conduct on Distribution and Use of Pesticides (Article 2). http://www.fao.org [Accessed 2 March 1999]
8. A Soft Approach to Cleaner Strawberry Fields. 1999. Novartis Crop Protection SC1328. NCP 801-00085-A 1/99
9. Harr, Jost. 1998. Research in NOVARTIS Crop Protection: Innovative Solutions for Healthy Crops. Novartis Crop Protection Internal Protected Intranet Site [Accessed 2 March 1999]
10. Fungicides. http://www.cp.novartis.com/a3_cont.htm [Accessed 2 March 1999]
11. Herbicides. http://www.cp.novartis.com/a1_cont.htm [Accessed 2 March 1999]
12. Fulfill™ Technical Bulletin. http://www.cp.us.novartis.com/products/cropprotection/brochures/fulfill_tech_bulletin/ [Accessed 2 March 1999]

13. Maxim® Fungicide.
http://www.cp.us.novartis.com/products_general_frame.html
[Accessed 2 March 1999]

14. Plant Activator Acts as "Vaccine", innovative compound activates plant immune system to fight disease.
http://www.cp.us.novartis.com/products/cropprotection/broc hures/growing/worldwide.shtml [Accessed 2 March 1999]

15. Novartis Seeds Innovation Drives Performance
http://www.nk.com/ [Accessed 2 March 1999]

16. Background Information: The Acuron™ Gene and PPO Herbicides.
http://www.cp.us.novartis.com/cropinfo/press/acuron_back.s html [Accessed 2 March 1999]

17. Flueckinger, Claude, 1999. Personal communication.

18. Responsible Product Use.
http://www.cp.us.novartis.com/news_general_frame.html
[Accessed 2 March 1999]

19. The Food Quality Protection Act (FQPA) of 1996.
http:www.epa.gov/oppfead1/fqpa/index.html [Accessed 2 March 1999]

Integrating New Fungicide Technologies in IPM

Eric Tedford and **Richard A. Brown**
Zeneca Ag Products, Western Research Center
Richmond, CA 94804-4610 USA

The definition of Integrated Pest Management (IPM) has gone through considerable revision since its inception by entomologists (37). The definition now includes concepts and terminology that are also relevant to plant pathologists and ecologists (17). In fact, the definition is now so inclusive that some prefer the term "ecologically based pest management" (32) and "biointensive IPM" (6). The current definition of IPM accepted by the National Coalition on Integrated Pest Management is as follows: "a sustainable approach to managing pests by combining biological, cultural, physical, and chemical tools in a way that minimizes economic, health and environmental risks." In general, this new definition changes the overall end-focus from one that attempted to least disruptively integrate chemical and biological control methods (40) to one of overall impact on economics, health, and safety.

The initiation of the *Food Quality Protection Act* (FQPA) in 1996 confirmed the government's commitment to reducing dietary risks of pesticide residues. As a direct consequence of implementing this act, changes in the agricultural chemical industry are essential. Provisions of the FQPA require that the Environmental Protection Agency (EPA) review and adjust tolerances for many of the currently registered agricultural chemicals. The result of this effort will, no doubt, alter use patterns for some compounds and eliminate the use of others altogether. Certainly, the hurdles for registration of any new agricultural chemicals have been raised. Only those compounds that meet the criteria of being "safer" products will proceed expeditiously through the registration process. The time has come for the agricultural chemical industry to focus proactively on developing and utilizing integrated new fungicide technologies in IPM.

Drivers of Change

North American agriculture is in the process of changing profoundly and irrevocably. The drivers for this change are economic (farm prices and processor/retailers), sociopolitical (environmental protection and food quality), and technological (information and biotechnology).

ECONOMIC DRIVERS

Farm prices are declining and will remain low, at least for the midterm. This has brought about a focus on economies of scale, which has resulted in farms becoming larger and more dependent on technology. Processors and retailers have merged to become powerful buyers. They have sought to increase their profitability through "branding" food products. Therefore, they are keen to protect their "brand equity." In order to do this, they will pay a premium price for agricultural commodities that are of a consistent predetermined quality. These buyers often seek to ensure quality by closely specifying the way in which the commodities are produced.

SOCIOPOLITICAL DRIVERS

Sociopolitical drivers result from a desire for environmental protection and concern over residues in food. They come from a relatively affluent and largely urban consumer base and are manifested in legislation such as the FQPA. Sociopolitical drivers have also created a viable organic food market. This will always remain small, because it is highly labor intensive. Supplying the entire food chain with organic foods would require a radical change in society, and it is highly unlikely that such change would be acceptable.

Advances in biotechnology have already changed U.S. agriculture substantially and will continue to do so. However, it is wrong to assume that biotechnology will quickly, if ever, provide "silver bullet" solutions to agricultural problems. Nonetheless, advances in other areas of biotechnology will enable the development of reduced-risk and more effective agrichemicals. For the foreseeable future, a mixed strategy involving new biotechnology and chemical products will be used, and information technology will be the key to advising how and where these products are to be used. Developments in precision farming will allow inputs to be optimized, which will benefit both the environment and the profitability of farming enterprises. Advances in the ability to devise and apply precise technological solutions to agricultural problems form the basis of a new, integrated business in which the information will have considerable value. This enhanced role of information, which can now be easily assigned a value as part of a package, will ensure that much of the original optimism for IPM will be realized. However, the costs of new research in this field and the difficulties in assuring a return on the investment in research have caused further consolidation in the agrichemical and seed industries, along with spectacular vertical and horizontal integration. Though many of the new technologies have been promoted as giving choice to farmers, there is a general fear among farmers that, in fact, choices are being removed.

All of these drivers have combined to force the greatest pace of change in the crop protection business since the introduction of synthetic pesticides. Though the exact details of the future are uncertain, it appears likely that IPM will play a pivotal role through the integration of new technologies and the ability to deliver specific solutions to individual geographic locations.

New Hurdles for Novel Fungicides

Along with the need to integrate fungicide technologies into IPM programs for disease management, there will always be a need for the discovery and development of new fungicides. Although the process

357

of discovering new fungicides is hardly new to the agrichemical industry, the rules have changed enough to warrant some discussion. First, the discovery process has changed from a traditional focus on in vivo screening assays (5,36) to high throughput *in vitro* screening assays similar to the automated screening methods that are common in the pharmaceutical industry (25). Second, new developments in automated combinatorial chemistry have provided for an exponentially larger number of candidate compounds to be evaluated as potential leads in a high throughput screen (4). Finally, new biochemical and molecular approaches that enable the identification of important novel target sites are now available (14,22,23). Therefore, instead of identifying potential fungicide leads based solely on efficacy, candidates can be selected based on their mode of action against target fungi. In addition to changing the overall discovery process, the downstream process of developing candidate fungicides has also changed. With the requirement for safety and reduced environmental impact, the toxicological and environmental profiles for candidate fungicides are evaluated much earlier in the development process than in the past (18). In general, the discovery process has changed to deal with the demand to find effective, safe, and environmentally friendly fungicides. The new hurdles for success have necessitated changes in the discovery process to ensure the continued ability to find needles in an ever-enlarging haystack.

Natural Compounds

An obvious benefit of fungicides that come from nature is that, by definition, they come with a "green image." There are several good examples of naturally occurring compounds that have been developed into fungicides for agricultural markets. These include fungicides that either came directly from fungi, bacteria, or higher plants such as blasticidin S, kasugamycin, or validamycin A (44), or fungicides that were derived from natural compounds, such as the phenylpyrroles (3) and the strobilurins (8).

Strobilurin A **Oudemansin A**

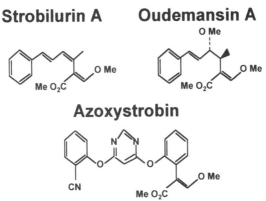

Figure 1. Photograph of *Oudemansiella mucida,* a basidiomycete that produces the natural strobilurin fungicide Oudemansin A (Top), and chemical structures of Strobilurin A, Oudemansin A, and Azoxystrobin (bottom).

STROBILURINS - IDEAL FUNGICIDES FOR IPM

Given that the hurdles that must be overcome in the development and registration of novel fungicides are greater today than in the past, we devote some discussion to a new class of fungicides that have a perfect fit in IPM systems. The discovery of the natural fungicidal activity of Strobilurin A and Oudemansin A, which came from the wood-rotting fungi *Strobilurus tenacellus* (2) and *Oudemansiella mucida* (28), respectively, led to the development of a new class of

359

agricultural fungicides, the 'strobilurins'. Zeneca developed azoxystrobin (15), a broad-spectrum fungicide containing a methyl β-methoxyacrylate toxophore similar to the natural strobilurins, Strobilurin A and Oudemansin A (Fig. 1). BASF and Novartis developed kresoxim-methyl (1) and trifloxystrobin (26), respectively, which are classified as oximinoacetates.

Azoxystrobin is ideal for use in IPM programs for the following reasons. First, azoxystrobin has a broad spectrum of activity against several important plant pathogenic fungi. Second, the use rate of active ingredient required to control these pathogens is substantially lower than that for most other fungicides. Therefore, lower chemical inputs are added to the environment. Third, azoxystrobin has an excellent environmental profile (Fig. 2). It degrades within the soil both photolytically and microbially, and therefore does not persist in the environment. Because azoxystrobin has low mobility in soil and is non-volatile, the potential for groundwater contamination is low. Finally, azoxystrobin has low toxicity to mammals, birds, honeybees, earthworms, and a variety of beneficial arthropods, and it presents only a low risk to fish and aquatic invertebrates. By definition, azoxystrobin is ideal for use in IPM programs.

Integration of New Fungicide Technologies in IPM

Throughout the history of plant pathology, there have been many attempts to modify the way that fungicides are applied in an effort to reduce the overall chemical inputs into the environment. We discuss several key fungicide technologies that relate to reducing fungicide inputs. These include technologies that have been around for a long time, such as biological control and disease forecasting systems based on weather data, as well as more novel technologies, such as remote sensing for disease forecasting, and integration of plant disease resistance into programs aimed at reducing fungicide inputs.

BIOLOGICAL CONTROL

Although the level of control provided by biological control programs seldom matches that provided by a good fungicide, biological control is an effective means of reducing the damage caused by some plant

Toxicity & Ecotoxicity Profile of Azoxystrobin

- Not a likely carcinogen
- Does not leach
- Rat LD_{50} > 5,000 mg/kg
- Bobwhite Quail LC_{50} > 2,000 mg/kg
- Earthworm LC_{50} 278 mg/kg
- Honey bee LD_{50} > 200 mg/bee
- Low aquatic risk
- Safety to all 12 IOBC spp tested in lab
- Safety to predatory mites in vines

Figure 2. Azoxystrobin has an excellent ecological profile that is ideal for use in IPM programs. Azoxystrobin has low toxicity to birds, bees, beneficial insects, earthworms, and soil micro-organisms. The active ingredient degrades photolytically and microbially, and the end product of this process is CO_2.

pathogens. Furthermore, use of biological control along with fungicides can reduce the overall level of fungicide inputs into the environment and, at the same time, provide an effective tool for fungicide resistance management (39). Some of the most promising biological agents that have been registered for use throughout the world are listed in Table 1.

The mechanisms by which biological control organisms control plant diseases include induced plant resistance, reduced virulence of aggressive pathogen strains via transmission of double-stranded RNA from a hypovirulent strain, competition between microorganisms for space and/or nutrients, production of antibiotics, or mycoparasitism. With the exception of mycoparasitism, in which the fungi that provide

Table 1. Examples of registered biological control agents, the diseases they control, and countries where they have been registered.

Biological control agent	Disease controlled, or effect on plant	Country of registration
Agrobacterium radiobacter	Crown gall	USA, Australia, NZ
Ampelomyces quisqualis	Powdery mildew	USA
Bacillus subtilis	Growth enhancement	USA
Pseudomonas fluorescens	Bacterial blotch	Australia
Pseudomonas fluorescens	Seedling diseases	USA
Peniophora gigantea	*Fommes annosus*	UK
Pythium oligandrum	*Pythium* spp.	USSR
Trichoderma viride	Timber pathogens	Europe
Trichoderma spp.	Root diseases	USA
Trichoderma harizianum	Root diseases	USA
Gliocladium virens	Seedling diseases	USA
Trichoderma harizianum	Wood decay	USA

Information from Elad and Chet (9).

biological control exist in intimate association with the fungi they control, most biological control agents must be able to competitively establish on the plant surface prior to, or coincidentally with the fungi they control. An inappropriately timed application of a fungicide could have deleterious effects on biological control organisms. Again, this illustrates the importance of using fungicides that have a minimal impact on beneficial organisms in IPM programs.

The development and use of biological control will likely be enhanced by the availability of new molecular approaches. For example, Staples et al. (38) transformed the entomophagous fungus *Metarhyzium anisopliae* to include a plasmid containing a gene for benomyl resistance. Genetic manipulations of biological control agents can possibly be used to enhance both the activity and spectrum of biological control agents, as well as to enhance the potential for successful integration of fungicides and biological control programs into IPM. Instead of focusing solely on efficacy, scientists can now view potential biological control agents as a gene pool for genetic manipulations. Plant pathologists can take an approach similar to that taken by entomologists who have transformed more than 50 plant crops to express *Bacillus thuringiensis* endotoxin, and by weed scientists who have transformed a multitude of crops to express numerous herbicide resistance genes (33).

DISEASE FORECASTING SYSTEMS THAT ARE BASED ON WEATHER DATA

Disease forecasting systems have been around since plant pathologists first practiced the principles of epidemiology. Some of the earliest disease forecasting systems were founded entirely on indirect assessments of the initial inoculum levels that were likely to be present, based on average winter, monthly temperatures (41). Other forecasting systems have been based on direct assessment of initial pathogen populations with little or no consideration of environmental conditions. Still other forecasting systems start with the assumption that initial inoculum will always be abundant, and then predict when infection periods will occur based on environmental conditions that would favor infection and disease development (27). Most of the forecasting systems that are popular today forecast initial fungicide applications based on the predicted occurrence of inoculum and then predict when subsequent applications should be made based on conditions favorable for secondary infections. BLITECAST is an excellent example (20). It is a forecasting program that was initiated in 1980 to determine fungicide application timings for the control of potato late blight, caused by *Phytophthora infestans*. The software for the BLITECAST program was modified a few years later in a new software package know as Potato Disease Management (PDM), which included forecasting for the control of early blight (*Alternaria solani*) as well as late blight (42). This program quickly became

363

popular in Wisconsin, where it was launched. In 1990, approximately 44% of the potato crop in that state was managed for disease control using this model (43). Growers who carefully collected environmental data, entered it into the program, and followed the recommendations for fungicide treatment reported saving up to two fungicide applications per season. Stevenson (43) reported cost savings exceeding $350,000 for Wisconsin, Minnesota, and Illinois potato growers who used the PDM program in 1987. In 1989, the Campbell Soup Company adopted the TOM-CAST disease forecasting system to predict fungicide application timings for tomato and carrot disease control (34). This forecasting system utilizes temperature and leaf wetness data to calculate disease severity values for various geographical zones. In this model, fungicide application timings are based on accumulation of disease severity values within the various geographical zones. During the 1992-1993 crop season, growers who utilized the PDM/TOM-CAST system reduced the number of fungicide applications by 54.4%, which provided an estimated cost savings of $168/ha over the conventional 10-day spray program (7). Another good example of an effective disease forecasting system is the UCIPM Pest Cast program that provides California grape growers with daily powdery mildew risk indices. As a component of this system, Terra Spase provides GIS maps that indicate the disease risk based on the model (Figure 3). Using this forecasting system, 2 to 3 fewer fungicide applications are typically applied, compared with a standard spray program (for detailed discussion of these results, see APSnet Web site **http://www.scisoc.org/feature/pmildew/Top. html**).

Although an annual saving on sprays may not be achieved during a year when risk of disease is high, a reduction in fungicide applications and a saving to the grower can be achieved over a number of years, regardless of the forecasting system used. The success of such IPM forecasting systems will pave the way for other specific model systems to follow.

REMOTE SENSING FOR DISEASE DETECTION

In the field of plant pathology, remote sensing and digital image analysis are rapidly gaining acceptance as tools to measure and interpret physiological changes in a plant canopy and, thus, to identify

364

disease development. Basically, remote sensing provides a means to detect and assess many of the physiological changes in plants that are associated with plant disease (examples include: changes in photosynthesis, evapotranspiration, and canopy temperature). Although there are many different pathogens and/or abiotic factors that may induce such physiological changes in plants and complicate a realistic interpretation of the actual disease pressure, remote sensing offers a relatively quick and non-visually biased methodology for capturing data. Until the day when diseases can be differentiated by specific remote signals, ground-truthing (confirmation by visual ground assessments) will be required to validate the biotic or abiotic factors responsible for creating the GIS map images that are generated via remote sensing. Although there are still many questions to answer and improvements to be made in remote sensing for disease detection, this area of technology has improved over the past several years (29). To date, remote sensing has been used to assess severity of Rhizoctonia blight on turf (16,35), dollar spot of turf (31), early blight of tomato (21), late leaf spot of peanut (30), and powdery mildew on barley (24). We are now beginning to see signs that remote sensing is being used for practical purposes in both disease forecasting programs (Fig. 3) and in basic field research programs (Fig. 4). Dr. W.R. Stevenson, plant pathologist at the University of Wisconsin-Madison has used remote sensing and GIS maps in his experimental-use, fungicide program (EUP) to evaluate fungicides for disease control in potatoes. In that progrtam, remote sensing technology provided a fairly rapid and unbiased assessment of vegetative vigor that had a strong correlation with visual assessment of disease severity obtained by standard ground-truthing.

MODELS FOR INTEGRATION OF PLANT RESISTANCE AND CHEMICALS

The most obvious means of reducing fungicide use is to grow varieties that have a high level of pathogen resistance. Although theoretically sound, many cultivated plants do not possess the appropriate genes, and the process of transferring resistance genes from related plant species by traditional breeding methods is time consuming and often difficult, if not impossible. Plant breeders who

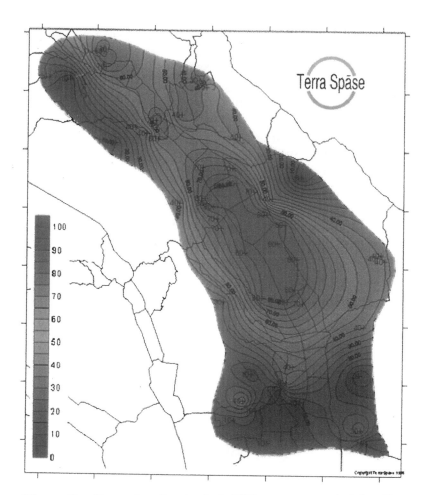

Figure 3. Example of a typical GIS map generated by Terra Spase illustrating the disease risk of grapevine powdery mildew in Napa County, California. Disease indices are calculated using the Pest Cast program, and these data are translated into color GIS maps that are easier to interpret than this black and white representative sample above. (Reprinted from Web page http://www.scisoc.org/feature/pmildew/Top.html by permission from Terra Spase).

366

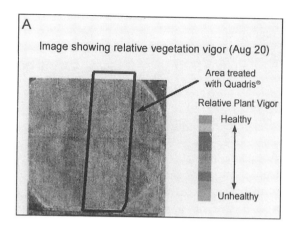

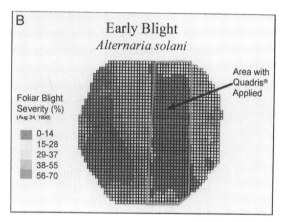

Figure 4. A. Remote sensed image of W.R. Stevenson's fungicide, experimental-use trial taken from a helicopter (image provided by T. Christensen, Zeneca Ag. Products). This image illustrates relative plant vigor. The area within the box was the section of the field that was treated with Quadris®. B. Map of percent early blight disease severity based on visual assessments on 0.404-ha grids with interpolation between points (data collected by G. Morgan and processed by T. Christensen).

successfully develop resistant varieties face the constant challenge of an ever-changing pathogen population.

The value of polygenic disease resistance on reducing fungicide usage can be illustrated by some of the pioneer research on late blight of potato (10,11,12). Fry (10,12) demonstrated that varietal differences among potatoes for late blight resistance could be integrated into programs that would reduce fungicide inputs, either reducing the rate of fungicide required, or extending the application interval. Although the use of lower-than-recommended rates in an application program is not desirable because it tends to enhance the development of fungicide resistant pathogens, Fry's research provides a monetary value for the contribution of individual components of disease control. Gans et. al (13) demonstrated the potential for extending the application interval of a fungicide (e.g., fluazinam) up to 3 weeks on a potato variety having a high level of disease resistance, without sacrificing excellent control of late blight under high disease pressure.

The potential to integrate plant disease resistance into IPM programs will be strengthened by new molecular tools that allow for the introduction of genes into germplasm lines that would be difficult or impossible to accomplish using traditional breeding methods. The future view of plant disease control is likely to be shaped by combinations of genetically modified crops, fungicides, and fungicide technologies, including those discussed in this chapter, as well as those that will be developed in the future.

Acknowledgments

We thank T. Sutton for the invitation to participate in this symposium; W.D. Gubler, M.R. Rademacher, S.J. Vasquez, C.S. Thomas and Terra Spase for allowing us the use of an image from their Web page for Figure 3; W.R. Stevenson for allowing us the use of a remote-sensed GIS map and data from his EUP field trial for Figure 4; H. Cox, K. Langan, and G. Olaya for quickly compiling a literature review; E. Franklin for assistance with graphic arts; and J. Frank, B. Lackey, N. Pain, and J. Godwin for helpful comments on this manuscript. Finally, we thank our many colleagues at Zeneca who have participated in the research and development of azoxystrobin.

Literature Cited

1. Ammerman, E., Lorenz, G., Schelberger, K., Wenderoth, B., Sauter, H., Rentzea, C. 1992. BAS 490 F- a broad-spectrum fungicide with a new mode of action. Brighton Crop Prot. Conf. Pests Dis. 1:403-410.

2. Anke, T., Oberwinkler, F., Steglich, W., Schramm, G. 1977. The strobilurins - new antifungal antibiotics from the Basidiomycete *Strobilurus tenacellus*. J. Antibiot. 30:806-810.

3. Arima, K., Imanaka, H., Kousaka, M., Fukuda, A., Tamura, G. 1964. Pyrrolnitrin, a new antibiotic substance, produced by Pseudomonas. Agric. Biol. Chem. 28:575-576.

4. Balkenhohl, F. von dem Bussche-Hunnefeld, C., Lansky, A., Zechel, C. 1996. Combinatorial synthesis of small organic molecules. Angew. Chem. Int. Ed. Engl. 35:2288-2337.

5. Bartley, M. R., and Youle, D. 1996. The use of *in vitro* and *in vivo* bioassays to screen for novel agrochemicals: approaches to experimental design. Pages 15-35 in: Bioactive Compound Design: Possibilities for Industrial Use. M. G. Ford, R. Greenwood, G. T. Brooks, and R. Franke, eds. Oxford: BIOS.

6. Benbrook, C. M., Groth, E., Halloran J. M., Hansen, M. K., Marquardt, S. 1996. Pest Management at the Crossroads. Yonkers: Consumers Union.

7. Bolkan, H. A., and Reinert, W. R. 1994. Developing and implementing IPM strategies to assist farmers: an industry approach. Plant Dis. 78:545-550.

8. Clough, J. M. 1993. The Strobilurins, Oudemansins, and Myxothiazols, Fungicidal Derivatives of β-Methoxyacryic Acid. Natural Product Reports 10:565-574.

9. Elad, Y. and Chet, I. 1995. Practical approaches for biocontrol implementation. Pages 323-328 in: Novel Approaches to Integrated Pest Management. R. Reuveni, ed. CRC Press, Boca Raton, FL.

10. Fry, W. E. 1975. Integrated effects of polygenic resistance and a protective fungicide on development of potato late blight. Phytopathology 65:908-911.
11. Fry, W. E. 1977. Integrated control of potato late blight - Effects of polygenic resistance and techniques of timing fungicide applications. Phytopathology 67:415-420.
12. Fry, W. E. 1978. Quantification of general resistance of potato cultivars and fungicide effects for integrated control of potato late blight. Phytopathology 68:1650-1655.
13. Gans, P., Carson, W. D., Pearson, N., and Owen, L. L. 1995. Exploiting cultivar resistance to control potato blight (*Phytophthora infestans*). Pages 345-356 in: *Phytophthora infestans*. L. J. Dowley, E. Bannon, L. R. Cooke, T. Keane, and E. O'Sullivan, eds. Boole Press Ltd, Dublin, Ireland.
14. Garber, R.C. 1991. Molecular approaches to the analysis of pathogenicity genes from fungi causing plant diseases. Pages 483-502 in: The Fungal Spore and Disease Initiation in Plants and Animals, G. T. Cole, and H. C. Hoch, eds. New York: Plenum.
15. Godwin, J. R., Anthony, V. M., Clough, J. M., Godfrey, C. R. A. 1992. ICIA5504: a novel, broad spectrum, systemic β-methoxyacrylate fungicide. Proc. Brighton Crop Prot. Conf. Pests Dis. 1:435-442.
16. Green D. E. II, Burpee, L. L., and Stevenson, K. L. 1998. Canopy reflectance as a measure of disease in tall fescue. Crop Sci. 38:1603-1613.
17. Jacobsen, B. J. 1997. Role of plant pathology in integrated pest management. Annu. Rev. Phytopathol. 35:373-391.
18. James, J. R., Tweedy, B. G., and Newby, L. C. 1993. Efforts by industry to improve the environmental safety of pesticides. Annu. Rev. Phytopathol. 31:423-439.
19. Knight, S. C., Anthony, V. M., Brady, A. M., Greenland, A. J., Heaney, S. P., Murray, D. C., Powell, K. A., Schulz, M. A., Spinks, C. A., Worthington, P. A., and Youle, D. 1997. Rationale and perspectives on the development of fungicides. Annu. Rev. Phytopathol. 35:349-372.
20. Krause, R. A., Massie, L. B., and Hyre, R. A. 1975. BLITECAST, a computerized forecast of potato late blight. Plant Dis. Rep. 59:95-98.

21. Lathrop, L. D., and Pennypacker, S. 1980. Spectral classification of tomato disease severity levels. Photogram. Eng. Remote Sens. 46:1433-1438.
22. Leong, S. A. 1988. Recombinant DNA research in phytopathogenic fungi. Advances Plant Pathol. 6:1-26.
23. Leong, S. A., and Holden, D. W. 1989. Molecular genetic approaches to the study of fungal pathogenesis. Annu. Rev. Phytopathol. 27:463-481.
24. Lorenzen, B., and Jensen, A. 1989. Changes in leaf spectral properties induced in barley by cereal powdery mildew. Remote Sens. Environ. 27:201-209.
25. Major, J. S. 1995. Current screening practices in the pharmaceutical industry. Proc. Brighton Crop Prot. Conf. - Weeds, 1995. 1:89-96.
26. Margot, P., Huggenberrger, F, and Amrein, J. 1998. CGA 279202: A new broad-spectrum strobilurin fungicide. Proc. Brighton Crop Prot. Conf. Pests Dis. 5A-8:375-382.
27. Mills, W. D. 1944. Efficient use of sulfur dusts and sprays during rain to control apple scab. Cornell Ext. Bull. 630.
28. Musilek, V., Cerna, J., Sasek, V., Semerdzieva, M., Vondracek, M. 1969. Antifungal antibiotic of the Basidiomycete *Oudemansiella mucida*. Folia Microbiol. (Prague) 14:377-387.
29. Nilsson, H. E. 1995. Remote sensing and image analysis in plant pathology. Annu. Rev. Phytopathol. 15:489-527.
30. Nutter, F. W., Jr. 1989. Detection and measurement of plant disease gradients in peanut with a multispectral radiometer. Phytopathology 79:958-963.
31. Nutter, F. W., Jr., Gleason, M. L., Jenco, J. H, and Christians, N. C. 1993. Assessing the accuracy, inter-rater repeatability and inter-rater reliability of disease assessment systems. Phytopathology 83:806-812.
32. Overton, J., ed. 1996. Ecologically Based Pest Management-New Solutions for a New Century. Washington, DC: Natl. Acad. Press.
33. Paoletti, M., and Pimental, D. 1996. Genetic engineering in agriculture and the environment. Bioscience 46: 665-673.
34. Pitblado, R. E. 1992. The development and implementation of TOM-CAST: A weather-timed fungicide spray program for field tomatoes. Ministry of Agriculture and Food, Ontario, Canada.

35. Raikes, C., and Burpee, L. L. 1998. Use of multispectral radiometry for assessment of Rhizoctonia blight in creeping bentgrass. Phytopathology 88:446-449.
36. Shephard, M. C. 1987. Screening for fungicides. Annu. Rev. Phytopathol. 25:189-206.
37. Smith, R. F., van den Bosch, R. 1967. Integrated control. Pages 295-340 in: Pest Control: Biological, Physical, and Selected Chemical Methods. W. Kilgore and R. Doutt, eds. New York. Academic Press.
38. Staples, R. C., Leger, R. J, Bhairi, S., Roberts, D. W. 1988. Biotechnology, biological pesticides, and novel plant-pest resistance for insect pest management. Pages 44-48 in: Strategies for genetic engineering of fungal entomo-pathogens. R. Grandos, ed. Ithaca (NY): Boyce Thomspson Institute, Cornell University.
39. Staub, T. 1991. Fungicide resistance: practical experience with antiresistance strategies and the role of integrated use. Annu. Rev. Phytopathol. 29:421-442.
40. Stern, V. M. , Smith, R. F., van den Bosch, R., and Hagen, K. S. 1959. The integrated control concept. Hilgardia 29:81-101.
41. Stevens, N. E. 1934. Stewart's disease in relation to winter temperatures. Plant Dis. Rep. 18:141-149.
42. Stevenson, W. R. 1983. An integrated program for managing potato late blight. Plant Dis. 67:1047-1048.
43. Stevenson, W. R. 1993. IPM for potatoes: A multifaceted approach to disease management and information delivery. Plant Dis. 77:309-311.
44. Worthington, P. A. 1988. Antibiotics with antifungal and anti-bacterial activity against plant diseases. Nat. Prod. Rep. 5:47-66.

Assessing Integrated Weed Management in Terms of Risk Management and Biological Time Constraints

Jeffrey L. Gunsolus, Thomas R. Hoverstad,
Bruce D. Potter, and Gregg A. Johnson
Department of Agronomy and Plant Genetics, University of
Minnesota, St. Paul, MN 55108 USA

Weed management decisions in commodity-based, large-area crops, such as corn (*Zea mays* L.) and soybean [*Glycine max* (L.) Merr.], are directly linked to time and labor management issues. The decline in the number of United States farm workers and the increase in farmland area per farm worker support this relationship.

The rapid and widespread adoption of genetically engineered row crops that are resistant to non-residual, broad-spectrum postemergence herbicides (e.g., Roundup Ready® and Liberty Link® crops) will change the way many row-crop producers manage weed populations. Herbicides such as glyphosate and glufosinate have little or no residual herbicidal activity in the soil and must be applied postemergence to the crop and weeds to be effective. A direct result of this weed management approach is that producers of large-area crops such as corn and soybean will need to become more aware of the periodicity of weed emergence, rate of weed growth and development, and time of weed removal to effectively use these herbicides.

Biological Time Constraints

Three groups of weeds must be considered in the context of a single application of a non-residual, broad-spectrum postemergence herbicide: those that emerge before the treatment and are controlled by it, those that emerge before the treatment but are not controlled by it, and those that emerge after the treatment. The relative importance of these groups of weeds to crop yield depends upon the time of weed emergence, the rate of weed growth, and the time of weed removal. These factors are referred to as biological time constraints. Understanding these biological time constraints can help a producer evaluate whether an integrated weed management system can be integrated into their crop production system.

PERIODICITY OF WEED EMERGENCE

The diversity of weed species in the soil-seed bank and the ensuing diversity of weed emergence patterns influence the number of applications of a non-residual, broad-spectrum postemergence herbicide that will be necessary for a given field. Research on annual weed emergence patterns in the midwestern United States (2) indicates that peak annual weed-emergence flushes may begin as early as mid-April (e.g., kochia [*Kochia scoparia* (L.) Schrad.] and wild mustard [*Sinapis arvensis* L.]) and as late as early July (e.g., common waterhemp [*Amaranthus rudis* Sauer]). It is not uncommon for early- and late-emerging weed species to exist in the same field. Herbicides applied early in the growing season may eliminate some early-emerging annual weed species but may leave late-emerging weeds that can still reduce crop yield or produce seeds that return to the seed bank. Over time, returning weed seed to the seed bank has the potential to shift the weed population to late-emerging annuals (e.g., common waterhemp), unless the weed management strategies are adequately diversified. Alternatively, delaying herbicide application to target late-emerging weeds may result in crop yield loss due to an extended period of competition with early-emerging weeds.

RATE OF WEED GROWTH AND DEVELOPMENT

The growth of weeds, as is the case for most biological organisms, is logarithmic. Many herbicide efficacy and crop-injury problems can be attributed to herbicide application to weeds and crops that are too large. Most postemergence herbicide labels list the ranges of weed and crop heights over which herbicide efficacy will be maximized and crop injury will be minimized. With the advent of herbicide-resistant crops, the concern over herbicide-induced injury has been reduced. However, the need for timely herbicide application to maximize efficacy and to minimize weed/crop interference is important. For example, in the Roundup Ready corn system, the glyphosate label states that the herbicide should first be applied when weeds are 10 cm in height. In the Roundup Ready soybean system, the glyphosate label states that weeds should be controlled in the 10- to 20-cm height range. Experience has taught us that translating weed height into units of time (i.e., days after crop planting) helps growers integrate weed and crop biology restrictions into their time and labor framework. Under most Minnesota spring and early-summer cropping conditions, it takes approximately four weeks for foxtail (*Setaria spp.*) to reach 10 cm in height and approximately six weeks for the foxtail spp. to reach 20 cm in height (2). Therefore, based on the glyphosate label, a Minnesota crop producer has an approximately two-week interval to complete a glyphosate weed management program in soybean, and a glyphosate weed management program in corn should be initiated approximately one month after corn planting. Disruption of the time of application framework could result in poor weed control or yield loss due to weed/crop interference.

CRITICAL PERIOD OF WEED CONTROL

The relative degree of competition between weeds and the crop depends upon the time of weed emergence and the time of weed removal, and is referred to as the critical period of weed control. Zimdahl (8) defines two key components of this critical period. The first is the length of time that weeds can remain in the crop before they interfere with crop growth. The second is the period of time over which weed control efforts must be maintained before the crop can effectively compete with weeds and prevent crop yield loss. Swanton and Weise (7) indicate that the critical period of weed control can be used in an integrated weed management program to optimize the timeliness of postemergence herbicide applications and to determine the timing of mechanical weed control practices.

In northern latitudes of the United States, the period of time that early-emerging weeds can remain in a crop before interfering with grain yield is approximately two to five weeks after corn emergence (4) and four to six weeks after soybean emergence (6). The time over which weed control efforts must be maintained before a crop can effectively compete with late-emerging weeds and prevent crop yield loss is approximately four to five weeks after crop emergence for corn and soybean. Therefore, the critical period of weed interference would indicate that postemergence weed control programs in corn are associated with more risk of yield loss due to untimely control of early-emerging weeds than are postemergence

375

weed control programs in soybean.

The critical-period concept is dependent upon the crop, geographic location, environmental conditions, and the weed species complex. Its usefulness to the producer is in the development of the timing of weed management strategies and in the determination of which crops are the most sensitive to key weed factors such as density and emergence patterns.

Density dependency. B. D. Potter et al. (unpublished) demonstrated the impact that weed density and cropping environment can have on the time of weed removal in a study using glyphosate in Roundup Ready soybean at three field locations in Minnesota (Table 1). Glyphosate provided excellent control of emerged weeds at all timings. All herbicide-treated fields yielded better than the weedy check. Soybean yield was not related to time of weed removal at the Morris, MN, site, where weed pressure was low and growing conditions were excellent. Yield loss due to weeds germinating after glyphosate application was observed at the Waseca, MN, and Lamberton, MN, sites with the one-pass postemergence application to weeds 5- to 10-cm in height. High weed density and low moisture combined to reduce yield from both early- (20 - 30 cm) and late-season (5 - 10 cm) weed competition at the Lamberton site. In the 25-cm row spacing used at Lamberton, yields were reduced in all treatments that allowed weeds to persist past 10 cm in height (data not shown). Glyphosate application to weeds 10 to 20 cm in height did not affect soybean yield with one exception: the 25-cm row spacing at Lamberton, where dry, early-season conditions combined with high weed density resulted in a row spacing by timing interaction. The results of this study illustrate the influence of weed density and environment (moisture stress) on risk associated with one-pass, postemergence, non-residual weed control, and illustrates the difficulty of designing a generic weed management strategy that will provide optimum economic benefit in all fields.

Time dependency. Berti et al. (1) developed a model that integrates weed/crop interference, periodicity of weed emergence, and economics in an effort to optimize when and how often to control weeds in order to maximize net profit margins. When they integrated weed/crop interference, weed biology, and economics, the net margins obtained for a specific weed and crop scenario were time-dependent. For example, in both corn and soybean, a one-pass, postemergence herbicide application was associated with a larger maximum net margin than was a sequential strategy involving a preemergence herbicide followed by a postemergence

376

herbicide. With the sequential strategy, however, the period of time over which the postemergence herbicide could be applied without a reduction in net margin was longer. Sequential herbicide strategies described by Berti et al. (1) extend the postemergence herbicide application period, because soil-applied herbicides reduce early-season weed densities, thereby reducing early-season weed/crop competition. Fields with low weed densities, like the Morris location in Table 1, are less likely to need timely postemergence weed control because of low weed populations.

These results also indicate that field size and variability of field working days (i.e., time and labor management issues) have a significant impact on the inherent risk associated with an integrated weed management program. Gunsolus and Buhler (3) have detailed the relationship of field size and field working days to timeliness of weed management operations.

Table 1. Soybean yield at three times of glyphosate postemergence (POST) application, averaged across 25- and 76-cm row spacings, 1998.

Treatment	Rate (kg/ha)	Lamberton	Morris	Waseca
		(Mg/ha)[a]		
POST (5-10 cm) glyphosate	0.84	3.46[b]	3.82	3.01[b]
POST (10-20 cm) glyphosate	0.84	3.71	4.09	3.52
POST (20-30 cm) glyphosate	0.84	3.42[b]	4.08	3.41
Check				
weed-free	---	3.86	4.00	3.49
weedy control	---	0.87[b]	1.09[b]	0.83[b]
LSD (0.10)		0.22	0.26	0.20

a/ Adjusted to 13% moisture.
b/ Yield significantly lower than the weed-free check.

Risk Management

Another method of evaluating integrated weed management systems is through a mean/variance economic analysis, in which the economic performance of weed management systems is evaluated and the economic returns from each system are compared to the standard error of the mean value of the economic return. This procedure provides a comparison of economic return and risk (i.e., the consistency of economic performance). Most corn and soybean producers are risk averse. That is, they are prepared to sacrifice some expected profit for more consistent crop yields and net returns over the growing seasons.

Olson and Eidman (5) have developed a model to explore the degree to which government policies could enhance the adoption of mechanical weed control systems over herbicide-based systems. They modeled a 162 ha farm in Minnesota that incorporates a risk coefficient. Their results indicated that with government incentives, risk-neutral producers would choose to produce crops using mechanical weed control practices. Risk-averse producers would choose to use herbicides even if the expected returns might be greater for the mechanical weed control system because of the increase in variability of returns when using the mechanical weed control system. Based on this research, the main value of herbicides to producers is their ability to reduce yield variability and, thus, income variability.

Since 1995, J. L. Gunsolus, T. R. Hoverstad, and G. A. Johnson (unpublished) have been developing a herbicide evaluation system that assesses the efficacy, cost, yield, and adjusted gross returns for various weed management systems currently available to Minnesota corn and soybean producers. The soybean weed management systems under evaluation (Tables 2 and 3) include soil-applied preplant-incorporated herbicides (PPI); preplant-incorporated herbicides followed by post-emergence herbicides (PPI / POST); preplant-incorporated herbicides followed by postemergence herbicides plus inter-row cultivation (PPI / POST / CULT); and postemergence herbicides (POST). The postemergence herbicide systems encompass both postemergence residual and non-residual herbicide options.

Tables 2 and 3 summarize the yield, weed management costs (Cost), adjusted gross returns (Income), and the "risk" associated with these returns for soybean weed management systems located in Lamberton and

378

Table 2. Risk efficiency for soybean weed management systems at Waseca, MN, 1996 - 1998.

Treatment	Rate (kg/ha)	Yield^x (Mg/ha)	Cost US$/ha	Income^{y,z} US$/ha	S.D. US$/ha
PPI					
Pendimethalin &					
Imazethapyr	1.05	2.76c	49.42	457.5b	114.9
PPI / POST					
Trifluralin /	0.84 /	3.50ab	83.49	559.3a	62.5
Imazethapyr	0.071				
Trifluralin /	0.84 /				
Bentazon &	1.03	2.48f	90.78	364.7c	85.0
Aciflurofen					
PPI / POST / CULT					
Trifluralin /	0.84 /				
Imazethapyr /	0.071 /	3.65a	95.85	574.5a	62.3
CULT	----				
Trifluralin /	0.84 /				
Bentazon &					
Aciflurofen /	1.03 /	3.10d	103.14	466.2ab	67.0
CULT	----				
POST					
Imazethapyr	0.071	3.30c	49.10	557.0 a	89.7
Bentazon&					
Aciflurofen &					
Sethoxydim +	1.35	3.35bc	81.84	533.2a	97.4
Chlorasulam					
Methyl	0.012				
Weedy control	----	0.71g	00.00	130.4d	46.7

x/ Yields with the same letter are not significantly different from one another at the p= 0.10 level.

y/ Incomes with the same letter are not significantly different from one another at the p= 0.10 level.

z/ Income = Yield (Mg/ha) x Price ($183.66/Mg) - Cost ($/ha)

Table 3. **Risk efficiency for soybean weed management systems at Lamberton, MN, 1996 - 1998.**

Treatment	Rate (kg/ha)	Yield[x] (Mg/ha)	Cost US$/ha	Income[y,z] US$/ha	S.D. US$/ha
PPI					
Pendimethalin &					
Imazethapyr	1.05	2.99c	49.42	499.7b	124.5
PPI / POST					
Trifluralin /	0.84 /	3.20ab	83.49	504.2ab	99.6
Imazethapyr	0.071				
Trifluralin /	0.84 /				
Bentazon &	1.03	2.80d	90.78	423.5d	133.9
Aciflurofen					
PPI / POST / CULT					
Trifluralin /	0.84 /				
Imazethapyr /	0.071 /	3.32a	95.85	513.9ab	71.4
CULT	----				
Trifluralin /	0.84 /				
Bentazon &					
Aciflurofen /	1.03 /	3.09bc	103.14	464.4c	105.5
CULT					

POST					
Imazethapyr	0.071	3.15b	49.10	529.4a	153.7
Bentazon&					
Aciflurofen &					
Sethoxydim +	1.35	2.82d	81.84	436.1cd	95.4
Chlorasulam					
Methyl	0.012				
Weedy control	----	0.93e	00.00	170.8c	146.5

x/ Yields with the same letter are not significantly different from one another at the p= 0.10 level.

y/ Incomes with the same letter are not significantly different from one another at the p= 0.10 level.

z/ Income = Yield (Mg/ha) x Price ($183.66/Mg) - Cost ($/ha)

Waseca, MN. Risk is expressed as the standard deviation of adjusted gross returns (S.D.). Lamberton is located in southwestern MN, and growing conditions tend to be warmer and dryer there than in Waseca, which is located in south-central MN. The predominant weed species at Lamberton, MN, are green foxtail [*Setaria viridis* (L.) Beauv.], yellow foxtail [*Setaria glauca* (L.) Beauv.], and common lambsquarters (*Chenopodium album* L.). The predominant weed species at Waseca, MN, are giant foxtail (*Setaria faberi* Herrm.), common lambsquarters (*Chenopodiuim album* L.) and common ragweed (*Ambrosia artemisiifolia* L.). Tables 2 and 3 contrast the use of the postemergence residual herbicide imazethapyr with the postemergence non-residual herbicides bentazon, aciflurofen, chloransulam-methyl, and sethoxydim.

At both locations, the residual PPI / POST, PPI / POST/ CULT, and POST treatments generated the greatest income. The only non-residual POST system that achieved a high income was at Waseca. At both locations, the more diversified residual PPI / POST / CULT and the PPI / POST weed management systems had a lower standard deviation of income over the 3-year study period than the residual one-pass PPI and POST systems, indicating a stable return, or, in other words, less risk. At Waseca, MN, the one-pass, postemergence residual and non-residual herbicide treatments (POST) provided the same income as the more diversified treatments, but their higher standard deviation would indicate more risk. This risk is likely associated with the inconsistent control of common ragweed by postemergence herbicide application (data not shown). At Waseca, MN, the one-pass, PPI treatment containing imazethapyr was associated with a significantly lower return than the other postemergence imazethapyr treatments, and was associated with the greatest risk. This greater risk is most likely due to the difficulty of consistently controlling large-seeded broadleaf weeds, such as common ragweed, with soil-applied herbicides (data not shown). At Lamberton, MN, the one-pass, soil-applied, postemergence residual herbicide treatment provided the same amount of revenue as the more diversified two- and three-pass residual weed management systems; however, their higher standard deviations would indicate an inconsistent income over the 3-year period. At Lamberton, MN, the one-pass, postemergence residual herbicide treatment appeared to be the riskiest. This risk may be associated with the effect that a drier environment can have on the consistency of postemergence herbicide efficacy (data not shown).

Berti et al. (1) found that, within a growing season, single postemergence herbicide applications in corn or soybean were associated with maximum economic returns, and sequential soil-applied, postemergence treatments were associated with lower-than-maximum economic returns. However, the period of time within the growing season that the sequential soil-applied, postemergence herbicides could be applied without any reduction in economic returns was longer than the optimum time period associated with single postemergence herbicide applications. The Lamberton and Waseca studies indicate that, over a 3-year period, sequential treatments can provide an income equal to or exceeding that of one-pass weed management strategies, and appear to provide a more stable ("less risky") income to the crop producer. Neither of these modeling approaches considered field size and field working days, but both models indicated that the early-season weed control associated with soil-applied herbicides gives the crop producer more time to apply the postemergence herbicides. Both studies highlight the need to understand and predict the impact of biological time constraints on risk management in attempts to improve decision-making capabilities and diversify integrated weed management strategies.

Conclusion

An understanding of biological time constraints and risk management analysis can help producers better carry out integrated weed management programs. It is the weed scientist's role to address biological time constraints and risk from the crop producer's perspective of time and labor constraints and to demonstrate how weed biology affects the economics of crop production.

Literature Cited

1. Berti, A., Dunan, C.,. Sattin, M., Zanin, G., and Westra, P. 1996. A new approach to determine when to control weeds. Weed Science 44:496-503.
2. Forcella, F. 1998. Real-time assessment of seed dormancy and seedling growth for weed management. Seed Science Research 8:201-209.
3. Gunsolus, J. L., and Buhler, D. D. 1999. A risk management perspective on integrated weed management. Journal of Crop

Production 2:167-187.

4. Hall, M. R., Swanton, C. J., and Anderson, G. W. 1992. The critical period of weed control in grain corn (*Zea mays*). Weed Science 40:441-447.

5. Olson, K. D., and Eidman, V. R. 1992. A farmer's choice of weed control method and the impacts of policy and risk. Review of Agricultural Economics 14:125-137.

6. Stoller, E. W., Harrison, S. K., Wax, L. M., Regnier, E. E.,. and Nafziger, E. D. 1987. Weed interference in soybeans (*Glycine max*). Reviews of Weed Science 3:155-181.

7. Swanton, C. J., and Weise, S. F. 1991. Integrated weed management: the rationale and approach. Weed Technology 5:657-663.

8. Zimdahl, R. L. 1988. The concept and application of the critical weed-free period. Pages 145-155 in: Weed Management in Agroecosystems: Ecological Approaches. M. A. Altieri and M. Liebman, eds. CRC Press, Boca Raton, FL..

Integrating New Insecticide Technologies in IPM

J. R. Bradley, Jr.
Department of Entomology
North Carolina State University, Raleigh, NC 27695 USA

Insecticides became the primary tools used against insect pests soon after World War II. During the immediate post-war era, insecticides appeared to most entomologists and others as the ultimate solution to insect problems. Insecticides were embraced, often to the detriment of other control methods (e.g., cultural) like no other technology before then (24). We reflect on that period in our history and typically criticize entomologists who preceded us for their lack of foresight into the pitfalls accompanying this unilateral approach to insect control. But how do we know that we would have had sufficient vision to predict the problems that would soon be so obvious? That was an era of rapid technological development, not that different from the present. How many of us have a crystal ball that will allow us to look into the future and predict the downside of the technologies we utilize today, much less place them in appropriate perspective relative to the benefits provided by these technologies?

The period from the late 1940s through the mid-1960s was considered by some to have been the "Dark Age" of insect control (20). Even though much of the insect control philosophy of that era has been widely condemned, it forced a paradigm of rapid improvement. Adverse human health and environmental effects, as well as the rapid development of insect strains that were resistant to insecticides, were obvious constraints to the continued use of insecticides as the sole means of achieving pest control. A new dawn of awareness, not only within the agricultural community, but also in the public in general, was ushered in by Rachel Carson's book, *Silent Spring* (2).

The concept of integrated pest management (IPM) was spawned more than 30 years ago in response to a need for changes in the way that mankind was combating insect pests. It had become very clear that the unilateral approach to insect control had failed, and a more ecologically based system of insect management had to be adopted. However, let's be fair in our assessment of what has been accomplished during the pest management era. Change has come about over the last three decades, remarkable change for the betterment of insect management and the environment. There has been a return to the reliance on cultural controls as the foundation for management of some of the most important insect pests, including pink bollworm, corn rootworms, Heliothines, and many others. Concepts such as the use of scouting techniques to monitor insect pests and their damage and the adoption of economic injury levels are now widely applied. In some crops such as cotton, the majority of hectares planted in most regions is scouted, and insecticide applications are made on the basis of perceived need. The private consultant is now a commonplace fixture on farms, whereas 30 years ago they were rare in most regions. Overall, the number of private consultants who make a living providing pest management advice to farmers has increased more than 10-fold since the late 1960s (1). There is widespread awareness among farmers that biological controls should be protected when possible and utilized in the management of insect pests. Insecticide selection and timing of applications are often conditioned by consideration for biological controls. Contemporary farmers are concerned about protecting their "beneficials," as they have come to recognize that these natural controls have an economic value. Admittedly, we have been unable to maximize the use of beneficial insects as population regulators of pest species. Our inability to achieve this objective has been due in part to our lack of knowledge, but also because there have been very few selective insecticides available to pest managers.

Status of new insecticide technology

Until very recently, practically all insecticides that had been developed and commercialized were broad-spectrum toxicants that did not discriminate between target pests and other arthropods in the system. Huffaker and Smith (12) clearly stated the problem:

385

"What we need is selective use of pesticides. All pesticides have some selectivity, but the degree of selectivity varies greatly. We have sought mainly materials of high toxicity to invertebrates and low toxicity to mammals. But many of our materials are simply biocides. We need differential toxicity within the insects, not necessarily species-specific specificity, but specificity for groups such as thrips, aphids, lepidopterous larvae, and so on."

Lack of selective insecticides has been an obstacle to full implementation of IPM principles, but this problem may be solved with the recent development of insecticides that are truly selective.

Also, the majority of insecticides available until recently were in four chemical classes: chlorinated hydrocarbons, organophosphates, carbamates and pyrethroids. With the narrow choice of modes of toxic action, there was a very high probability that resistant strains of insect pests would develop rapidly, and they did. It is now estimated that more than 500 arthropod species have developed resistance to one or more insecticides (7). Many of the older insecticides have either been banned or have been voluntarily withdrawn from the market. Use of those that remain will gradually decline as a function of insect adaptation and preferences for the newer, more environmentally friendly insecticides.

There has been a renewed effort to develop insecticides during the past decade because of environmental problems associated with older compounds, insecticide resistance, and the need for more selective insecticides. The U.S. Environmental Protection Agency enhanced development of more environmentally compatible insecticides by announcing that such compounds would receive preference in the registration process and that all compounds must meet stricter environmental criteria in order to be labeled. The agrichemical industry has responded to the challenge by intensifying the effort to discover and bring to the market insecticides that possess new or under-exploited modes of action. This group of new insecticides has greater diversity in chemistry and mode of action than ever before (Table 1). With few exceptions, these new insecticides have vastly improved environmental profiles and possess a level of species selectivity that make them ideal candidates for inclusion in IPM programs. The following is a brief overview of the most prominent of the new insecticides that have recently received full EPA labels or have labels pending.

Emamectin benzoate (Denim®) is a new, semi-synthetic avermectin insecticide derived from the fermentation product avermectin B1 (abamectin) (15). Emamectin benzoate acts through increasing the potential of neurotransmitters like glutamate and gamma-amino butyric acid (GABA) to stimulate an influx of chloride ions into nerve cells. This causes loss of cell function and disruption of nerve impulses. This insecticide is effective against a broad range of lepidopteran larvae primarily when ingested as a stomach poison causing larvae to stop feeding and become irreversibly paralyzed shortly after exposure. Due to emamectin benzoate's rapid photodegradation from foliage surfaces and the feeding habits of most predatory and parasitic arthropods, it is considered to be minimally disruptive to biological controls and should be very well suited for use in IPM programs. Emamectin benzoate is initially targeted for use against caterpillar pests such as Heliothines and *Spodoptera* spp. (5).

Imidacloprid (Admire®, Gaucho®, Provado®) is a neo-nicotinoid that acts on the nicotinic receptor and appears to function as an acetylcholine agonist (19). Imidacloprid is very effective against sucking insect pests and may be used as a seed treatment or as a foliar systemic providing control for several weeks. A similar neo-nicotinoid insecticide being developed by Novartis Crop Protection, Inc., is thiamethoxam. It is a thianicotinyl insecticide that controls a wide spectrum of sucking and chewing insects through contact and stomach activity. Thiamethoxam (Actara®, Adage®, Cruiser®, Platinum®) may be applied to seeds, soil, and foliage, and exhibits rapid plant uptake and is xylem-transported. Due to the methods of application (e.g., seed treatments) and its systemic nature, it is considered to have minimal impact on beneficial species and should be well suited for use in IPM programs (11,16). .

Indoxacarb, the pure active ingredient in the insecticide Steward® (DPX-MP062), represents a new class of insecticidal chemistry, the oxadiazines. The mode of action appears to be interference with a group of ion channels by inhibiting the flow of sodium into nerve cells, causing pest paralysis and death (29). Indoxacarb has demonstrated efficacy against a broad range of lepidopterous pests, and appears to possess an excellent environmental profile (9, 22, 27). Thus, Steward has the species specificity profile that is highly desirable in IPM programs.

The phenylpyrazole, fipronil (Regent®, etc.), also represents a new class of insecticides that has recent EPA registrations for use against a wide range of insect pests (8). The mode of action is through the blocking of the gamma-aminobutyric acid (GABA)-regulated chloride ion channel

Table 1. Mode of action of selected new classes of insecticides that meet IPM needs (adapted from Sparks, (25)).

Insecticide	Target Site	Action	System
Avermectins	GABA gated Cl^G channel	Increase GABA binding	Nervous
Neo-Nicotinoids	Nicotinic ACh receptor	Receptor agonist	Nervous
Oxadiazines	Voltage gated Na^+ channel	Na^+ channel blocker	Nervous
Phenylpyrazoles	GABA gated Cl^G channel	Block channel	Nervous
Pyridine azomethine	Insect CNS	Neural inhibition of feeding	Nervous
Spinosyns	Nicotinic (GABA) receptors	Alter function	Nervous
Juvenoids	JH receptors	Mimic JH	Endocrine
Diacylhydrazines	Ecdysone receptors	Ecdysone agonist	Endocrine

Abbreviations used: ACh - acetylcholine; GABA - gamma-aminobutyric acid; JH - juvenile hormone; Na^+ - sodium; Cl^G – chloride

in the insect central nervous system (18). It has registrations, or pending registrations, for use on crop, urban, and animal insect pests. Fipronil has a desirable environmental profile, but the impact on beneficial insects varies with species.

Pymetrozine (Fulfill®) is a representative of a novel new chemistry that exhibits a unique mode of action that can be characterized as neural inhibition of feeding behavior. It does not have a general toxic or paralyzing effect on insects, but selectively interferes with normal feeding activities by affecting neural regulation of fluid intake in sucking insects, particularly aphids. Fulfill is a highly selective, systemic insecticide with excellent efficacy against a wide range of aphid species on a variety of cropping systems (10,13). Pymetrozine has a very favorable ecotoxicology profile, and is an exceptional fit for IPM programs because of its selective toxicology to sucking insect pests.

The spinosyns are a new class of fermentation-derived tetracyclic macrolides (25) that act via the insect nervous system and are especially active against lepidopterous pests (3, 26). The suggested mode of action involves alteration of nicotinic receptor function, and perhaps GABA-gated chloride channel function as well, in a unique manner (23). Spinosad (Tracer®, the initial commercial product of the class) is a mixture of spinosyns A and D, has very favorable mammalian toxicity and enviromental profiles, and should be a candidate for inclusion into a variety of IPM programs.

Juvenoids are compounds that mimic the action of juvenile hormone, thereby disrupting the process of metamorphosis and leading to a variety of deleterious effects (4). Despite a great amount of effort into synthesis and testing of insect juvenile hormone analogues or juvenoids, as a class they have had very little impact on IPM. However, the recent introduction of new, more active and more photostable juvenoids, such as pyriproxyfen (Knack®, Esteem®, Distance®), may elevate the juvenoids to a position of prominence among the insecticides suited for inclusion into IPM programs. Pyriproxyfen acts by suppressing embryogenesis within the insect egg and by inhibiting metamorphosis and adult emergence of target insects. Pyriproxyfen has proven to be highly effective against whitefly species (silverleaf, sweetpotato, greenhouse) on cotton, shrubs and ornamentals, and various vegetable crops, with minimal adverse impacts on beneficial arthropods. Tree fruit insect pests, including pear psylla, codling moth, and scale insects, are also targeted by this compound. In addition, pyriproxyfen appears to be an ideal candidate insecticide for use against imported fire ant.

The diacylhydrazides are a recent and unique class of IGRs typified by tebufenozide (Mimic®, Confirm®) and methoxyfenozide (Intrepid®, Runner®) that have been classified as MAC (Molting Accelerating Compound) mode-of-action insecticides (4). As MAC insecticides, tebufenozide and methoxyfenozide act as mimics of the natural insect molting hormone, 20-hydroxyecdysone, inducing a premature lethal molt of the larvae within hours of ingestion of treated plant tissue. Their applications extend to a wide range of lepidopterous pests of field, vegetable, and tree fruit crops, as well as forestry. The specificity to lepidopterous pests, together with favorable mammalian toxicity and ecotoxicity profiles, make these diacylhydrazides highly desirable IPM tools.

Thus, insecticide discovery and development have evolved to the point that biorational insecticides may soon be the rule rather than the exception. It is apparent that soon there will be an array of insecticides available that will provide us with the capability to reduce an insect pest population to an acceptable level with very little ecological disruption. Perhaps for the first time, we will have the insecticide tools that will allow us to maximize the effectiveness of indigenous natural enemies in many cropping systems.

New Wave IPM

Despite the widespread adoption of the IPM concept and general implementation during the past few decades, critics point out that the rate of IPM adoption has been too slow and that there still remain too many cases where insecticides are the primary management tactics. In response to these concerns, Frisbee and Smith (6) proposed the transition to "biologically intensive" IPM systems. These biointensive IPM systems would emphasize host-plant resistance, cultural controls, and biological controls, which would be supplemented by biorational insecticides. There was nothing new about this concept, since it had been advanced nearly three decades ago by the originators of the IPM concept (14). The problem then and until recently was that there were very few biorational insecticides with adequate effectiveness for inclusion in most IPM programs. The recent developments in insecticide technology may facilitate the transition to more biologically intensive IPM systems at the beginning of the new millennium.

The condemnation of IPM as it is generally practiced achieved a new level of absurdity with two recent publications. The first was a report by a special committee of the National Research Council (21) in which a new paradigm, termed "ecologically based pest management," was proposed. This "new" ecologically based pest management concept was proposed to be cost effective and easy to adopt and integrate with other production practices; safe to growers, farm workers, and consumers; durable rather than ephemeral; without adverse environmental, economic, or safety

390

effects; and with an ecological focus at the ecosystem level. How could anyone disagree with such an altruistic goal? The second scenario (17) proposed "a total system approach to sustainable pest management." These authors opined that therapeutics of any type are not the answer, and that long-term, sustainable solutions may be achieved only through restructuring agroecosystems. Both of these concepts are simply rebirths of the original IPM concept and as such provide nothing new conceptually (14); however, they do serve as a reminder of the breadth of the IPM concept as it was originally proposed. These are ideological concepts whose application appears as unlikely today as it has been for the past 30 years.

Constraints to the full implementation of IPM concepts

Conceptualization has always been easier than implementation, even though implementation proceeds as new technological developments are proven to be effective and economical. The constraints that have prevented implementation of more ecologically based insect pest management systems for the past 30 years largely remain today. The primary constraints now facing us are: 1) lack of sufficient funding needed to support research essential to generate the necessary knowledge base, and 2) agroecosystem structure and dynamics that are outside the influence of pest managers. However, the constraint posed by lack of selectivity in insecticides is rapidly being solved through the unprecedented development of new classes of insecticides.

It is convenient for us to sit in conference and develop theories about how things should be, when we have little or no responsibility for making the critical decisions at the point where the "rubber meets the road." Let us not forget that we as entomologists are small "cogs" in a large machine. We don't build the system; we simply design ways to protect it from insect pests. Farmers and their advisors, as well as financiers and landowners, make technology decisions almost totally from the perspective of how a return on investment may be procured. Financial constraints demand that most farmers have a short-term rather than long-term perspective. The present economic health of farming is tentative at best, and simple survival is the immediate goal of most farmers.

Agroecosystems are more dynamic today than ever before; their structure is affected more by international economics and consumer demands than perhaps any other factors. Modern farmers and their advisors are in tune with commodity markets and are prepared to act in response to market volatility. As commodity prices change, farmers may decide which crops to plant and the number of hectares to be allotted to each. At times, entire crops are destroyed prior to harvest and others are planted in their place in response to market vagaries. Whether we like it or not, the modern world of farming has a level of fluidity never before experienced. Modern communication systems have allowed farmers to keep their fingers on the pulse of commodity markets as never before in history, and "freedom to farm" has provided the flexibility to change farm structure almost overnight. Vertical integration with maximum flexibility is the template for the future. The numbers of full-time farmers continues to decline, and farm sizes are increasing. Supply and demand are no longer local issues in the new world economy. Our lack of control of the agricultural systems we wish to manage reminds me of the following lines from the movie, "Forrest Gump":

"Life is like a box of chocolates;
it's all sweet, so you might as well eat.
And if you don't like coconut,
get over it."

Admittedly, many insect pest problems occur because modern farms consist of disturbed habitats that are planted to monocultures of introduced crop plants where environmental resistance is minimal. Also, a large proportion of insect pests are foreign, introduced species that are not in tune with native biological controls. The boll weevil is an example of a pest for which serious doubt exists that it would ever be effectively managed over most of the U.S. cotton belt through biological and cultural means alone. Where would we be in our quest to eliminate the boll weevil as a pest without insecticides? These characteristics of agroecosystems, as well as their dynamic state, ensure our continued dependence on insecticides or some other remedial-action technologies. It is unlikely that entomologists will be able to substantially alter the structure of whole agricultural systems in a way that will eliminate insect pest problems, even if we knew how. We may nibble at the system and recommend minor structural changes (e.g. planting geometry), but wholesale changes will occur only in response to economic and other factors. I submit that these factors are beyond our control and are likely to remain outside our sphere of influence. Human nature in the developed world is to move forward, not backward; thus, our dependence on new technological advances will continue ad infinitum. My point is that for the foreseeable future, IPM will require tools that will prevent a pest problem from developing (e.g., Bt crops) or tools which may be immediately applied to relieve a pest problem (e.g., insecticides) when thresholds are met or exceeded.

Insecticide Resistance Management

The insecticides discussed earlier are just the first to be developed from a diversity of new modes of action or underutilized modes of action. We now have more insecticide tools available to use against insect pests than at any other time. It is imperative that these new insecticides be used wisely and judiciously within the context of IPM systems; otherwise, they too may be ephemeral, as were many of their predecessors. The selectivity profiles characteristic of the newly discovered insecticides should be a factor working against insect pest adaptation, since species that are not targeted will not be inadvertently selected. Also, biological controls should play a greater role than in the past, thus reducing dependence on insecticides. The economic incentive to prolong the useful life of newly developed insecticides is obvious to the agrichemical industry, since recouping investments made toward discovery and development take time. Insecticide resistance management guidelines are among the initial considerations for commercialization of new insecticides, and cooperation by all parties involved is required to make these products more sustainable. The fact that there are several new modes of action included in the new insecticides should minimize selection for insect pest adaptation. Also, the planting of Bt crops should delay resistance development in caterpillar pests to the new insecticides and vice versa. Obviously, resistance management must be a factor considered when integrating new insecticide technologies in IPM.

Novel Insecticide Delivery Systems

A previous chapter by Frank Hall focused on delivering new crop protection agents within an IPM environment. Nevertheless, I would be remiss without brief mention of some of the novel insecticide delivery systems, because delivery is an essential requirement for the successful integration of new insecticide technologies in IPM. Technology development toward novel insecticide delivery systems has accelerated recently, particularly with the development of transgenic plants that produce their own Bt endotoxins. Bt plants may simply be considered as unique insecticide delivery systems that have recently been added to the pest managers' toolbox. Where the transgenic plant technology will go is anyone's guess. Obviously, many more toxicants will be delivered to insect pests through transgenic plants in the future. In that the subject of genetic engineering for IPM is addressed elsewhere in these Proceedings, I will only mention that these technologies must be integrated into IPM systems along with traditional insecticides on a foundation of cultural and biological controls.

Another novel insecticide delivery system is the attract and kill (attracticide) technique that incorporates a pheromone or plant attractant with small amounts of an insecticide designed to kill adult insects before reproduction. This technique is most appropriate for use in wide-area insect management programs. For example, promising results with the use of the semiochemical insecticide-bait SLAM® have been obtained recently in several midwestern states for control of corn rootworms (28). SLAM is a commercial bait formulation that combines a powdered cucurbit (buffalo gourd) and carbaryl insecticide. Adult rootworm control with the semiochemical bait system may require as little as 5% as much insecticide as traditional insecticide application systems. The attracticide technique has been successfully used against the pink bollworm in California for a decade or more and is having some success in boll weevil management. The appeal of the attracticide control technique is that the very small amounts of insecticide required have minimal adverse environmental impact and do not disrupt biological controls of other pests. Newly developed insecticides may be appropriate candidates for this method of insecticide delivery, particularly if cost of new insecticides is a constraint to traditional methods of application. Novel insecticide delivery systems must go hand-in-glove with development of novel insecticides for maximum utility of the technology.

Limitations to Integrating New Insecticide Technologies in IPM

Optimum benefits from new insecticide technologies may be realized only through intelligent use. The experiences from the past have prepared pest managers to recommend more efficient systems for utilizing insecticides. However, there are inherent limitations for the new insecticides that involve their cost, species selectivity, stage-specific selectivity, and residuality profiles. Also, there are external factors that must be given careful consideration when designing patterns for use of new insecticides.

Cost of the newly developed insecticides is a major concern of users. It is obvious that new insecticides must be more expensive on a per application basis, because the expenses of discovery, development, and manufacture are substantially greater than for insecticides that have been on the market for some time. Let's face the facts: there will never be a class of insecticides as cost effective as the pyrethroids. Twenty years of pyrethroid use have not prepared farmers for the prices they will have to pay for new insecticides. From a pest management point of view, this may not be so bad, because higher cost will very likely lead to more judicious use.

Selectivity is a very desirable trait from the IPM perspective, but it may be a two-edged sword. We anticipate the ability to reach into a system and impact a single pest or closely related pest group, but in so doing, pests that are unaffected by the selective insecticide may then achieve their biological potential. The extent to which this manifests itself as a problem will vary with the pest and the degree to which other controls exert their influence on the non-targeted pest population. In a well-designed management program, biological and cultural controls will typically keep potential insect pests from reaching economically damaging levels. Thus, the pest managers' knowledge of the systems and insects involved becomes much more important than it had been previously, when broad-spectrum insecticides were used. Knowledgeable crop consultants will become more essential for optimum use of selective insecticides.

Some of the new insecticides possess stage-specific selectivity, and none are active against as broad a range of life stages as are the older insecticides, such as the pyrethroids. Many of the new insecticides are active only against larvae and do not affect egg and adult stages, whereas the pyrethroids kill all three stages. Also, many of the new insecticides act as stomach poisons and must be ingested to be toxic. This results in some damage occurring before the insect pest is actually killed. The variation in feeding behavior of specific insects on different crop hosts or on different plant parts will affect efficacy. For example, heliothines are much more likely to acquire a lethal insecticide dose when feeding on foliage than when feeding on fruit.

Registration requirements have generally resulted in new insecticides that have reduced residuality in comparison to insecticides developed earlier. Short residual insecticides are desirable when residues on foods are considered, but oftentimes, insect management may be compromised. Reduced residual profiles place a premium on scouting efficiency and may necessitate more total annual applications on a crop, resulting in higher costs. The influence of biological controls, which should be minimally affected by the new insecticides, will determine the extent to which reduced residuality becomes a problem.

Conclusion

Recently discovered insecticides with new modes of action possess the environmental profiles needed for IPM programs. For the first time, the insect manager should have insecticides that may be effectively combined with biological and cultural controls to achieve truly integrated control of arthropod pests. New insecticide delivery systems may allow reductions in the amounts of insecticide required for the management of insect pests. New insecticides will require an advanced level of management expertise to optimize their use because of cost and selectivity characteristics.

Acknowledgments

T. C. Sparks provided valuable information for this manuscript.

Literature Cited

1. Allen, W. A., and Rojotte, E. G. 1990. The changing role of extension entomology in the IPM era. Annu. Rev. Entomol. 35: 379-397.
2. Carson, R. 1962. Silent Spring. Houghton Mifflin. Boston, MA.
3. DeAmicis, C. V., Dripps, J. E., Hatton, C. J., and Karr, L. L. 1997. Physical and biological properties of the spinosyns: Novel macrolide pest control agents from fermentation. Pages 144-154 in: Photochemical for Pest Control. P. A. Hedin, R. Hollingworth, E. P. Masler, J. Miyamoto, D. Thompson, eds. American Chemical

Society, Washington, DC.

4. Dhadialla, T. S., Carlson, G. R., and Le, D. P. 1998. New insecticides with ecdysteroidal and juvenile hormone activity. Annu. Rev. Entomol. 43:545-569.

5. Dunbar, D. M., Lawson, D. S., White, S. M., and Ngo, N. 1998. Emamectin benzoate: control of the Heliothine complex and impact on beneficial arthropods. Pages 1116-1119 in: Proc. 1998 Beltwide Cotton Conf., P. Dugger and D. Richter, eds. National Cotton Council of America, Memphis, TN.

6. Frisbie, R. E., and Smith, J. W., Jr. 1991. Biologically intensive integrative pest management: the future. Pages 151-164 in: Progress and Perspectives for the 21st Century. Entomol. Soc. Am. Cent. Symp. J. J. Menn and A. L. Steinhauer, eds. Entomol. Soc. Am., Lanham, MD.

7. Georghiou, G. P. 1990. Overview of insecticide resistance. Pages 18-41 in: Managing Resistance to Agrochemicals. M. B. Green, H.M. LeBaron, and W. K. Moberg, eds. American Chemical Society, Washington, DC.

8. Hamon, N., Shaw, R., and Yang, H. 1996. Worldwide development of fipronil insecticide. Pages 759-765 in: Proc. 1996 Beltwide Cotton Production Conf., P. Dugger and D. Richter, eds. National Cotton Council of America, Memphis, TN.

9. Harder, H. H., Riley, S. L., and McCann, S. F. 1997. DPX- MP062: a novel broad-spectrum, environmentally soft, insect control compound. Pages 48-50 in: Proc. 1997 Beltwide Cotton Conf., P. Dugger and D. Richter, eds. National Cotton Council of America, Memphis, TN.

10. Harrewijn, P., and Kayser, H. 1997. Pymetrozine, a fast-acting and selective inhibitor of aphid feeding. *In situ* studies with electronic monitoring of feeding behavior. Pestic. Sci. 49:130-140.

11. Hofer, D., and Brandl, F. 1999. Cruiser®/Adage® performance features of thiamethoxam as a seed treatment in worldwide cotton. Pages 1101-1104 in: Proc. 1999 Beltwide Cotton Conf., P. Dugger and D. Richter, eds. National Cotton Council of America, Memphis, TN.

12. Huffaker, C. B., and Smith, R. F. 1980. Rationale, organization, and development of a national integrated pest management project. Pages 1-24 in: New Technology of Pest Control. C. B. Huffaker, ed. John Wiley and Sons, Inc. New York.

13. Koenig, J. P., Lawson, S. D., White, S. M., and Dunbar, D. M. 1997.

Utility of Fulfill 50WG for aphid and whitefly management in cotton. Pages 997-999 in: Proc. 1997 Beltwide Cotton Conf., P. Dugger and D. Richter, eds. National Cotton Council of America, Memphis, TN.

14. Kogan, M. 1998. Integrated pest management: historical perspectives and contemporary developments. Annu. Rev. Entomol. 43:243-270.

15. Lasota, J. A., and Dybas, R. A. 1991. Avermectins, a novel class of compounds: Implications for use in arthropod pest control. Annu. Rev. Entomol. 36:91-117.

16. Lawson, D. S., Dunbar, D. M., White, S. M., and Ngo, N. 1999. Actara® 25 WG: Control of cotton pests with a new neonicotinoid insecticide, thiamethoxam. Pages 1106-1109 in: Proc. 1999 Beltwide Cotton Conf., P. Dugger and D. Richter, eds. National Cotton Council of America, Memphis, TN.

17. Lewis, W. J., van Lenteren, J. C., Phatak, S. C., and Tumlinson, J. H. 1997. A total system approach to sustainable pest management. Proc. Natl. Acad. Sci. 94:12243-12248.

18. Moffat, A. S. 1993. New chemicals seek to outwit insect pests. Science 261:550-551.

19. Mullins, J. W. 1992. Imidacloprid: A new nitroguanidine insecticide. Pages 183-198 in: Pest Control with Enhanced Environmental Safety. S.O. Duke, J. J. Menn, and J. R. Plimmer, eds. Am. Chem. Soc., Washington, DC.

20. Newsom, L. D. 1980. The next rung up the integrated pest management ladder. Bull. Entomol. Soc. Am. 26:369-374.

21. National Research Council. 1996. Ecologically Based Pest Management: New Solutions for a New Century. Natl. Acad., Washington, DC.

22. Ruberson, J. R., and Tillman, P. G. 1999. Effects of selected insecticides on natural enemies in cotton: laboratory studies. Pages 1210-1213 in: Proc. 1999 Beltwide Cotton Conf., P. Dugger and D. Richter, eds. National Cotton Council of America, Memphis, TN.

23. Salgado, V. L., Watson, G. B., and Sheets, J. J. 1997. Studies on the mode of action of spinosad, the active ingredient in Tracer® insect control. Pages 1082-1086 in: Proc. 1997 Beltwide Cotton Conf., P. Dugger and D. Richter, eds. National Cotton Council of America, Memphis, TN.

24. Smith, R. F. 1970. Pesticides - their use and limitations in pest management. Pages 103-118 in: Concepts of Pest Management. R. L. Rabb and F. E. Guthrie, eds. N. C. State Univ. Press., Raleigh, NC.

25. Sparks, T. C. 1998. New insecticides and acaricides: mode of action and potential role in IPM. Pages 123-130 in: Proc. Sixth Australasian Applied Entomological Research Conf. Vol.2. M. P. Zalucki, R. A. Drew, and G.G. White, eds. Brisbane, Austral.

26. Sparks, T. C., Thompson, G. D., Larson, L. L., Kirst, H. A., Jantz, O. K., Worden, T. V., Hertlein, M. B., and Busacca, J. D. 1995. Biological characteristics of the spinosyns: A new class of naturally derived insect control agents. Pages 903-907 in: Proc. 1995 Beltwide Cotton Conf., P. Dugger and D. Richter, eds. National Cotton Council of America, Memphis, TN.

27. Tillman, P. G., Mulrooney, J. E., and Mitchell, W. 1998. Susceptibility of selected beneficial insects to DPX-MP062. Pages 1112-1114 in: Proc. 1998 Beltwide Cotton Conf., P. Dugger and D. Richter, eds., National Cotton Council of America, Memphis, TN.

28. Tollefson, J. J. 1998. Rootworm areawide management program in Iowa. J. Agri. Entomol. 15:351-357.

29. Wing, K. D., Schnee, M. E., Sacher, M., and Connair, M. 1998. A novel oxadiazine insecticide is bioactivated in lepidopteran larvae. Archives of Insect Biochem. Physio. 37:91-103.

Fulfilling the Role of Resistance Management to Preserve Effectiveness of New Insecticide Technologies

T. J. Dennehy
University of Arizona, Department of Entomology,
Extension Arthropod Resistance Management Laboratory,
Tucson, AZ 85721 USA

The disruptive influence that pesticides can have on IPM programs has been reviewed extensively (e.g., 11,17) and has changed little since the conference, Concepts of Pest Management, was held at NC State University in the 1970s (14). What has changed since the 1970s, and largely for the worse, I believe, is the degree to which many states have fallen behind in meeting stated goals in developing and implementing IPM, especially in intensive agricultural production systems. Intensive agricultural systems typically have high crop value and/or long growing seasons, coupled with a complex of pests that are limiting factors in production. Pesticides in such systems frequently account for as much as 30% of variable production costs, i.e., those costs over which the grower has some management control during the season.

In this paper, I will discuss the role of chemical management in intensive agricultural systems and serious hurdles that are limiting efforts in this regard. I will then narrow the discussion to resistance management, provide examples of the enhanced role that Cooperative Extension is serving in resistance management in Arizona, and discuss related challenges and impediments. My take-home message is that, in these times of increasing public concern about insecticide use in agriculture, land grant universities, Cooperative Extension, and counterpart organizations in the USDA should be increasing, not shying away from, commitments to chemical management and resistance management. To do otherwise is to imperil hard-fought advances in IPM programs and to risk accelerated losses of new technologies.

Continued Importance of Insecticides in Intensive Agricultural Production Systems

Impressive advancements have been made in the chemical control component of IPM programs since the 1970s. As detailed in numerous passages elsewhere in this volume, insecticides are generally safer to workers and the environment, and gains have been made in the manner in which they are selected and used in production agriculture. Equally clear is the fact that pesticides have been, and will for the foreseeable future continue to be, essential tools for maintaining economically competitive production in many, if not most, intensive agricultural systems in the USA. However, the term "insecticide" now includes chemicals that are safer and more sophisticated in mode of action than in the past. Pheromones, insect growth regulators, chloronicotinal insecticides, and transgenic crops expressing Bt toxins are but a few examples of new, safer insecticidal products that have gained widespread acceptance in Arizona and elsewhere in the USA.

Pivotal Role of Chemical Management in IPM Programs

Improper selection and use of insecticides causes well-documented problems with destruction of biological control and the resulting resurgence of primary pests, outbreaks of secondary pests, and resistance treadmills (17). This hurts growers economically, often most seriously in arid regions. It is also a matter of public concern, because it can greatly increase insecticide load in the environment and exposure of workers and the public to insecticides. It is the goal of chemical management to limit such problems in IPM.

Resistance management is logically a subset of chemical management in IPM, but in practice, the activities of resistance managers are often quite distinct from the latter. Chemical management programs typically focus on: understanding which chemicals provide the best control of key pests, selectivity to natural enemies and other aspects of the integration of chemical and non-

401

chemical tactics, cost of control regimes, application methods, action thresholds for triggering chemical treatments, etc. Resistance management workers strive to keep producers informed of the resistance status of key pests (monitoring), of the best response to existing resistance problems (isolation and characterization of resistance, cross-resistance, stability of resistance), and how to formulate strategies to thwart resistance (strategy development and evaluation).

General agreement has been reached over the conceptual framework for judicious use of pesticides in IPM programs (Table 1). Most definitions of IPM call for pesticides to be used only when economically justified, and to assume that non-chemical alternatives have been given appropriate consideration. Therefore, it is fair to say that we pretty much know what is necessary to manage chemicals in IPM systems. However, a number of problems are increasingly keeping us from achieving the desired goals in this critical area of IPM and show signs of getting more serious.

MARKETING OF PESTICIDES

What information does the grower need to have in order to allow objective decisions to be made regarding insecticide selection and use? Producers are exposed to an abundance of marketing information at meetings, in trade journals, and from sales representatives. It is commonplace for trade journals that target growers to publish technical articles that are underwritten by chemical producers and that are highly biased in favor of specific pesticides. Third-party evaluations of chemicals, such as those routinely conducted in university and USDA field trials (often with partial funding from chemical producers), are essential for preserving growers' ability to evaluate claims of pesticide producers. There is a greater need than ever for such objective input regarding selection and use of insecticides.

INCREASED NEED / FEWER WORKERS

In many areas of the United States, there are increasing deficiencies in university and extension infrastructure, resources, and personnel devoted to IPM. University budget cuts and losses to inflation have resulted in a decaying infrastructure for field stations and laboratories in many states. Extension workers now must rely on

Table 1. Fundamental Principals of Insecticide Management

1. *Limit use.* Treat only when pest damage economically justifies intervention (action threshold).
2. *Proper selection.* Choose chemicals that are most compatible with other components of the IPM program (safety, selectivity) and the environment.
3. *Proper application.* Use application methodology that takes into account worker safety, drift, and selectivity to natural enemies.
4. *Appropriate rate.* Use lowest rate of insecticide that provides needed degree of control (few exceptions).

outside funding, often exclusively, for hiring summer helpers, for leasing vehicles, and in some cases, even for supporting IPM educational programs. Funding is difficult to sustain, often only year-to-year, and tied to specific products. Nowhere are these problems more obvious than in the management of the chemical component of IPM programs.

In extreme cases, deficiencies in the size of the work force have made the IPM movement appear farcical in the eyes of producers. Extension educational programs repeatedly tell growers that they should avoid using hard pesticides and that they should not treat pest populations until they reach critical levels. However, they frequently cannot provide producers with research-based action thresholds or alternatives. As a result, many growers view IPM as a rhetoric-driven initiative, rather than something based on replicated science.

Recent waves of Executive Branch and consumer interest in increased use of IPM are in striking discordance with this reality of fewer IPM professionals covering more problems. Producers of many crops continue to experience problems with resistance, resurgence, and secondary pest problems stemming from mismanagement of insecticides. Meanwhile, many older pesticides are not being re-registered and are being replaced with alternatives that are less familiar to producers, are often more difficult to use, and are more expensive. We are expecting increased sophistication in the manner that growers select and use insecticides at the same time that it has become harder for them to obtain impartial, research-based pest management information and relevant educational programs.

If we indeed agree that success of IPM programs can be greatly influenced by the quality and timeliness of information that producers are provided regarding chemical management, then why is the public sector of the pest management establishment moving in a direction away from meeting this need? To a very large degree, we know what needs to be done to achieve integrated management of agricultural pests. At least in the Southwest, the often unspoken question is, Who is going to do it?

UNIVERSITY AND EXTENSION FACULTY

University Faculty. University faculty have contributed to the current deficiencies in chemical and resistance management. Not infrequently, campus-based academic faculty have low regard for research involving pesticides, and this can be reflected in the peer review process. It is not uncommon for faculty to believe that: 1) pesticides are going to be eliminated from use in agriculture; 2) investigations of pesticides promote their increased use; 3) pesticide research is not scientifically rigorous; 4) attention of research gets focused unduly on chemicals or problems for which there is chemical-industry funding; and 5) time and money would be better invested in alternatives to pesticides.

I do not wish to refute these views, because some have a basis in fact. However, as long as IPM programs, field workers, and the environment continue to be endangered by the current use of pesticides in agriculture, it is socially irresponsible for academic departments to expect such problems to go away by ignoring them or by simply advocating that pesticides not be used.

Extension's Identity Crisis. Cooperative Extension is evolving in response to deficiencies in IPM and other areas of applied research. Applied research on land grant campuses was previously done principally by academic faculty. In many states, this has changed during the past two decades as shorter-term applied research, including most chemical management investigations, has shifted to the domain of extension faculty. This has set the stage for what has very aptly been called a "clash of cultures" (D. Byrne, personal communication) between the long-standing majority in extension who believe that their mission should primarily be that of conducting educational programs, versus department-based extension specialists who were hired to do problem-solving research as well as education.

The reality on most land grant campuses is that if extension faculty do not address the most pressing short-term research problems of their constituents, many will go unanswered. Unsolved pressing problems, in turn, leave extension education programs appearing out of touch and out of date. Indeed, this reality has been the driving force behind the trend toward hiring experienced research scientists in extension specialist positions. At the University of Arizona, extension leadership has adjusted to these new realities by formally adding problem-solving research to the long-standing mission of extension to conduct educational outreach.

Resistance Management Programs are Needed to Stabilize IPM in Intensive Agricultural Systems

The 1980s represented a period of awakening of academics and practitioners to the need for incorporating resistance management into IPM programs of intensive agricultural systems (9,11,12,16). This dawning of applied resistance management was prompted by the compounding effects of mounting resistance problems and the concomitant slowing of the rate at which new insecticides were being registered (11). Moreover, increased producer reliance on monitoring and thresholds in IPM has made it even more important that insecticides work as expected when thresholds are exceeded.

Simply put, it became clear in the 1980s that resistance is the Achilles' heel of IPM programs and that measures must be taken to keep hard-fought gains in IPM programs from falling victim to mismanagement of pesticides. This awareness was reflected in a series of recommendations stemming from a meeting of resistance experts convened by The National Research Council in 1985 (12). One of the recommendations contained the rationale for extension's future role in resistance management.

"Resistance detection, monitoring and management organizations should be formed at the local or regional level and assume responsibility for education, coordination and implementation of activities to deal with resistance problems" (12).

To a very large degree, we have learned during the past 20 years what is necessary to combat resistance on a practical basis (1,12,15). Unfortunately, like most other advancements in IPM, it involves investments in field research and strong extension education efforts. We first *formulate* a best-guess resistance management strategy. The strategy should be nested within the recommendations of the IPM program with which it is associated. It should strive to limit chemical use, diversify chemical modes of action utilized, and harmonize chemical use with other crops in the agricultural ecosystem. Secondly, we *implement* the resistance management strategy and intensively educate users regarding its use and benefits. This step involves conventional extension programming but is most effectively done in partnership with grower groups and chemical producers (see below). Third, we *monitor* success of the strategy on a routine and systematic basis. This allows continual evaluation of the strategy. Lastly, we *modify* the recommendations based on new findings from monitoring and other research.

Once resistance is detected in the field or selected in the laboratory, information from studies of cross-resistance, stability, and genetics of resistance may be used to modify the initial strategy. A critical issue is often that of incorporating into the resistance management strategy limitations on the number of times per season a given chemical or group of related chemicals may be used. The U.S. Environmental Protection Agency has in recent years demonstrated a willingness to consider such limitations when granting Section 18 Emergency Exemptions. In the case of whitefly control in Arizona cotton (detailed below), strict limitations placed on the use of two new insect growth regulators have been instrumental to success of the strategy.

The ultimate objective of resistance management is to sustain the efficacy of insecticides to help stabilize IPM programs. In reality, we know that resistance development is often inevitable under the conditions of intensive agriculture, irrespective of limitations placed on chemical use. However, as a practical criterion, I submit that resistance management efforts may be considered successful if efficacy against key pests can be sustained for at least 10 years. Limiting overall insecticide use by employing IPM is the first and most powerful principle of resistance management. Reductions of even one or two insecticide applications per season can greatly improve the chances of sustaining the efficacy of a chemical-use regime.

It has been argued that we have not really managed resistance if we simply extend the life of a chemical by, for example, limiting its use to once per year for 9 years, rather than burning it up in 3 years by using it three times per year. While it is true that in this hypothetical case, we did not extend the number of uses obtained from the product, this perspective does not take into account the value of stability to IPM programs. Each resistance episode has victims. Growers experience reduced profit when they use ineffective insecticides and have to increase the rates and frequencies of treatments. It often takes 1 or 2 years to modify IPM recommendations to account for changes brought about by introduction of a new pesticide.

ARIZONA'S EXTENSION-BASED RESISTANCE MANAGEMENT PROGRAM

Arizona's extension-based resistance management program was established in 1994 with the formation of the Extension Arthropod Resistance Management Laboratory (EARML). Now in its fifth year, EARML strives to empower producers to make timely decisions regarding chemical selection and use that will thwart resistance development and assist in combating existing resistance problems. Brief summaries of three recent EARML programs are described below.

The Whitefly Crisis in Arizona Cotton: A Success in Resistance Management. The silverleaf whitefly (*Bemisia argentifolii*) is a major pest of irrigated desert agricultural systems of the southwestern USA. It infests winter vegetables, spring and fall melons, cotton, and alfalfa. Adult whiteflies move between a succession of crops grown throughout the year on the low desert, providing uninterrupted pest development (8).

Arizona cotton producers experienced severe whitefly resistance problems two times in the past decade. This fact is revealed by changes in the average number of insecticide treatments made for whiteflies in cotton from 1990-98 (Fig. 1). Resistance episodes in 1992 and 1995 resulted in four to six times as many whitefly treatments being made in cotton as in years when insecticide efficacy against whiteflies was not compromised (8). Monitoring of whitefly resistance to synergized pyrethroid insecticides by EARML in 1994 provided early warning of the resistance problem that occurred in 1995 (7) and prompted rapid action in 1995 by a multi-agency resistance working group to formulate a resistance management strategy that would reduce reliance on synergized pyrethroid insecticides in cotton.

In the areas most severely affected by resistance in 1995, growers made as many as 12 treatments of mixtures of two to five insecticides and still sustained severe contamination of the crop from whitefly honeydew (5). Indirect costs of the Arizona whitefly resistance problem stemming from honeydew contamination of cotton were estimated to be 2 to 4 cents per pound over the years 1996-98, summing to losses of tens of millions of dollars. Moreover, lygus bugs exposed to the high numbers of whitefly treatments became more difficult to control with insecticides (4), adding more costs to the resistance treadmill on which producers were trapped.

A three-phase resistance management program was implemented in 1996 (6), centered around once-per-season use of each of two new insect growth regulators, buprofezin (a chitin biosynthesis inhibitor) and pyriproxyfen (a juvenile hormone mimic). Older, non-pyrethroid (conventional) insecticides were recommended for use in the second phase of the strategy, and synergized pyrethroids were reserved for phase three, during the late season. Seminars to teach growers about the new strategy and to reinforce the fundamentals of using whitefly sampling procedures and thresholds were conducted by Cooperative Extension around the state. These were sponsored by the Arizona Cotton Growers Association and Cotton Incorporated, and EPA mandated those wishing to use the IGRs, under the provisions of the newly obtained Section 18 Exemption, to attend the educational program. Deployment of the resistance management strategy has coincided with sharply reduced numbers of whitefly treatments in 1996-98 (Fig. 1) and striking reductions in whitefly resistance to synergized pyrethroid (Fig. 2) and some non-pyrethroid insecticides (3).

The insect growth regulators buprofezin and pyriproxyfen have been major factors in this success in resistance management. Preventing recurrences of whitefly control problems in Arizona cotton will require that we sustain the effectiveness of these new technologies. To date, there have been no confirmed failures of IGRs in cotton. However, statewide monitoring has revealed unexpected reductions in susceptibility of whiteflies to buprofezin from 1996-98

408

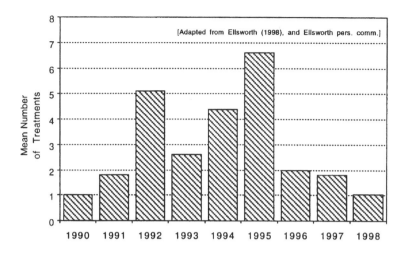

Figure 1. Estimated numbers of insecticide treatments targeted against whiteflies in Arizona cotton from 1990-98.

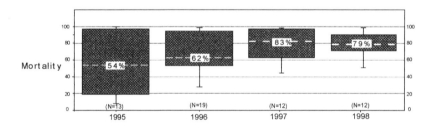

Figure 2. Box plots illustrating the significant increase in susceptibility of whiteflies collected throughout Arizona to mixtures of fenpropathrin + acephate from 1995 through 1998. Shown is mortality in bioassays of 10 µg/ml fenpropathrin + 1000 µg/ml acephate.

(2), despite its use in cotton being limited to one application per season. EARML efforts continue to focus on sustaining this fragile success.

Chloronicotinyl Insecticides in Arizona Vegetables and Melons: An Impending Problem. Imidacloprid is an insecticide of the chloronicotinyl group that exhibits both systemic and contact activity.

Admire® is a systemic formulation of imidacloprid that, since 1993, has provided exceptional control of whiteflies in Arizona vegetables and melons (20). The extreme effectiveness of imidacloprid in vegetables and melons has meant that fewer whiteflies move into cotton in the spring. For this reason, its efficacy in vegetables is very much a concern to cotton growers in Arizona. Imidacloprid is used on a very high proportion of Arizona winter vegetable crops that are whitefly hosts (18). This extensive use of the insecticide, coupled with the rapid life cycle of whiteflies and capacity to disperse between crops, puts imidacloprid at considerable risk for resistance development (19). Moreover, we anticipate, in the near future, the registration in Arizona of two additional chloronicotinyl insecticides, acetamiprid and thiamethoxam, for control of whiteflies and other pests.

Owing to its importance to whitefly management in Arizona, we have monitored resistance to imidacloprid on a statewide basis since 1995 (2,18,19,20). Statewide collections have revealed significant reductions in whitefly susceptibility to imidacloprid each of the past 3 years (Fig. 3) and a striking drop from 1997 to 1998 (2). Moreover, individual populations of whiteflies from cotton, melons, and greenhouse poinsettias were considerably more resistant than were the averages from statewide collections (Fig. 3).

There are no verified cases of failures of imidacloprid in Arizona melon or vegetable fields, and we cannot predict when failures will begin to occur. However, given the striking increases we documented in resistance to imidacloprid in 1998, the Southwest Whitefly Resistance Working Group has begun to formulate contingency plans for responding to this eventuality. Of greatest concern initially is correlating field failures with specified levels of resistance as measured in bioassays, and determining the impact that resistance to imidacloprid will have on the other chloronicotinyl insecticides under development.

Transgenic Cotton Expressing a Bt Toxin: Can it be sustained? Transgenic cotton derived from *Bacillus thuringiensis* represents the biggest technological breakthrough in cotton insect management in many decades. In Arizona, it has been extremely effective at controlling pink bollworm (PBW) (*Pectinophora gossypiella*). Control of this pest was estimated to cost cotton producers in central and northwestern Arizona $44-207 per hectare in 1997 (13). The benefits of using Bt cotton include reduced environmental and worker exposure to conventional insecticides, reduced selection for resistance to conventional insecticides, and improved conservation of natural enemies. Bt cotton in Arizona was grown on 60% to 70% of the land planted to Upland cotton in 1997 and 70% in 1998 (13).

Bt transgenic crops are widely agreed to be at high risk of resistance development owing to the five to seven months that pests are exposed to Bt toxins in the plants. Collaborative efforts underway at EARML to monitor and manage pink bollworm resistance to Bt cotton include: 1) statewide monitoring of susceptibility to Cry1Ac endotoxin; 2) selection and characterization of resistance in the laboratory and greenhouse; and 3) a rapid response team for responding jointly with the Arizona Cotton Research and Protection Council to reports of unusual survival of pink bollworm on Bt cotton.

Statewide monitoring of pink bollworm susceptibility to Bt toxin was conducted in Arizona in 1997 (Fig. 4) and 1998 (13) and revealed

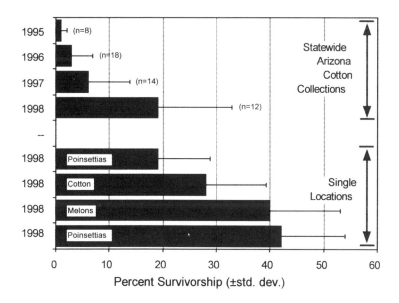

Figure 3. Reductions in average susceptibility to Admire® (imidacloprid) of whiteflies collected from Arizona cotton in 1995-98, and responses of four of the less susceptible single populations evaluated in 1998. Values shown are percent survivorship in replicated bioassays of 1000 μg/ml imidacloprid.

no reductions in susceptibility to Cry1Ac. However, selection in the laboratory produced a 100- to 460-fold resistance to Cry1Ac (13). Preliminary results from greenhouse evaluations of survival of laboratory-selected strains on Bt cotton showed that approximately 30% were able to complete development in bolls of Bt cotton, pupate, and successfully reproduce (10). We concluded from these preliminary results that the type of resistance we have isolated to Bt cotton in the laboratory is likely to negatively impact field performance of Bt cotton in Arizona in the future. However, intensive and extensive monitoring of fields by the ACRPC revealed no evidence of diminished performance of Bt cotton in 1998 (13). We will continue to maintain very active monitoring and collaborative research efforts in attempting to extend the useful life of transgenic Bt technology in Arizona cotton.

IMPEDIMENTS AND MISCONCEPTIONS AFFECTING THE FUTURE SUCCESS OF RESISTANCE MANAGEMENT

Pro-activity Versus Crisis Mentality. For resistance management actions to be most effective, they should be implemented prior to the onset of resistance problems in a region. Yet, it is very difficult to interest chemical manufacturers, crop producers, and extension personnel in taking action prior to the confirmed widespread failure

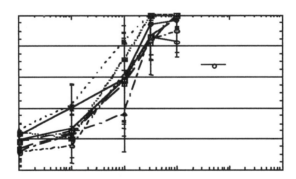

Figure 4. Baseline susceptibility of Arizona pink bollworm populations to Bt toxin as estimated in diet bioassays. Selection of these populations in the laboratory resulted in a 100 to 460-fold resistance to Cry1Ac.

of an insecticide in the field. As a result, resistance monitoring and research to isolate and characterize resistance typically has been initiated well after problems first are experienced in the field. This usually means that there is a lag of 1 to 3 years before producers receive research-based recommendations for mitigating the problem. The obvious solution to this problem is routine monitoring of resistance in key pests and to investigate resistance to new insecticides before problems occur in the field. Because extension workers and grower groups alike often have a crisis mentality that has them accustomed to responding to the year's most pressing problems, it can be very difficult to sustain interest and support for pro-active monitoring and research. As a result, some commodities go from one crisis in chemical control to another every few years. Effective resistance management efforts cannot be started and stopped in response to such crises; to be successful, they must be made long-term entities of production systems.

Industry Should Solve Resistance Problems. During the past two decades, we have seen the chemical industry serving an increasingly active and positive role in addressing resistance problems in agriculture. Indeed, our colleagues in industry have been instrumental in formulating a number of resistance management strategies and in contributing to the research on which such programs were based. However, it is unrealistic to expect chemical company personnel to be able to investigate and report on resistance problems impartially. The chemical industry is highly competitive, and good biological judgement often gives way to the imperatives of competition, marketing, and concerns about political fallout of resistance.

IPM programs rely on the chemical industry to generate innovative new technologies, so their economic survival is very relevant to the future of IPM. However, the first line of responsibility for minimizing resistance-related losses to growers and associated damage to IPM programs logically must reside in the public sector, with parties that do not directly profit from sales of the products involved.

Resistance Monitoring is a Waste of Time and Resources. Typically this sentiment is propounded by academics who are aware of a monitoring program with results that were inadequately translated into action (i.e., nothing was done with the data). Monitoring that does not impact chemical management decisions *is* a waste of time and resources. However, this is a problem with how information is being processed and used rather than with the rationale for monitoring. A related misconception is that by the time we can measure resistance in the field, it is too late to do anything about it. As shown by the aforementioned example of regained susceptibility of Arizona whiteflies to pyrethroids, this cannot be assumed *a priori*, though it may be true in some cases. Moreover, this misses the point that resistance monitoring helps to limit economic losses experienced by growers and the environmental damage commonly associated with resistance episodes.

CRITICAL ROLE OF EDUCATION AND PARTNERSHIP

A key goal of EARML programs is to increase through education the awareness of Arizona producers regarding arthropod resistance and especially the high costs they incur from resistance problems. As with all extension educational campaigns, these goals are achieved by providing timely, relevant research findings, consistency of message, and multiple years of delivering it to producers. We feel we are making some inroads in resistance education in Arizona. New information is routinely presented at county meetings throughout the state and is disseminated throughout the year via a variety of written and verbal formats, including those sponsored by the Arizona Cotton Growers Association and Cotton Incorporated.

Sustaining long-term resistance management efforts requires partnership and vision on the part of producer groups and the chemical industry. In our case, it has also been necessary for the university to be willing to dedicate considerable laboratory and greenhouse space to EARML. Furthermore, experienced technicians are required to do complex bioassay procedures, and this means maintaining continuity of staff over time. Few agricultural commodities have the resources and vision necessary to sustain such programs. However, the Arizona Cotton Growers Association and Cotton Incorporated have done so since the inception of EARML. They have served critical roles in formulation, development, and delivery of results of the programs I previously described in Arizona cotton.

Conclusion

IPM, chemical management, and resistance management share the common imperative of limiting chemical use and helping producers to make decisions that will sustain profitability of crop production. As illustrated in examples from Arizona agriculture, we have remarkable, new insecticide technologies for use in IPM programs. These tools have been very costly to develop, serve critical roles in IPM programs, and need to be managed carefully to sustain their effectiveness. Chemical and resistance management will be critical to accomplish this end. Yet, the future of such programs is tenuous at best.

Acknowledgments

Many individuals and organizations have contributed to the collaborative programs summarized herein. I wish to acknowledge the contributions of L. Antilla, I. Denholm, R. Horowitz, P. Ellsworth, R. Nichols, J. Palumbo, A. Patin, and B. Tabashnik toward program design, execution and implementation. I thank X. Li, M. Sims, and M. Wigert of EARML for outstanding technical support. Lastly, I thank the Arizona Cotton Growers Association, Cotton Incorporated, the Arizona Cotton Research and Protection Council, Monsanto, and the University of Arizona for providing financial and physical support of the programs.

Literature Cited

1. Denholm, I., and Rowland, M. W. 1992. Tactics for managing pesticide resistance in arthropods: theory and practice. Annu. Rev. Entomol. 37:91-112.
2. Dennehy, T. J., Wigert, M., Li, X., and Williams, L., III. 1999. Arizona whitefly susceptibility to insect growth regulators and chloronicotinyl insecticides: 1998 season summary. 1999 University of Arizona Cotton Report. University of Arizona Cooperative Extension (in press).
3. Dennehy, T. J., Li, X., and Wigert, M. 1999. Successful management of whitefly resistance to pyrethroid insecticides in

Arizona cotton: a four-year retrospective. 1999 University of Arizona Cotton Report. University of Arizona Cooperative Extension (in press).

4. Dennehy, T. J., Russell, J. E., Antilla, L., and Whitlow, M. 1998. New insights regarding estimating lygus susceptibility to insecticides. Pages 1255-1261 in: Proc. 1998 Beltwide Cotton Conferences.

5. Dennehy, T. J., and Williams, L., III. 1997. Management of resistance in *Bemisia* in Arizona cotton. Pesticide Science 51:398-406.

6. Dennehy, T. J., Ellsworth, P. C., and Nichols, R. L. 1996. The 1996 whitefly resistance management program for Arizona cotton. University of Arizona IPM Series No. 8.

7. Dennehy, T. J., Simmons, A., Russell, J., and Akey, D. 1995. Establishment of a whitefly resistance documentation and management program in Arizona. Pages 287-297 in: Cotton Report. University of Arizona College of Agriculture.

8. Ellsworth, P. C. 1998. Whitefly management in Arizona: looking at the whole system. Proc. Beltwide Cotton Conferences. National Cotton Council.

9. Georghiou, G. P., and Saito, T. 1983. Pest Resistance to Pesticides. New York: Plenum.

10. Liu, Y. B., Tabashnik, B. E., Dennehy, T. J., Patin, A. L., and Bartlett, A. C. 1999. Development time and resistance to Bt crops. Nature 400:519.

11. Metcalf, R. L. 1980. Changing role of insecticides in crop protection. Ann. Rev. Entomol. 25:219-256.

12. National Research Council. 1986. Pesticide Resistance: Strategies and Tactics for Management. National Research Council, National Academy Press, Washington, DC.

13. Patin, A. L., Dennehy, T. J., Sims, M. A., Tabashnik, B. E., Liu, Y. B., Antilla, L., Gouge, D., Henneberry, T. J., and Staten R. 1999. Status of pink bollworm susceptibility to Bt in Arizona. Pages 991-996 in: Proc. Beltwide Cotton Conferences. National Cotton Council.

14. Rabb, R. L., and Guthrie, F. E., eds. 1970. Concepts of Pest Management. N.C. State University, Raleigh.

15. Roush, R. T., and Tabashnik, B. E., eds. 1990. Pesticide Resistance in Arthropods. Chapman and Hall, London.

16. Sawicki, R. M. 1987. Definitions, detection and documentation of insecticide resistance. in: Combating Resistance to

416

Xenobiotics. M. G. Ford, D. W. Holloman, B. P. S Khambay, and R. M. Sawicki, eds. Horwood, Chichester (England).

17. Smith, R. F. 1970. Pesticides: their use and limitations in pest management. Pages 103-112 in: Concepts of Pest Management, R. L. Rabb and F. E. Guthrie, eds. N.C. State University. Raleigh.

18. Williams, L, III, Dennehy, T. J., and Palumbo, J. C. 1998. Can resistance to chloronicotinyl insecticides be averted in Arizona field crops? Proc. Beltwide Cotton Conferences. National Cotton Council.

19. Williams, L., III, Dennehy, T. J., and Palumbo, J. C. 1997. Defining the risk of resistance to imidacloprid in Arizona populations of whitefly. Proc. Beltwide Cotton Conferences. National Cotton Council.

20. Williams, L., III, Dennehy, T. J., and Palumbo, J. C. 1996. Whitefly control in Arizona: development of a resistance management program for imidacloprid. Proc. Beltwide Cotton Conferences. National Cotton Council.

Section 6

Geographical Information Systems and Global Positioning Systems

Spatial and temporal variation in occurrence of plant pathogens, plant-pathogenic nematodes, weeds, and insects, and in such soil characteristics as moisture level, fertility, and pH have long posed difficulties for both farmers and crop protection specialists. Recognition that great efficiencies could be gained by matching crop inputs to patterns of spatial variation has long existed, but the technology necessary to accomplish this was lacking. Recent advances in computing and communications technology and infrastructure, sensors, and multispectral imaging, in conjunction with the development of powerful procedures for analyzing spatial information, hold great promise for enabling us to realize these efficiencies through the use of geographical information and global positioning systems (GIS/GPS) to achieve precision agriculture.

In this section, Michael Ellsbury et al. provide an overview of the application and potential of GIS/GPS in insect, weed, and plant disease management, and discuss some of the major constraints to the adoption of emerging GIS/GPS technologies for IPM. Douglas Marin et al. then describe the current use of GIS/GPS in banana production in Costa Rica.

Use of GIS/GPS Systems in IPM: Progress and Reality

Michael M. Ellsbury, USDA, ARS, Northern Grain Insects Research
Laboratory, Brookings, SD 57006 USA
Sharon A. Clay, Plant Science Department, South Dakota State
University, Brookings, SD 57007 USA
Shelby J. Fleischer, Department of Entomology, Pennsylvania State
University, University Park, PA 16803 USA
Laurence D. Chandler, USDA, ARS, Northern Grain Insects
Research Laboratory, Brookings, SD 57006 USA
S. M. Schneider, USDA, ARS, Horticultural Crops Research
Laboratory, Fresno, CA 93727 USA

Geographical information systems (GIS) have evolved as integrated computerized tools for mapping and spatial analysis of georeferenced information. Capabilities of contemporary GIS include assemblage, storage, manipulation, retrieval, and graphic display of the volume of information about attributes, georeferenced position, and the spatial interrelation of layered features in a geographical context. The advent in recent years of economical, commercially available global positioning systems (GPS) for civilian use (34) has greatly improved the utility of GIS for general agricultural applications. Availability of advanced GPS technology has eliminated much of the imprecision in the georeferencing of spatial data, considered a serious shortcoming in early GIS technology (5). The ability of GPS to provide real-time positioning is critical, because variable rate applications of agricultural inputs are adjusted during field operations (34). Rapid adoption of GIS/GPS technologies is occurring for monitoring of crop yields, describing agronomic field features, grid sampling of soil properties, and site-specific soil nutrient management. Likewise, GIS/GPS technology, including geostatistical methods, mapping software, and remote sensing, are being increasingly applied to integrated pest management.

419

GIS/GPS and Spatial Variability of Pests

A Paradigm Shift?

For management purposes, distributions of pest populations usually are considered in terms of a mean density per unit area fluctuating through time. In reality, the actual distributions for weeds (6,12), plant pathogens (10,35), and insects (3,15,27) frequently are heterogeneous or nonrandom because of environmental influences that vary over the landscape. The result is patchy distributions that make reliable and economical sampling, modeling, or management strategies difficult to develop (6).

The tools of GIS/GPS, including remote sensing technology and geostatistics, offer alternative approaches to the characterization,

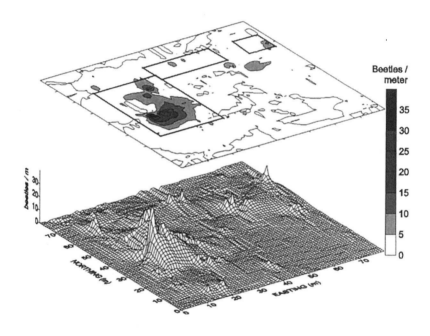

Figure 1. Interpolated densities for combined counts of adults and 3rd and 4th instar larvae of *Leptinotarsa decemlineata* in an ~80 by 80 m potato plot. Rectangles indicate areas targeted for insecticide application. (Figure by Paul Blom)

sampling, and management of pest populations that engender a fundamental paradigm shift in the way integrated pest management is likely to be accomplished in the future. That is, we now have the option and capability to make pest management decisions on the basis of knowledge about spatial variation in distributions of pests visualized as maps of populations or disease incidence that vary in space as well as in time (15,25). These maps can be used to determine where control measures can be applied for maximum effectiveness using precision or variable-rate application technology.

INSECTS

There are many examples of GIS/GPS application for the mapping of spatial variation of insect pest populations. Several recent reviews of these applications are available (2,15,27,28,29). Among the earliest of these were assessments of spatial variability in insect populations for predictive purposes in forest and rangeland ecosystems (20). More recently, successes have been reported in use of GIS/GPS technologies for integrated pest management of public health and stored-product insects (2). Routine use of GIS/GPS technologies in fully operative integrated pest management programs for field crops is yet to be attained. A promising possibility is a site-specific approach to potato pest IPM (38) that maps Colorado potato beetle distributions (Fig. 1) sampled visually using an efficient distance-walk sampling method. Crop host maps have been developed to predict future infestation risk of potato plants (39) and to characterize temporal dynamics in the spatial continuity of potato pest density (Fleischer, unpublished). Site-specific control measures also may help slow development of insecticide resistance by creating temporally dynamic refuge areas within a managed field (22).

Georeferenced data for northern corn rootworms (13) show spatial variation in grid-sampled egg distributions (Fig. 2a) and adult emergence densities (Fig. 2b). The lowest emergence densities usually occurred in wetter, low-lying areas or on ridge tops, and the highest densities were found in better-drained soils (13). Larval distributions (Fig. 2c) exhibited spatial correlation with ratings of root injury (Fig. 2d). Because survival of these soil-dwelling insects is determined, at least in part, by the soil conditions they experience, we hypothesize that information about soil physical properties gained through grid sampling and geostatistics should be useful for decision-making regarding management of the larval stages

421

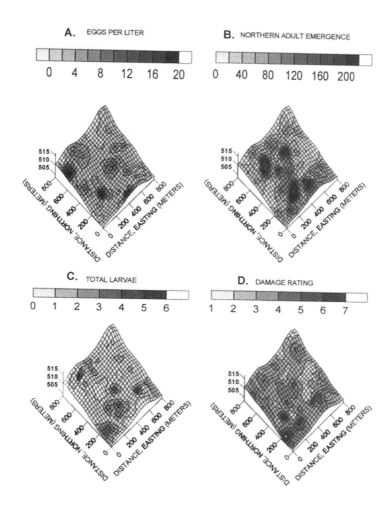

A. EGGS PER LITER

0   4   8   12   16   20

B. NORTHERN ADULT EMERGENCE

0   40   80   120   160   200

C. TOTAL LARVAE

0   1   2   3   4   5   6

D. DAMAGE RATING

1   2   3   4   5   6   7

Figure 2. Interpolated maps of northern corn rootworms for spring egg densities (A), numbers of emerging adults (B), numbers of larvae in a 4-plant sample (C), and root injury ratings (D). (Map by Sharon Nichols)

of corn rootworm. Soil characteristics and other ecological data may provide surrogate measures of the risk of insect infestation to guide allocation of sampling effort.

Targeted Sampling. Because insect populations frequently occur in spatial distributions that are aggregated and nonrandom, there usually is not enough prior information about these distributions to support decisions on directed sampling of areas where insect numbers are highest (32). Use of remote-sensed images displayed through GIS technology offer the opportunity for stratified or targeted sampling directed to areas where anomalies in plant spectral characteristics or soil and terrain features can be identified (23). For plant bugs (*Lygus* spp.) in cotton, anomalous areas, where plant bug abundance and associated variability were highest, corresponded with regions where cotton plants were the greenest and most vigorous (J. L. Willers, unpublished). Remote-sensed imagery has been used to produce host plant resource maps for cotton fields (3). Landform features on these maps were correlated with whitefly density and were used to produce maps of whitefly density and as inputs for spatio-temporal models of whitefly population dynamics. In these examples, use of GIS technology to process aerial images allowed identification of variation in crop growth associated with higher densities or spatial variation in insect populations. In the case of plant bugs, the result was more-efficient sampling, because scouting routes and time allocations could be planned using the remote-sensed images before the fields are actually sampled.

Area-wide Management. Perhaps the greatest utility of GIS/GPS to pest technologies will be seen in informationally intensive area-wide approaches to pest management. Rossi et al. (28) estimated northern corn rootworm populations and evaluated economic risk for treatment decisions on a regional scale using geostatistical techniques with data taken from northwestern Iowa. Applications of GIS/GPS technologies also are ongoing in area-wide management programs for the corn rootworm complex (9) and pink bollworm (36). The extensive geographical coverage necessary for sampling in cooperators' fields during area-wide programs is well suited to the application of GIS technology. GIS/GPS technologies enable the compilation of geo-referenced layers of information about land use, topography, crop history, pest monitoring, and control activities that are needed for area-wide management decisions. Using GIS/GPS technology, a thematic map of the South Dakota management area shows classification of program fields by crop rotation status (Fig. 3).

First-year fields were those that had been out of corn production for at least a year. Continuous fields were planted to corn for a minimum of 2 years, and mixed fields were planted to two or more crops, e.g.,

423

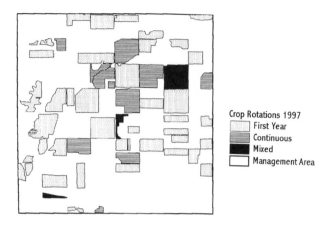

Figure 3. GIS-based map, produced in ArcView (ESRI, Redlands, CA), showing cropping status of producers' fields during 1997 in the South Dakota corn rootworm area-wide management zone. (Map by David Woodson)

corn and soybeans, during the previous year. The unique spatial perspective of this map shows that almost all first-year and mixed fields were within 1.6 km of continuous corn that will probably have an established western corn rootworm population. This information would not have been available without the benefit of GIS for information management.

WEEDS

Recent developments in GIS/GPS technologies also have stimulated interest in site-specific weed management. Weeds are ideal for site-specific management because they are not uniformly distributed in fields. Weed distributions are species-dependent (19,33), and up to 80% of a field may be free of a given weed species. The problem thus becomes one of obtaining reliable and accurate maps to identify target areas for site-specific management.

Weed Mapping. Weed management requires knowledge of weed species, density, and location in the field. The site-specific management of weeds has several benefits, including better weed control, reduction in excessive or inappropriate treatment, increased yield by better targeting specific weed species, and increased environmental safety. Site-specific herbicide technology includes controllers and software for application based on recommendations that vary the herbicide applied and/or the amount applied. The goal for weed mapping is to improve management decisions by matching recommendations to real, rather than perceived, needs. Only the necessary amount of chemical, applied only where needed, will be used. This, in turn, will help to maximize economic return.

Precision herbicide applications will be only as good as the maps that are provided. Weed complexes at field edges may have the most influence on decisions about herbicide treatments, but they may not give an accurate depiction of the center of the field. Scouting during harvest in the fall or during early spring will improve decisions, but weed infestations and their locations may still be missed. Sampling on a 1.0 to 1.6 ha grid (as done in soil sampling for nutrients) is often too coarse a scale to provide accurate weed maps for site-specific applications (19). Grid sampling at finer scales improves map quality but is too time-consuming and labor intensive to be cost effective.

Remote Sensing, Aerial, and Satellite Imagery. To solve the problem of obtaining site-specific weed information, it has been proposed that weed sensors mounted on spray booms be used to control applicators. The advantage with this approach is that extensive field scouting is not required. The major problem with this approach is that the sensors are extremely sensitive to changes in light irradiance levels and to soil and plant surface characteristics, which makes real-time weed identification difficult (4). Plant stress also has been studied using spectral reflectance (7). Spectral wavelengths that can be used for remote sensing and spectral imaging of several weed species have been reported (4,14,21). Spectral differences between weed patches, corn, and soil (or residue) are distinguishable using still video cameras at heights of 10 to 600 m above the soil (4), with the best separation occurring between classes at higher altitudes.

Aerial images photographed using multispectral band-detectors (red, blue, green, and near-infrared) promise to be a valuable scouting tool for mapping weed distributions. Multispectral remote sensing offers the potential to identify, categorize, and determine differences and similarities over a wide range of crop conditions. Multidate, synoptic views of fields, farms, and watersheds can be collected routinely by remote sensors. Sensors can record data beyond the reflectance part of the electromagnetic spectrum. For example, thermal and microwave sensors have shown potential for indicating crop conditions and soil-water status (16,37). The establishment of crop spectral signatures across a crop calendar is a necessary basis for the detection of anomalies, or departures from normal (1). The causes of anomalies are many and may vary from soil-water and fertility stress to weed, insect, or disease problems. Obviously, field knowledge gained through observation and awareness of management decisions is needed to best utilize the remote-sensed information.

Aerial imagery also provides a spatial perspective that is not possible during systematic ground-level scouting. Areas that are anomalous (having irregular patterns or colors, or both) can be quickly located in the image, and the image then becomes a map to direct or target field scouting efforts. Healthy plants look different on multispectral imagery when compared with stressed plants. Aerial photographs filtered at critical wavelengths reveal patterns related to stress in crop plants. Spectrophotometric data have been used to identify diseased and rust-resisting cereal crops (18), and spectral data in the 745 to 785 nm range can be used to classify the physiological maturity of wheat and grain sorghum (11).

Reliable information also may be obtained from multispectral satellite imagery, but there are several limitations to such imagery. First, it has not been available in a timely manner. The spectral bands in which the images are taken are limited, and although over-flights may occur frequently, weather conditions ideal for obtaining information cannot be guaranteed.

While researchers have been developing protocols for agricultural use of multispectral band information for agriculture, the everyday practitioner has used little, if any, of the data. Although attempts at remote-sensing applications are showing economic promise, the challenge now is to make this potential a reality. The development of improved spectral sensors and high-speed, low-cost computers will do much to bring aerial imagery and remote sensing into the realm of everyday agricultural use.

Multispectral Imagery for Weeds in South Dakota. Our project is exploring the use of multispectral imagery obtained by airborne remote sensing as a guide to mapping weed infestations in production-sized fields. Near-infrared aerial images were obtained to evaluate weed distributions in a 65 ha field in 1996. During the 2 previous years, the field had been intensively sampled on a 15 by 30 m grid (about 1300 data points each year) to define temporal and spatial weed distribution patterns. Both the species present and the georeferenced samples of

426

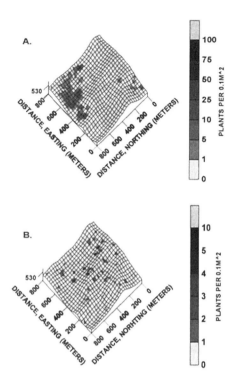

Figure 4. Maps of weed densities interpolated from grid samples of common ragweed (A) and Canada thistle (B) from a field in Moody Co., South Dakota, during 1996. (Maps by Scott Christopherson)

population densities were entered on a field map, and contour maps were produced through the use of kriging to estimate weed densities in un-sampled areas. Common ragweed occurred in two poorly drained areas of the field (Fig. 4a) but was rare throughout the remainder of the field. Canada thistle occurred in drier hilltop or side-slope areas but rarely in the toe-slope (Fig. 4b). Near-infrared digital images taken in mid-September 1996 and superimposed on a topographical map (Fig. 5) had several anomalies that corresponded to high weed populations. For example, many of the Canada thistle patches show up as dark spots on the aerial image in the center of the field, whereas larger anomalies on the image match large patches of common ragweed. Areas of the image that show anomalies not matched to ragweed or Canada thistle were determined to be quackgrass, barnyardgrass, and yellow foxtail. Similarities and differences between the weed map and aerial image clearly demonstrate that valuable information can be obtained through aerial images, but ground-truthed data still are required for verification.

Multispectral images provide a map allowing rapid identification of irregularities in the field that can be scouted with a minimum of time and effort. The change from a "normal" to an "abnormal" spectral signature

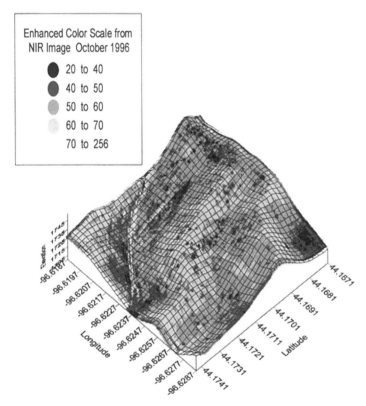

Figure 5. Gray-scale rendering of a color-enhanced, near-infrared, digital image of a field in Moody Co., South Dakota. (Figure by Cheryl Reese and Scott Christopherson)

provides indications of problems and the need for on-site observations. In the early spring, soil-water content, drainage patterns, and residue cover are easily observed. In mid to late June, aerial imagery can be used to identify anomalous areas that correspond to weed infestations. In late July and August, greenness indices can be calculated, and areas in the field where the crop is stressed can be determined. Images from the fall months show both stressed areas and areas that remain green; these images have a high correlation with weed infestations.

Factors to Consider. An understanding of the relationships between spectral, spatial, and time variables is important in utilizing remote-sensed information for assessing crop conditions. Spatial variables relate to both the pixel and minimum object/feature size to be seen by remote sensing. A 1 m^2 resolution integrates the reflectance in the size of about a tabletop, and gives one value for the area. At 5 m^2 resolution, the integrated area is approximately the size of an average bedroom, and at 30 m^2, the integrated area is about the size of six school buses parked together. The object or feature size must also be considered. If the feature occupies a small point in space, very fine resolution is needed. As the size increases, coarser resolution can be used. In Figure 4b, the minimum Canada thistle patch size that could easily be distinguished at 1 m^2 resolution was about 30 m^2 that contained about 11 shoots/m^2. Airborne platforms will normally have between 1 and 5 m^2 resolution, depending on the height of the overflight. Today's satellite imagery has a resolution that is about 30 m^2. Generally, remote-sensed information must be rectified to the proper geometry to allow integration with other data sources in GIS.

Although the use of aerial imagery in agriculture is still in the developmental stages, it is likely that valuable information can be gained about production-sized fields by using aerial imagery. In the future, aerial imagery of weed infestations will be one of the many tools that practitioners will use routinely for site-specific management.

PLANT DISEASES

Spatial variability. Plant disease incidence and severity seldom are distributed either uniformly or randomly over a field. Although patchy distribution of diseases and nematodes is the rule rather than the exception, relatively few studies have applied GIS/GPS technology to the detection of spatial variability in plant pathogens. An example of spatial variability of nematodes in potato is shown in Figure 6. Reviews of the early application of geostatistics to the description of spatial variability in plant diseases and nematodes are available (10,37). Applications of GIS for management of plant virus disease on a regional scale also have been proposed (25).

Microclimates. Use of GIS/GPS to identify microclimates within a field, based on topography, windflow patterns, or soil drainage, can be an important step in site-specific management for plant diseases. Specific areas with environments favorable for disease development can be managed with preventive treatments or more intensive scouting through adaptive or targeted sampling for signs of disease development. Intensive sampling of entire fields can be economically infeasible. Targeting limited sampling resources in those areas of the field at greatest risk of disease development due to host susceptibility, microclimate, or inoculum potential can obtain information for sound decisions without unacceptably high costs. Early treatment of only those areas in the field with environmental conditions favorable to disease, rather than treatment of the entire field, will reduce the cost of treatment without increasing the risk of disease development. Expensive, but effective, treatment might be economically acceptable if applied only to those areas of the field at greatest risk for disease development. A series of management options, based on risk of disease development, can be tailored to microclimates present in each field.

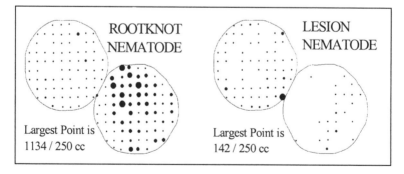

Figure 6. Spatial distribution of nematodes in a commercial potato field based on 1.6-ha grid samples. The density of the nematode population at a sample point is proportional to the size of the circles.

430

Soil Spatial Variability. Spatial variability of soil factors, such as soil texture, water-holding capacity, pH, and nutrient status, impacts not only the incidence and severity of disease development, but also the efficacy of disease management strategies. In two 32-ha commercial potato fields that shared a common boundary, pH values measured on a 0.4-ha grid ranged from a low of 4.5 to a high of 8.5. It would be unlikely that a chemical or biological agent would perform the same at pH 4.5 as it would at pH 8.5. Inconsistent results in field tests of soilborne disease-control strategies might be due to differences in effects of the soil environment on the chemical or biological control agent rather than on any inherent variability or unreliability in the agent itself. Characterizing the spatial variability of important soil characteristics and selecting disease management strategies that are best suited to each combination of soil factors could result in more consistent, and therefore less risky, disease management. Management decisions should be made after considering the environmental conditions surrounding the disease agents and the control agents.

Considerations for IPM of Plant Diseases. Integrated site-specific pest management will facilitate the inclusion of biological control agents into the total pest management strategy. Use of biological control agents would be less risky if stronger measures, such as a chemical, could be applied in a site-specific manner only to those areas of the field where pest populations exceed the action threshold. Because chemical control is applied only in specific areas of the field, biological agents in the rest of the field would not be harmed and could quickly recolonize the chemically treated area. Another advantage to this approach is that growers might be more willing to try biological control agents if a safety net existed to control hot spots.

Economic Thresholds. The concept of economic threshold will change with site-specific management, because treatment is focused only on those areas that need treatment at the minimal effective rates of application. Treating large areas of the field with "softer" treatments and reserving the "harder" treatments for hot spots might mean that more costly, but more effective, treatments would be economically justified in the hot spots. Treatment might begin sooner, since a smaller area would be treated, or later, if site-specific microclimate considerations predict a slowing of disease development. Small hot spots (and resulting yield loss), which might be ignored if the only option is to treat the entire field, could be treated if site-specific technologies were available.

Factors Constraining Adoption of Emerging Technologies in IPM

The National Research Council (24) recently completed an assessment and overview of technological advances in spatial and temporal approaches to agricultural information management that form the basis for the concept of precision agriculture. That report enumerated several factors that were considered as constraints to the adoption and diffusion of precision agriculture, particularly where public institutions are involved. These points will serve as a point of departure for the present discussion (reproduced with permission):

- a need for new measurement and analysis tools
- an unbiased approach to evaluation of precision agriculture
- new approaches to research on complex systems
- training and education targeted to needs of precision agriculture
- a need for rapid information transfer technologies at the rural level

MEASUREMENT AND ANALYSIS

The cost of manual sampling and scouting in terms of labor and time is a serious constraint to obtaining data at the intensity necessary for accurate characterization of spatial variability in pest populations. Ideally, a capability for real-time in-field decision-making would allow immediate implementation of timely management practices. Remote-sensing technologies linked to GIS/GPS technology offer great promise for real-time monitoring of pest populations to facilitate map-driven application technology. Public and private research efforts are being invested in sensor development, but progress in this area lags behind other aspects of site-specific agriculture (30). The necessary GIS/GPS capabilities are available but have not been effectively combined into platforms capable of real-time monitoring and mapping of pest or disease variability. Optical sensors are being developed that will detect weed seedlings in real-time (17), but consistent performance of these systems is hampered by effects of variable light and soil-surface conditions. It has been suggested (15) that optical sensors might be applied to detection of canopy-dwelling insect pests. Targeted sampling can be directed by

432

analysis of remote-sensed aerial images that identify anomalous areas indicative of severe disease incidence or pest infestations (Willers, unpublished), provided that the cost of the imagery can be kept at reasonable levels and still provide rapid turn-around.

UNBIASED EVALUATION

The National Research Council report (24) suggested that public institutions and agencies could play an important and critical role in providing unbiased and objective evaluation of precision farming technology. The report also warned that meaningful cooperation between private concerns and public agencies, and among public agencies themselves, may be difficult to bring about. Particularly important will be unbiased comparisons between precision farming approaches and more traditional agricultural methods. Such comparisons must account for regionally variable conditions in order for producers to adapt them to their own unique circumstances. It was also noted that the technology of site-specific agriculture is changing so rapidly that evaluation of these technologies will be difficult without collaboration between commercial concerns and public agencies.

NEW RESEARCH APPROACHES

Precision agriculture seeks to apply information technologies to decision-making and development of an understanding of interactions between complex and interrelated components of agroecosystems. These components range from integrated pest management in its broadest sense (insects, weeds, and diseases, inclusively) to agronomic factors, such as yield history, land use, and soil nutrient management, approached on a whole-farm or regional basis. Traditional research approaches using small plot techniques and controlled experiments are not well suited to the layered, integrative, and holistic approach involved in site-specific management. Cooperation between researchers from public institutions, producer-cooperators, and private firms will be critical to successfully bring about meaningful on-farm research in production settings that cannot be duplicated in small plot studies or on institutional research farms. A critical issue will be removal of constraints to sharing and managing data between researchers and cooperators in a manner that will make the best use of large databases for decision-making purposes.

Precision agriculture and site-specific integrated pest management are by definition information-intensive undertakings. Accessibility to computer-based information technology, i.e., GIS/GPS capability, is essential and assumed in the precision-farming approach. Computer-based information management, training, and education will be necessary for farmers, crop production professionals, and researchers in multi-disciplinary precision-farming projects. Some commercial firms already are offering short courses in site-specific agriculture. This has occurred partly out of necessity, because agricultural professionals must have trained employees to stay competitive with the growing demand from producers for site-specific farming input (31). The need for education and training in site-specific farming has motivated modifications in course offerings and teaching methods at public institutions, attuned to the multidisciplinary concepts engendered in precision agriculture.

RURAL-LEVEL INFORMATION TRANSFER

The effectiveness of site-specific approaches to integrated pest management will likely be dependent on accessibility to the data generated through precision-agriculture technologies. Network capability for information exchange, such as that possible through Internet access, will be critical. Information exchange must be routinely accomplished for full benefit to be attained in rural areas from the potential for regional accumulation of layered, spatially referenced data that is now possible through use of GIS/GPS technologies. This information can be used to drive pest management decision-making and predictive-modeling efforts (26).

Literature Cited

1. Bauer, M. 1995. Spectral inputs to crop identification and condition assessment. Proc. IEEE 73:1071-1085.
2. Brenner, R. J., Focks, D. A., Arbogast, R. T., Weaver, D. K., and Shuman, D. 1998. Practical use of spatial analysis in precision targeting for integrated pest management. Amer. Entomol. 44:79-101.

3. Brewster, C. C., and Allen, J. C. 1997. Spatiotemporal model for studying insect dynamics in large-scale cropping systems. Environ. Entomol. 26:473-482.

4. Brown, R. B., Steckler, J. P., and Anderson, G.W. 1994. Remote sensing for identification of weeds in no-till corn. Trans ASAE. 37:297-302.

5. Burrough, P. A. 1986. Principles of geographical information systems for land resources assessment. Monogr. on Soil and Resources, Survey No. 12, Oxford Univ. Press, New York.

6. Cardina, J., Sparrow, D. H., and McCoy, E. L. 1995. Analysis of spatial distribution of common lambsquarters (*Chenopodium album*) in no-till soybean (*Glycine max*). Weed Sci. 43:258-268.

7. Carter, G. A. 1993. Responses of leaf spectral reflectance to plant stress. Am. J. Bot. 80:239-243.

8. Chandler, L. D., Woodson, W. D., and Ellsbury, M. M. 1998. Corn rootworm areawide IPM: Implementation and information management with GIS. Pages 129-143 in: Proc. 52nd Ann. Corn Sorgh. Res. Conf., Dec. 10-11, 1997, Chicago.

9. Chellemi, D. O., Rohrbach, K. G., Yost, R. S., and Sonoda, R. M. 1988. Analysis of the spatial pattern of plant pathogens and diseased plants using geostatistics. Phytopathology 78:221-226.

10. Collins, W. 1978. Remote sensing of crop type and maturity. Photogrammetric Engineering and Remote Sensing. 44:43-55.

11. Donald, W. W. 1994. Geostatistics for mapping weeds, with a Canada thistle (*Cirsium arvense*) patch as a case study. Weed Sci. 42:648-657.

12. Ellsbury, M. M., Woodson, W. D., Clay, S. A., Malo, D., Schumacher, J., Clay, D. E., and Carlson, C. G. 1998. Geostatistical characterization of the spatial distribution of adult corn rootworm (Coleoptera: Chrysomelidae) emergence. Environ. Entomol. 27: 910-917.

13. Everitt, J. H., Pettit, R. D., and Alaniz, M. A. 1987. Remote sensing of broom snakeweed (*Gutierrezia sarothrae)* and spiny aster (*Aster spinosus*). Weed Sci. 35:295-302.

14. Fleischer, S. J., Weisz, R., Smilowitz, Z., and Midgarden, D. 1997. Spatial variation in insect populations and site-specific integrated pest management. Pages 101-130 in: The state of site-specific management for agriculture. F. J. Pierce, and E. J. Sadler, eds. ASA, CSSA, and SSSA, Madison, WI.

15. Heilman, J. L., Heilman, W. E., and Moore, D. G. 1981. Remote

sensing of canopy temperature in complete cover. Agron. J. 73:403-406.

16. Johnson, G. A., Cardina, J., and Mortensen, D. A. 1997. Site-specific weed management: Current and future directions. Pages 131-147 in: The state of site-specific management for agriculture. F. J. Pierce, and E. J. Sadler, eds. ASA, CSSA, and SSSA, Madison, WI.

17. Keegan, H. J., Schleter, J. C., Hall, W. A., Jr., and Haas, G. M. 1956. Spectrophotometric and colorimetric study of diseased and rust resisting cereal crops. Natl. Bur. Stds. Rept. 4591.

18. Lems, G. J. 1998. Weed spatial variability and management on a field-wide scale. M.S. Thesis. South Dakota State University.

19. Liebhold, A. M., Rossi, R. E., and Kemp, W. P. 1993. Geostatistics and geographic information systems in applied insect ecology. Annu. Rev. Entomol. 38:303-327.

20. Menges, R. M., Nixon, P. R., and Richardson, A. J. 1985. Light reflectance and remote sensing of weeds in agronomic and horticultural crops. Weed Sci. 33:569-581.

21. Midgarden, D. M., Fleischer, S. J., Weisz, R., and Smilowitz, Z. 1997. Impact of site-specific IPM on the development of esfenvalerate resistance in Colorado potato beetle (Coleoptera: Chrysomelidae) and on population densities of natural enemies. J. Econ. Entomol. 90: 855-867.

22. Mulla, D. J. 1997. Geostatistics, remote sensing and precision farming. Pages 100-119 in: Precision agriculture: Spatial and temporal variability of environmental quality. J. V. Lake, G. R. Bock, and J. A. Goode, eds. Ciba Foundation Symposium 210, January 21-23, 1997, Wageningen, The Netherlands. Wiley, New York.

23. National Research Council. 1997. Precision Agriculture in the 21st Century: Geospatial and Information Technologies in Crop Management. National Academy Press, Washington, D.C.

24. Nelson, M. R., Felix-Gastelum, R., Orum, T. V., Stowell, L. J., and Myers, D. E. 1994. Geographic information systems and geo-statistics in the design and validation of regional plant virus management programs. Phytopathology 84:898-905.

25. Parker, W. E., and Turner, S. T. D. 1996. Application of GIS modelling to pest forecasting and pest distribution studies at different spatial scales. Aspects Appl. Biol. 46:223-230.

26. Roberts, E. A., Ravlin, F. W., and Fleischer, S. J. 1993. Spatial data representation for integrated pest management programs. Amer.

Entomol. 39:92-107.

27. Rossi, R. R., Borth, P. W., and Tollefson, J. J. 1993. Stochastic simulation for characterizing ecological spatial patterns and appraising risk. Ecol. Applic. 3:719-735.

28. Rossi, R. E., Mulla, D. J., Journel, A. G., and Franz, E. H. 1992. Geostatistical tools for modeling and interpreting ecological spatial dependence. Ecol. Monogr. 62:277-314.

29. Sudduth, K. A., Hummel, J. W., and Birrell, S. J. 1997. Sensors for site-specific management. Pages 183-210 in: The state of site-specific management for agriculture. F. J. Pierce, and E. J. Sadler, eds.. ASA, CSSA, and SSSA, Madison, WI.

30. Swinton, S. M., and Lowenberg-DeBoer, J. 1998. Evaluating the profitability of site-specific farming. J. Prod. Agric. 11:439-446.

31. Thompson, S. K. 1997. Spatial sampling. Pages 161-172 in: Precision agriculture: Spatial and temporal variability of environmental quality. J. V. Lake, G. R. Bock, and J. A. Goode, eds. Ciba Foundation Symposium 210, January 21-23, 1997, Wageningen, The Netherlands. Wiley, New York.

32. Thompson, J. F., Stafford, J. V., and Miller, P. H. C. 1990. Selective application of herbicides to UK cereal crops. ASAE Paper No. 90-1629 ASAE, St. Joseph, MO.

33. Tyler, D. A., Roberts, D. W., and Nielsen, G. A. 1997. Location and guidance for site-specific management. Pages 161-181 in: The state of site-specific management for agriculture. F. J. Pierce, and E. J. Sadler, eds. ASA, CSSA, and SSSA, Madison, WI.

34. Wallace, M. K., and Hawkins, D. M. 1994. Applications of geo-statistics in plant nematology. Suppl. J. Nematol. 26:626-634.

35. Walters, M. L., Sequeira, R. A., and Staten, R. T. 1998. GIS use in area-wide pink bollworm management. Pagess 314-319 in: Proc. 1st International Conference on Geospatial Information in Agriculture and Forestry, Vol. II, June 1-3, 1998, Lake Buena Vista, Florida. ERIM International, Ann Arbor, Michigan.

36. Waring, R. H., Way, J. B., Hunt, E. R., Jr., Morrissey, L., Ranson, K. J., Weishampel, F. F., Oren, R., and Franklin, S. E. 1995. Imaging radar for ecosystem studies. BioSci. 45:715-723.

37. Weisz, R., Fleischer, S. J., and Smilowitz, Z. 1996. Site-specific integrated pest management for high-value crops: impact on potato pest management. J. Econ. Entomol. 89:501-509

38. Weisz, R., Smilowitz, Z., and Fleischer, S. J. 1996. Evaluating risk of Colorado potato beetle (Coleoptera: Chrysomelidae) infestation across a landscape as a function of migratory distance. J. Econ. Entomol. 90:855-867.

Geographical Information Systems and Global Positioning Systems for Optimizing Banana Production

Douglas H. Marin[1], **José Eduardo Soto**[2], and **Roberto Valenciano**[1]

[1]Del Monte Agricultural Development Corporation, Banana Division (BANDECO), San José, Costa Rica, and [2]National Banana Corporation (CORBANA), San José, Costa Rica.

"Though of ancient origin, the banana has become a staple food of modern times. Probably few of the millions who today enjoy this fruit know of its colorful and varied history, or realize the wide range of human endeavor involved in its cultivation and in its long but rapid journey from the tropical plantation to the consumer's table." (17)

Bananas *(Musa* AAA) and plantains (*Musa* AAB) are important components of the human diet in almost every country of the world, either as a cooked food or fresh fruit. They are an excellent food source, and in some countries of the world (e.g., parts of Africa) are the principal components of the diet. Bananas provide a source of fiber, are low in sodium, and are a rich source of vitamin B6 and potassium (2). They are the most widely consumed fresh fruit in the United States and are found year-round in supermarkets in the largest cities and the smallest towns. The average American citizen eats more than 10 kg of bananas yearly (9).

Over 76 million metric tons of bananas and plantains are produced worldwide each year, accounting for over 1.5 million jobs (16). Tropical areas that produce dessert fruit for export are, in order of importance, Ecuador, Costa Rica, Colombia, The Philippines, Honduras, and Panama. However, production in these countries amounts to only 14% of all *Musa* production worldwide. Other tropical production areas are located in the Caribbean (Windward Islands, French West Indies) and West Africa (Ivory Coast,

Cameroon). Their export volumes are considerably smaller than those from Latin America (18). Most of the bananas commercially produced today belong to the "Cavendish" subgroup. In Costa Rica, banana exports accounted for more than half a billion dollars in revenue, second only to tourism. Costa Rica produces about 20% of the world's premium bananas.

Bananas are affected by more than a dozen major pests and diseases that can ruin fruit quality and lower production. Potential losses caused by pests and diseases are estimated as 10-20% by foliage feeders, 1-15% by fruit insects, 20-50% by plant parasitic nematodes, 5-10% due to weed competition, and 50-100% by black Sigatoka (14). Many years of research have been devoted to producing disease-resistant cultivars; however, there are currently no disease-resistant banana cultivars that produce commercially acceptable fruit. Bananas have no seed and therefore traditional plant breeding methods are extremely difficult.

Modern crop protection chemicals make it possible to manage pest and disease problems and produce the quality and quantity of fruit that consumers worldwide demand. Crop protection chemicals are used only when needed. Continuous research by the banana industry, backed by quality assurance programs and government testing, assure that the fruit meet established safety and quality standards. The banana industry in Costa Rica spends millions of dollars annually on research to protect the environment and improve fruit quality and pest control. Integrated pest management (IPM) is widely practiced to limit the use of crop protection chemicals.

Understanding the spatial and temporal variation within management areas at the plantation level opens the possibility of rational agrichemical use to maximize yield with a minimum environmental impact. This approach, combined with software specially designed for this purpose, would provide a diagnostic tool to identify problematic areas for further analysis (25). Banana Management (BanMan) software, developed by J. J. Stoorvogel from the Agricultural University in Wageningen, The Netherlands, and supported by the Fundación para el Desarrollo de la Investigación y la Agricultura de Precisión en el Cultivo de Banano (FUNDIAP, Foundation for the Development of Research and Precision Agriculture in Bananas), is the first step toward implementing precision agriculture in the banana plantations of Costa Rica.

FUNDIAP is a non-profit foundation whose objectives are the promotion and development of applied research and precision agriculture for the cultivation of bananas (8).

The small number of agronomic and scientific alternatives for low-input banana cultivation using the current cultivars in conjunction with declining productivity, and more stringent environmental constraints are the driving forces for implementing precision agriculture for banana production in the humid tropics (10,21,24).

Global Positioning Systems for Black Sigatoka Control

Control of Sigatoka leaf spot diseases is one of the most difficult challenges facing the production of bananas, plantains, and other cooking bananas. There are two Sigatoka diseases: yellow Sigatoka (YS), caused by *Mycosphaerella musicola* Leach and previously known as Sigatoka disease, and black Sigatoka (BS), caused by *Mycosphaerella fijiensis* Morelet (synonym, black leaf streak) (5,13,15). The first reports of BS in Fiji emphasized the increased destructiveness of the disease compared to YS and the speed with which it spreads and develops to epidemic proportions where climatic factors are favorable and large areas of bananas are grown.

In Central America, BS is the most damaging and costly disease; its control accounts for 20-30% of all production costs (11). *M. fijiensis* attacks the leaves of banana plants, destroying the foliage very rapidly if control measures are not applied. It affects plant growth and yield because of the reduction in the photosynthetic area. The most important cause of losses is probably the premature ripening of fruit, which can occur in the field and/or during transport and storage (12).

Chemical control of BS is the only available option in export situations because of the susceptibility of the commercial Cavendish cultivars. Chemical control consists of applications of a fungicide plus mineral oil. The protectant fungicides (mancozeb and chlorothalonil) and the systemics (benomyl and other benzimidazols, tridemorph, and several DMI fungicides) are widely used. Resistance to benzimidazols and DMI's is widespread. Recently, azoxystrobin, a strobilurin fungicide, has been incorporated in black Sigatoka management programs. Timing of applications is mainly based on the residual effect and eradicant action of the fungicides, as well as knowledge of the climatic conditions and the fluctuation of the disease level during the year. Low-volume sprays are generally applied from the air or, less often, from the ground. Aerial application is commonly performed with turbo-prop airplanes, flying at speeds ranging 224-256 km/h at 15-17m from the ground (5,6,12). Until recently, aerial applications were guided by flagging personnel on the ground, carrying "flags" through lines traced in the field. These lines were usually located from 500-700 m away to facilitate the airplane orientation (12).

Electronic guidance systems were adopted by the agricultural aviation industry because of safety, legal, labor, and time issues, among other factors (1). The first electronic flagging system was used commercially in Costa Rica in June 1994; today, most companies have replaced flagging personnel with these sophisticated devices. The use of the differential Global Positioning Satellite System (GPS) in the banana industry in Costa Rica has proved to be advantageous over other guidance systems. Global Positioning Systems are not affected by land irregularity and protection areas, which can interrupt the GPS signals, deciding factors in selecting one system over another.

Global positioning systems depend upon a receiver that receives satellite-transmitted data and determines the receiver's location. The U.S. government uses selective availability (SA) and scrambles the satellite-transmitted data. Thus, to eliminate SA, there must be a second GPS receiver at a known location, usually a tower with an UHF radio frequency. The receiver at the known location calculates the difference between the information it is receiving from the satellites and its actual location, correcting the satellites' inaccuracies. This difference is transmitted to the aircraft and thus eliminates SA (3).

The use of GPS has improved the quality of aerial applications and eliminated exposure of flagging personnel to the fungicides. Exposure of field workers has also been reduced by special programs, such as night flying.

Figures 1 and 2 illustrate flight tracks of aerial applications made with flagging personnel in the field and with GPS equipment. In spite of the fact that "flagging" lines were topographically measured and established in the field, human errors resulted in irregular patterns of

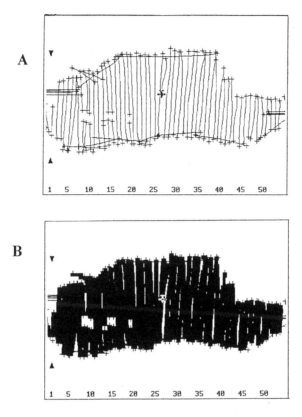

Figure 1. Flight tracks of an aerial application of a banana farm using flagging personnel in the field. A) Flight tracks, B) Theoretical coverage based on used swath. Drawings produced by Del Norte Flying Flagman® (Flite-Trac™).

application (Fig. 1 A). These patterns resulted in non-protected areas (Fig. 1B), which became evident when the application was analyzed using the GPS software, Flite-Trac™ (Del Norte Technologies, TX). In contrast, an application made using the electronic flagging system followed a regular pattern and provided complete spray coverage (Fig. 2).

GPS and other technologies have greatly changed the application of fungicides for black Sigatoka control in the banana industry

443

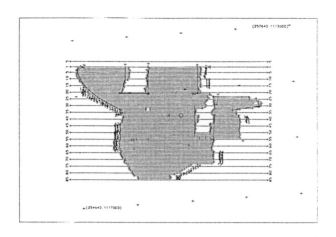

Figure 2. Flight tracks of an aerial application of a banana farm using GPS equipment (Del Norte Technologies). Drawing produced by Del Norte Flying Flagman® (Flite-Track™).

during the last decade. The use of turbo-prop aircraft, CP-nozzles (constant-pressure), and electronic flagging systems has improved quality and satisfied needs for increased productivity to meet a demanding market.

Geographical Information Systems for Optimizing Banana Production

Precision agriculture applies principles and practices of crop management to relatively small areas according to their specific conditions and needs. Geographical information systems are cornerstones for the establishment of precision agriculture. The systems are useful to manage, analyze, and visually present data and georeferenced images (20,22). In Costa Rica, particularly in bananas, GIS are being developed to optimize crop production. Initially, GIS was used to analyze productivity, but it is beginning to be used to better manage pests and diseases.

For the establishment of a GIS, soil characterization of the banana farm is required. In addition to the location of the farm

444

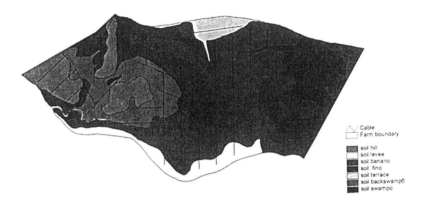

Figure 3. Map of soils classified according to their suitability for banana cultivation.

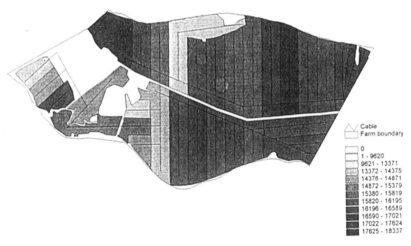

Figure 4. Biomass production (kg of fruit/ha) per cable of a banana plantation.

boundaries, cables and drainages must be located (Fig. 3). Information on production (i.e., bunch weight) is collected using an electronic scale connected to a computer. Further analyses can be done according to needs, which may vary from whole cables to specific map units (i.e., 2000 sq. m.). From the practical point of view,

445

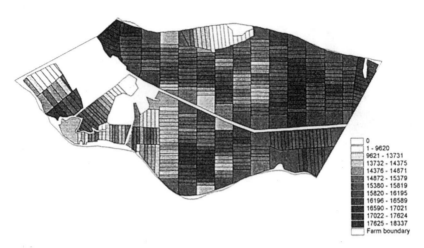

Figure 5. Biomass production (kg of fruit/ha) per unit of 2000 sq. m. of cultivated bananas.

analysis by cable (Fig. 4) is operationally sound; however, technical decisions are usually made at a more detailed level. Units from 0.2 to 1.0 ha are probably optimal for evaluating the productivity of a banana farm. Specific units (i.e. 2000 sq. m., Fig. 5) may give a better idea of the performance of an area, especially because of the soil heterogeneity of banana plantations.

Spatial analyses can help separate the effect of differential productivity due to soil type. If the proportion of each soil type in each unit is accounted for in an analysis, then the resulting information can be related to the mean production and used to estimate more objectively the condition of an area. Further observation under field conditions is required to determine the cause of low productivity. Figure 6 provides an example of this type of analysis. The clear areas represent specific units that are producing yields that are within 10 percent of mean yield for a specific soil type or its combination. The shaded areas show sites producing yields below and above (at 10% intervals) the mean yield.

GIS has begun to be used to evaluate the spatial distribution of diseases such as black Sigatoka (Fig. 7 and 8). The extent of black Sigatoka infection is usually measured by the youngest leaf infected

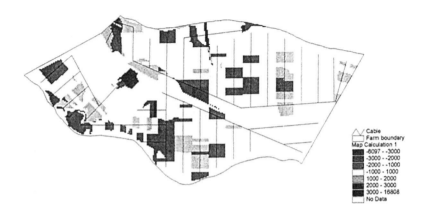

Figure 6. Estimation of mean biomass production according to soil type.

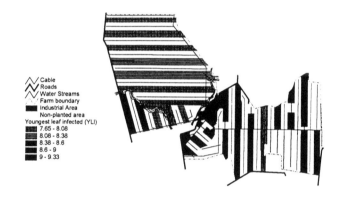

Figure 7. Black Sigatoka infection on a banana farm. Youngest leaf infected.

(YLI) and the youngest leaf spotted (YLS), which gives an idea of the status and development of the disease. The analysis of black Sigatoka infection is mostly based on the mean of the farm, and the distribution in the field is rarely considered. However, a spatial analysis such as the one presented in Figures 7 and 8 (YLI and YLS, respectively) may provide a useful tool to understand the development of the infection and to establish strategies for control.

Plant-parasitic nematodes are also widespread and are the most damaging soilborne pests of all bananas, wherever they are grown. Nematodes cause severe crop losses in commercial Cavendish plantations and seriously limit the productivity of non-export dessert and cooking bananas (7,19,23). Successful control of nematodes in established banana plantations relies on the use of granular nematicides (7,23), which are applied to the plants up to three times a year in Costa Rica. Nonetheless, nematicides are usually applied without considering nematode distribution in the field. A GPS–based decision system would help to improve nematode control strategies; however, basic research on temporal and spatial distribution of pest nematodes is still required before a GIS can be used.

GPS are fully established in most banana-growing areas of Central America, and are in their implementation phase in other banana areas. Use of GPS has improved fungicide applications and reduced the risk of worker exposure to these products. GIS are being developed for optimizing banana production. Although most efforts currently are focused on improving productivity, crop production is not an isolated event, and banana pests must be included for banana production systems to become more sustainable in time and space.

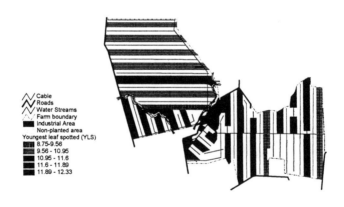

Figure 8. Black Sigatoka infection on a banana farm. Youngest leaf spotted.

Acknowledgments

We would like to thank FUNDIAP and Del Monte for their support to this project, and J. C. Madrigal, A. Obando, and H. Fonseca for their technical assistance. Critical review of manuscript by T. B. Sutton was highly appreciated.

Literature Cited

1. Anonymous. 1993. Electronic flagging systems. Choices and confusion. Agricultural Aviation 20(4):10.
2. Chandler, S. 1995. The nutritional value of bananas. Pages 468-480 in: Bananas and Plantains. S. Gowen, ed. Chapman & Hall, London.
3. Denton, J. 1994. GPS Systems. How they work and more. Agricultural Aviation 21(2):11.
4. FAO. 1992. Production Yearbook 1991. Vol. 45. Food and Agricultural Organization, Rome, Italy.
5. Fullerton, R. A. 1994. Sigatoka leaf diseases. Pages 12-14 in: Compendium of Tropical Fruit Diseases. R. C. Ploetz, G. A. Zentmeyer, W. T. Nishijima, K. G. Rohrbach, and H. D. Ohr, eds. APS Press, St. Paul.
6. González, M. 1987. Enfermedades del cultivo del banano. Oficina de Publicaciones de la Universidad de Costa Rica, San José.
7. Gowen, S., and Queneherve, P. 1990. Nematode parasites of bananas, plantains, and abaca. Pages 431-460 in: Plant Parasitic Nematodes in Subtropical and Tropical Agriculture. M. Luc, R.A. Sikora, and J. Bridge, eds. CAB International, Wallingford, U.K.
8. Guzmán, M., and Marín, D. H. 1998. La creación de FUNDIAP. Investigaciones al día 2(5):1.
9. Hallam, D. 1995. The world banana economy. Pages 509-533 in: Bananas and Plaintains. S. Gowen, ed. Chapman & Hall, London.
10. Hartshorn, G., Hartshorn, A., Atmella, A., Gomez, L. D., Mata, A., Mata, L., Morales, R., Ocampo, R., Pool, D., Pool, C., Quesada, C., Solera, C., Solorzano, R., Stiles, G., Toshi, J.,

Umaña, A., Villalobos, C., and Wells, R. 1982. Costa Rica country and environmental profile. A field study. Tropical Science Center, San José, Costa Rica.

11. Jacome, L. H., and Schuh, W. 1992. Effects of leaf wetness duration and temperature on development of black Sigatoka disease on banana infected by *Mycosphaerella fijiensis* var. *difformis*. Phytopathology 82:515-520.

12. Marín, V. D., and Romero C. R. 1992. El combate de la Sigatoka negra. Boletín No.4. Corporación Bananera Nacional, San José, Costa Rica.

13. Meredith, D. S. 1970. Banana leaf spot disease (Sigatoka) caused by *Mycosphaerella musicola* Leach. Phytopathological Papers, No. 11. Commonwealth Mycological Institute, England.

14. Mirenda, J. 1998. Bananos y ambiente. CORBANA 23(49): 101-108.

15. Mourichon, X., and Fullerton, R. A. 1990. Geographical distribution of the two species *Mycosphaerella musicola* Leach (*Cercospora musae*) and *M. fijiensis* Morelet (*C. fijiensis)*, respectively agents of Sigatoka Disease and Black Leaf Streak Disease in bananas and plantains. Fruits 45:213-218.

16. Price, N. S. 1995. The origin and development of banana and plantain cultivation. Pages 1-13 in: Bananas and Plantains. S. Gowen, ed. Chapman & Hall, London.

17. Reynolds, P. K. 1927. The Banana: Its History, Cultivation and Place Among Staple Foods. Houghton Mifflin Co., Cambridge, MA.

18. Robinson, J. C. 1996. Bananas and Plantains. CAB International, Wallingford, England.

19. Sarah, J. L. 1989. Banana nematodes and their control in Africa. Nematropica. 19:199-216.

20. Schuiling, R., and Soto, E. 1998. Sistema de información geográfica (SIG). Pages 145-146 in: Informe Annual 1997. Corporación Bananera Nacional, Departamento de Investigaciones y Asistencia Técnica, San José, Costa Rica.

21. Serrano, E., and Marín, D. H. 1998. Disminución de la productividad bananera en Costa Rica. CORBANA 23:85-96.

22. Soto, J. E. 1998. ¿Qué es agricultura de precisión? Investigaciones al día 2:1.

23. Stover, R. H., and Simmonds, N. W. 1987. Bananas. 3rd ed. Longman Scientific & Technical.
24. Stoorvogel, J. J. 1995. Geographical information systems as a tool to explore land characteristics and land use with reference to Costa Rica. Doctoral Thesis. Wageningen Agricultural University, Wageningen, The Netherlands.
25. Stoorvogel, J. J., and Vargas, R. 1998. La agricultura de precisión en banano. Resúmenes del Taller Internacional sobre Producción de Banano Orgánico y/o Ambientalmente Amigable. EARTH-INIBAP, Guácimo, Costa Rica.

Section 7

Information Processing and Delivery

IPM has always been an information-intensive activity. However, only within the last decade have information technologies suitable for use in IPM involved both prediction and decision-support computer software and rapid, inexpensive communication. Fast, accurate, and complex decision-support systems have moved from theory to practice with the advent and continuing evolution of high-speed personal computers. Because of rapid technological advances in the ability to process and deliver information, information technology will play an ever-increasing role in pest management decision-making.

Timely and reliable access to accurate weather information has long limited our ability to make full use of our knowledge of basic pest and pathogen biology in the implementation of efficient and reliable pest management. In the first chapter of this section, Joseph Russo describes recent advances in the kinds, scale, and providers of weather information for IPM and the impact those are having on the ability to use IPM models. He then describes the implications for IPM of improvements in weather information that are likely to be achieved in the near future.

The emerging concepts for information processing and delivery are being built around information and database access via the World Wide Web. In the second chapter of this section, Ron Stinner provides a brief history of the World Wide Web and discusses developments underway that are likely to enhance the flow of information for IPM.

Weather Forecasting for IPM

Joseph M. Russo
SkyBit, Inc., Boalsburg, PA 16827-0010 USA

Weather is an important concern in any integrated pest management (IPM) program. It determines the timing and intensity of pest and host development, and it influences management decisions. In recent years, technical advances have affected the kinds, scale, and providers of weather information. This paper discusses these advances in the context of IPM programs. In particular, it focuses on the impact that recent innovations in weather forecasting are having on IPM models.

The paper is written from the perspective of an active participant in the development of commercial "information technology" products for agriculture, products based on decades of research in both agriculture and meteorology. The paper does not give adequate recognition to those scientists, both past and present, who have contributed to this research. However, it does, hopefully, convey the potential of these products to be an important source of data for IPM and other programs supporting management decision-making in crop production.

Background

The use of weather information in the study of pests and their management has a long and rich history. However, when one mentions weather and IPM, it is usually in reference to crop and pest models. Since the early1970s, agricultural meteorologists have worked closely with horticulturists, agronomists, plant pathologists, and entomologists in developing IPM models. Over the years, these models evolved from simple "rule-of-thumb" guidelines (14) to more sophisticated computer simulations (10). The goal of these models was to track crop and pest development in order to identify the best time, or "window," to optimize control tactics. This optimization has been shown to minimize, and sometimes eliminate, unnecessary pesticide applications without sacrificing crop quality and earnings.

Early pest models employed weather data from a nearby weather station, which, in most cases, was part of the federal cooperative network. Stations in the national network were equipped with manual instruments, which were read daily and the observations stored as paper records. The cooperative network varied in density from state to state, with many of the stations being distant from agricultural areas.

Beginning in the late 1970s, statewide networks began to supplement the federal networks. The new state networks began to favor electronic instruments over manual designs. The electronic instruments permitted weather data to be stored upon observation and read at a later time. Weather data from these electronic instruments, coupled with observations from traditional cooperative stations, became the main source of input into IPM models. Federal and state agencies and universities were responsible for collecting data, running the models, and delivering the results to the agricultural community.

By the mid-1980s, automated weather stations began to become a competitive alternative to network stations. Automated stations had a compact assembly of sensors powered by a battery. In early designs, electronically stored data could be viewed in a digital display by keying in a variable. In later designs, observed data could be "downloaded" to a portable computer or electronically sent over telephone lines via modems to a government or university processing center. Like the earlier electronic networks, the processed data from an automated station served as input into IPM models. Model output, which was in the form of tables, graphs, and crude maps, was transmitted to extension agents and other specialists for interpretation. Their interpretation included alerts to growers about imminent pest outbreaks. While attractive in design, price prevented automated weather stations from replacing manual instruments on the farm.

By the early 1990s, there was a marked drop in price and an improvement in the sensors and performance of automated weather stations. Growers could choose an affordable station from a number of manufacturers. The automated stations could be either a stand-alone unit or one linked by cable to a desktop computer. IPM models began to be distributed on floppy disks to extension personnel, growers, and consultants. In some cases, the same models were being programmed directly into automated stations.

The trend of lower price, more choice, and improved performance in automated weather stations continues even today. At one extreme, a single, enclosed temperature sensor can be hung in a canopy. It is capable of storing hourly electronic observations for months. The stored records can be "downloaded" to a laptop computer. At the other extreme, an automated weather station can consist of a framework of sensors feeding weather data to a microprocessor, which is powered by solar panels. The stored data are entered into built-in models and their results transmitted by radio to a nearby farm. The received model results can be displayed in various forms on a monitor and converted to hardcopy using a printer. Analysis software and expert system programs can help a grower interpret model results and make weather-sensitive decisions about growing a crop or controlling a pest.

The use of local weather equipment and models by a grower can be referred to as a "plug" technology; a grower purchases a self-contained unit, complete with sensors and models, and "plugs" it in for local use. As sophisticated as automated weather stations have become, they do not provide forecasts of pest development. In IPM programs, the term "prediction" is often confused with the term "forecast." To meteorologists, prediction and forecast are synonymous. They both mean a future weather state. In pest modeling, "prediction" refers to a model result, which is based on observed weather data as input. To avoid this confusion, it is probably best to speak of "real-time" simulations, or "now-casting," when referring to a modeler's definition of prediction. Favoring the meteorologist's bias, true pest "predictions" would be based on future weather conditions, or "forecasts."

Weather Forecasts and Data Scales

Pest predictions using weather forecasts have historically been at the "synoptic," or weather-system, scale. Meteorologists define weather phenomena according to their geographic extent and lifetimes.

Larger phenomena, such as long waves associated with the jet stream, can straddle a continent and last weeks, while smaller phenomena, such as dust devils, are about as wide as a road and last for seconds. Weather phenomena and the numerical models used to predict them are commonly organized along spatial and temporal scales, as shown in Figure 1. As depicted in the figure, "synoptic" models predict the future positions of weather systems, such as lows and highs. A forecaster trained in agricultural meteorology can interpret the surface weather conditions associated with the future position of these systems. He or she is able to predict the impact these conditions will have on pest development. Because of the relatively crude synoptic scale, these predictions of pest development have been only general in nature and for a large region.

Even today, many IPM programs rely on the forecast abilities of a synoptic meteorologist, whether he or she is in government or private industry. The synoptic or regional predictions of pest development provided by a forecaster are in sharp contrast to the local monitoring provided by an automated weather station. This difference in scale is depicted in Figure 1. The synoptic models that guide a forecaster are valid for spatial scales between 100 and 1000 km, while automated weather stations, which are the source of data for local monitoring, represent scales between 1 and 10 m. Even a traditional network of weather stations, whose records are the basis for 30-year and shorter-period climatologies, represents an area on the order of 100m^2. The "farm" scale, where a grower must manage pests in a field, sits about equidistant between the synoptic and the near-canopy scale of an automated weather station (Fig. 1). This in-between position means that a forecaster must, as he or she traditionally has done, predict surface weather conditions at the farm scale based on the movement of weather systems at the synoptic scale. Conversely, weather data recorded by an automated station at a point near or within a canopy must be interpreted upward for the farm scale. For highly variable terrain and different land covers, this "upscaling" may be difficult and, in the end, require additional instrumentation. Even if the data collected from local automated stations were successfully interpreted to track pest behavior at the farm scale, they still are not forecasts.

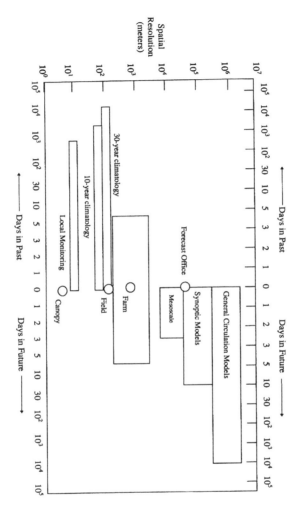

Figure 1. Spatial and temporal scales of models and data.

457

A forecaster trained in agricultural meteorology can certainly interpret weather at the farm level from future positions of weather systems. When making predictions, he or she can be guided by numerical weather predictions, such as from general circulation, synoptic and mesoscale models, and other forecasting tools, such as upper-air soundings and radar. However, given the sheer volume of data and the time required to make a forecast for even one farm, it is impractical to rely on human skill for site-specific weather information. Even if people were available to produce farm-scale forecasts, the cost for the information would be too prohibitive for a grower. While traditional practices are clearly untenable economically, new modeling approaches, which take advantage of surface data bases and the continual increase in computer power, offer an alternative path for realizing site-specific weather forecasts.

Site-Specific Forecasts

It is very clear from Figure 1 that there is an increase in detail of at least two orders of magnitude as one moves from the scales of forecast models down to site-specific predictions for a farm. The only time-sensitive and economically feasible way to move from the cruder scales of traditional forecasts to the scale demanded by IPM programs is to use models and other computer-supported tools to bridge the gap. SkyBit, Inc., has pioneered the development of an automated, electronic weather service by developing interpolation schemes and finer-scale forecast models to bring synoptic-scale observations and forecasts down to the farm scale (11,13). They call this service "E-Weather." Like all commercial weather firms, their source of weather data is from the National Weather Service (NWS), but, unlike other companies, they use computer technology and models to mimic the skill of a forecaster. In some sense, SkyBit has created a "virtual" meteorologist who can demonstrate skills comparable to those of a human meteorologist when making site-specific forecasts.

The flow of weather information in SkyBit's electronic, or "E-Weather," service is outlined in Figure 2. Beginning at the top of the figure, data from different sources, including surface weather stations (SRF), satellite (SAT), radar (RAD), and radiosonde balloons (SONDE) are collected by the National Weather Service (NWS) and

458

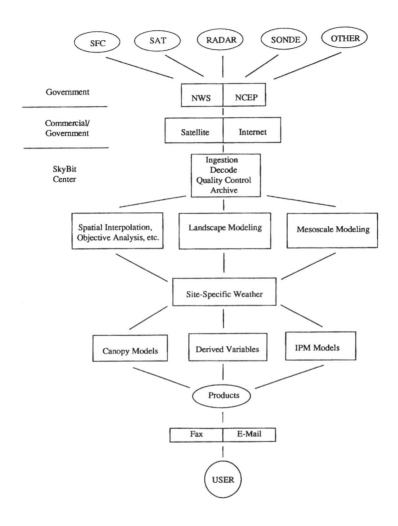

Figure 2. Flow of weather information from government to user.

shared with the National Center for Environmental Prediction (NCEP). NCEP is responsible for the public numerical weather prediction models. There are other (OTHER) sources, such as severe weather observations, river levels, and surface indices. The data managed by the federal government are delivered by satellite (NOAAPORT), dedicated data lines (Family of Services), and the Internet to other agencies and commercial entities. A few of the larger weather companies redistribute the "raw" data from the NWS to smaller businesses and individuals via satellite, telephone lines, cable, and the Internet. SkyBit receives NWS data via government and commercial channels.

After raw weather data are "ingested" at the SkyBit Center, they are decoded, subjected to quality control, archived, and reorganized in preparation for further processing by spatial interpolation and objective analysis schemes and by landscape and mesoscale models (Fig. 2). The resulting output from the analysis schemes and models are site-specific, simulated observations and forecasts for weather and weather-dependent soil and crop variables. The simulated observations and forecasts from these variables are organized electronically into single-sheet formats that form the core of "Tier 1" products (Fig. 2). An example of a Tier 1 product is presented in Figure 3. It is called the Combination Forecast and Summary, or "Combo" product for short. The product example is for a rural site in central Pennsylvania. The Combo product combines a short-term forecast out to 2 days in 3- or 6-hour timesteps, a medium-range forecast out to 7 days in 1-day timesteps, a long-range outlook beginning at 8 days and going out to 10 days in 1-day timesteps, and simulated daily observations for the past week. The Combo product is a good example of the large selection of variables available to the agricultural community. It includes custom variables, (e.g., the drying and spraying indices) that are designed for management decision-making. These indices, along with a number of other variables listed in the product, are not available from public sources.

The simulated observations and forecast data used to create Tier 1 products also serve as input into "application" models. These application models generate simulations and forecasts of soil moisture balances, pest development, and extreme weather conditions in a canopy. Output from application models are the source of data for

460

Figure 3. Example of E-Weather Combination Forecast and Summary (or Combo product) for a Site in Central Pennsylvania.

```
E-WEATHER FORECAST AND SUMMARY
For: PA-CENTRE COUNTY-ROCKSPRING
Date: THU MAR 18, 1999
```

<-------------- 0-48 HOUR FORECAST -------------->

DATE	Mar 18						Mar 19							
HOUR (EST)	7a	10a	1p	4p	7p	10p	1a	4a	7a	10a	1p	4p	7p	10p
TEMP (F)	42	41	42	40	36	31	27	26	25	28	34	35	31	27
2"- SOIL TEMP (F)	31	35	39	41	40	36	32	26	27	30	33	34	34	31
REL HUM (%)	58	63	61	63	67	74	80	80	75	66	53	49	55	64
6HR PRECIP(in)	.00/		.00/		.00/		.00/		.00/		.00/		.00/	
6HR PRECIP PROB(%)	4/		5/		6/		0/		1/		0/		2/	
3HR EVAP (in)	.01	.02	.03	.02	.01	.01	.00	.00	.00	.02	.03	.03	.01	.01
3HR WETNESS (hrs)	0	0	0	0	0	0	0	0	0	0	0	0	0	0
WIND DIR (pt)	W	W	WNW	WNW	WNW	WNW	WNW	WNW	WNW	WNW	WNW	WNW	WNW	NW
WIND SPEED (mph)	16	22	25	26	22	20	17	15	15	20	19	19	16	11
CLOUD COVER	SCT	BKN	BKN	BKN	BKN	OVC	OVC	OVC	BKN	SCT	BKN	BKN	BKN	BKN
3HR RADIATION (ly)	3	77	138	113	23	0	0	0	3	80	153	115	24	0
PCT RADIATION (%)	70	75	70	71	70				66	76	77	70	70	
DRYING (key)	6	6	7	7	6	5	4	4	6	7	7	7	1	5
SPRAYING (key)	0	0	0	0	0	0	0	0	1	0	0	0	1	3

	<-------	1-7	DAY	FORECAST	-------	--->	<8-10D	Outlook>	
DATE	Mar 18	19	20	21	22	23	24	25	26	27
DAY	THU	FRI	SAT	SUN	MON	TUE	WED	THU	FRI	SAT
MAX AIR TEMP (F)	44	36	46	47	41	45	46	48	48	49
MIN AIR TEMP (F)	27	24	21	28	30	27	27	28	28	29
PRECIP PROB (%)	15	6	5	63	75	43	38	43	42	41
AVG DAILY RH (%)	68	66	63	80	80	69	68	67	67	67
AVG WND SPD (mph)	20	17	14	7	11	12	8	9	9	8
DRYING (key)	6	6	6	3	4	5	5	5	5	5
SPRAYING (key)	0	0	2	6	3	3	5	5	5	5

	<-------	-------	-------	1-7 DAY SUMMARY	-------	-------	--->
DATE	Mar 11	Mar 12	Mar 13	Mar 14	Mar 15	Mar 16	Mar 17
DAY	THU	FRI	SAT	SUN	MON	TUE	WED
MAX AIR TEMP (F)	38	38	42	32	41	47	62
MIN AIR TEMP (F)	18	24	23	27	29	25	39
MAX 2" SOIL TMP(F)	37	37	41	32	41	46	61
MIN 2" SOIL TMP(F)	19	25	24	29	30	31	44
AVG DAILY RH (%)	59	54	56	79	60	49	42
PRECIP (in)	0.11	0.00	0.00	0.64	0.21	0.00	0.00

EVAP (in)	0.11	0.13	0.10	0.04	0.12	0.15	0.14
WETNESS (hrs)	13	0	0	12	23	0	0
WIND SPEED (mph)	12	13	6	6	11	10	8
RADIATION (ly)	434	437	444	225	435	495	438
PCT RADIATION (%)	93	92	93	47	90	100	89
DRYING (key)	6	7	6	3	6	7	8
SPRAYING (key)	3	2	7	7	4	5	6

key: 0..1..2..3..4..5..6..7..8..9..10
    LESS FAVORABLE   MORE FAVORABLE

Note: Key for drying and spraying assumes no precipitation.
      Local precipitation may result in unfavorable conditions.

Phone: 1-800-454-2266

Copyright(c) 1999 SkyBit, Inc.

463

"Tier 2" products. The application models, which are used to create the Tier 2 IPM products for SkyBit's E-Weather Service, have a legacy going back to the traditional disease and insect models created under university IPM research projects. However, unlike these earlier IPM pest models, SkyBit's models use hourly instead of daily weather data as input. While E-Weather IPM products were under development, university researchers and extension specialists provided invaluable assistance in bringing to light both data and publications on pest behavior. They also play an important role as outside evaluators of the accuracy and usefulness of the products (1,5,6,7,8,9,12).

An example of SkyBit's Tier 2 IPM product is the "IPM Apple Disease," which is shown in Figure 4. The product example depicts the simulated observations and forecasts of apple scab, fire blight, and sooty blotch for a rural site in central Pennsylvania during the 1998 growing season. Individual models for each disease simulate and predict, day by day, the accumulated wetness (AW) and temperature (TW) of a potential disease event and a pest "window" (PW). The pest window, through the use of symbols, informs the user of days when a disease is expected to be either not active (-), active but not infectious (+), or active and infectious (++). The pest models in SkyBit's IPM products determine if the weather conditions at a particular site are conducive to disease and insect development. If conditions are favorable, on-site scouting must still be carried out to determine the presence and population of a simulated disease or insect in a canopy.

SkyBit's E-Weather Service products are delivered either by electronic mail (e-mail) or by telephone facsimile (fax) to customers as end-users (Fig. 2). Customers provide either latitude and longitude coordinates or a geographic description of their farms. They pay a monthly subscription according to their choices of Tier 1 and Tier 2 products. The subscription includes frost alerts and other informational notices. As a season progresses, a customer has the option of adding or deleting products from his or her service and is billed accordingly. The product choices under Tier 1 apply to all crops, while the choices under Tier 2 are specific to a particular commodity. A summary of the present products offered through SkyBit's E-Weather Service is given in Figure 5. It is organized by

Figure 4. Example of E-Weather IPM Apple Disease Product for a Site in Central Pennsylvania During the 1998 Growing Season.

```
E-WEATHER SERVICE AGWEATHER IPM APPLE DISEASE PRODUCT
For: PA-CENTRE COUNTY-ROCKSPRING          Date: THU Jul  9, 1998
```

	WEATHER					APPLE SCAB 980324				FIRE BLIGHT 980331				SOOTY BLOTCH 980504	
Date	TMX F	TMN F	PREC in	ARH %	LW hr	ASM %	AW hr	TW F	PW	ADH 65F	AW hr	TW F	PW	ALW hr	PW
BASED ON OBSERVATIONS															
0701	79	59	0.02	62	9	100	30	68	++	225	30	68	++	424	++
0702	80	54	0.00	64	0	100	0	-	+	225	0	-	-	424	++
0703	82	55	0.00	68	7	100	7	60	++	225	7	60	+	431	++
0704	75	59	0.21	81	18	100	12	68	++	225	12	68	++	449	++
0705	78	57	0.00	69	9	100	21	65	++	225	21	65	++	458	++
0706	80	52	0.00	71	8	100	8	58	++	225	8	58	+	466	++
0707	80	60	0.42	79	14	100	8	74	++	225	8	74	++	480	++
0708	70	62	0.96	92	24	100	32	68	++	225	32	68	++	504	++
BASED ON FORECASTS															
0709	80	62	0.00	74	10	100	10	67	++	225	10	67	++	514	++
0710	77	60	0.00	62	0	100	0	-	+	225	0	-	-	514	++

465

```
0711  79  55  ----  76   0  100   0   -  +  225   0   -   -  514  ++
0712  77  54  ----  80  13  100  13  62  ++ 225  13  62  ++ 527  ++
0713  81  62  ----  68   2  100   2  65  +  225   2  65  ++ 529  ++
0714  84  63  ----  71   0  100   0   -  +  225   0   -   -  529  ++
0715  85  64  ----  77   8  100   8  70  ++ 225   8  70  ++ 537  ++
=================================================================
```

********** IMPORTANT: Check the dates at the top of each column. **********

```
Green Tip Date   - is used for Apple Scab
Blossom Date     - is used for Fire Blight
Petal Fall Date  - is used for Sooty Blotch
ASM = Apple Scab Maturity Percentage
ADH = Accumulated degree-hours from blossom date up to a max of 225.
AlW = Accumulated leaf wetness hours from petal fall date.
AW  = Accumulated wetness hours for the most severe event.
TW  = Average temperature during the most severe event.
PW  = Pest Wait/Watch/Warning:  - = not active
      + = active but no infection  ++ = possible infection & damage
=================================================================
```

Copyright (c) 1998 SkyBit, Inc.   Phone: 1-800-454-2266

```
=================================================================
```

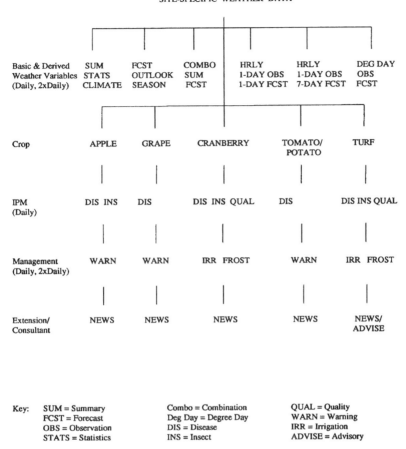

Figure 5. E-Weather products generated from site-specific weather data.

467

type of product and, where applicable among the products, by commodity.

SkyBit's E-Weather Service has several advantages over weather information provided by the NWS, state agencies, and universities, and by data that are locally collected with automated instrumentation. First, SkyBit, like the NWS, draws upon national networks of surface station, satellite and radar observations, and upper-air soundings. Unlike the NWS, SkyBit personnel can focus their energy in developing products for specific industries, such as agriculture. SkyBit can improve the spatial resolution of numerical output from NCEP synoptic models by applying its own landscape and mesoscale models. The finer-scale data serve as site-specific input into application models, such as those used to generate IPM products. Second, SkyBit, unlike state agencies, can design and evaluate products for national distribution. By working with universities and agencies, SkyBit can bring the best available weather data and the latest research on pest and IPM models to growers in their respective states. Third, growers as end-users do not have to make any investment in special equipment in order to receive E-Weather products. They simply need either a fax machine or a computer with Internet access. For growers and other end-users who have an automated weather station, SkyBit data can be used to track "drift" in sensor readings and can substitute for missing records.

Looking to the Future

SkyBit's E-Weather Service is in its infancy. It is part of an information technology era that will revolutionize how people in industry will make decisions. The power of the computer, coupled with analysis and modeling, permits an unprecedented processing of large amounts of data and the timely presentation of that data to end-users. The presentation of data is in an easy-to-read, accessible format and allows users to make informed and cost-effective decisions. As computers and models continue to improve, so will E-Weather products. In the not-too-distant future, technology will not be the limiting factor; it will be our lack of knowledge of how weather influences pest behavior at finer spatial and temporal scales.

What does the immediate future hold? Since SkyBit's data are the result of the further processing of NCEP's numerical predictions, improvement in government models translates into improvement in E-Weather products. There should be an incremental improvement in the short-to-medium range (1 to 10 days) forecasts provided by NCEP models. However, there should be a substantial improvement in the long-range (30 day to 1 year) forecasts. The improvement in the longer-range forecasts is due to the gradual maturity of global climate models and the better understanding of the role oceans play in long-term weather patterns. Another improvement to forecasts in the immediate future is the inclusion of a "risk" measure with a datum. Presently, only the 6-hour precipitation has a risk factor associated with it. The "probability" of precipitation is reported, along with a precipitation total on a forecast product. The expanded use of risk measures for other variables will identify times when a forecast is expected to be good or poor based on past model performance for similar weather conditions.

A future improvement in site-specific forecasts means an improvement in IPM models that use the fine-scale predictions as input. Besides the expected increased accuracy in the input, there very likely will be an improvement in the design of IPM models, especially those simulating pest development. University researchers, who become accustomed to using hourly weather data, will be able to design experiments and conduct surveys to collect pest data at a comparable temporal resolution. The results derived from this field research can be incorporated into future IPM model designs. In addition to improving the pest development component, new features will be added to IPM models. Some of these features will be oriented toward management decisions, such as the best time to spray given the forecasted appearance of a potentially harmful pest. Other features will be related to data collected under "precision agriculture" programs. Under the precision agriculture banner, global positioning systems (GPS) are being used to pinpoint geographically the position of crops and pests in a field. Frequent sampling with GPS will provide high-resolution data sets on crop and pest development during a growing season. These data sets will be invaluable in future designs of IPM models, especially when combined with site-specific weather forecasts.

What are the future roles of industry, government, and university? Government will continue to play an important support role in the creation and distribution of national data sets. Businesses will play an increasingly important role in the further processing of government data for industry-specific applications, such as IPM models for agriculture. Entrepreneurs will develop new automated electronic services similar to SkyBit's E-Weather Service. These new services will offer a wide selection of information technology products at competitive prices to growers. Not only will the selection of products improve, but the means to deliver them will, as well. Cable, satellite, and telephone lines will offer faster and less expensive transmission rates. Growers will increasingly have the option of "downloading," or "pulling," data products from Web sites available through the Internet. In fact, among agricultural weather products, there are three competing ways for a grower to retrieve data. The first, which was mentioned earlier, is the "plug" technology. A grower can "plug" in an automated weather station and locally download data to a desktop or laptop computer. The second is the "pull" technology. A grower can use an Internet connection to "pull" products from a government, university, or private Web site. The third competing technology is "push." SkyBit's E-Weather Service is a good example of "push" technology. SkyBit automatically delivers, or "pushes," products to a grower through his or her e-mail address or fax number. There is only a minimum effort on the part of a grower to receive products.

Another important group under the industry label that will play an increasingly important role in the information technology era is "consultants." The traditional crop consultant will grow in sophistication after he or she begins to work with site-specific, precision agricultural and other similar data-intensive field-level products. The future crop consultant will be less involved in data collection and more focused on data "interpretation." The consultant will be responsible for interpreting site-specific data in the context of planning a seasonal IPM strategy and in the recommendation of day-to-day tactics in the management and marketing of a client's crop. The grower will still be making the final decisions; the consultant will keep him or her informed of different production options based on field observations and forecasts.

The university has probably the most important role in the information technology era. Information is the business of universities. They will be responsible for educating the future consultants, researchers, and entrepreneurs as providers of the new information technology products and for educating the growers as users of these products. University researchers and extension specialists will play an increasingly visible role as independent evaluators of information technology products. Their evaluations will help growers pick the right products for their management practices.

Conclusion

Site-specific forecasts represent a true innovation that will help promote the acceptance of IPM programs among growers. In the recent past, growers were responsible for either locally monitoring or securing meteorological data in order to make weather-dependent management decisions. With the automated delivery of electronic products, such as through SkyBit's E-Weather Service, this data responsibility is shifted from grower to industry, allowing more attention to be paid to the decision-making. Grower acceptance of site-specific forecasts in support of their decision-making has been documented in surveys as part of a 3-year orchard study conducted in Pennsylvania (2,3,4). It is clear from their responses that forecast products, like those delivered by the E-Weather Service, are becoming another important information source in the management of their crops.

Literature Cited

1. Felland, C. M., Travis, J. W., Russo, J. M., Kleiner, W. C., and Rajotte, E. G. 1997. Validation of site-specific weather data for insect phenology and disease development in Pennsylvania apple orchards. Pennsylvania Fruit News 77:10-17.
2. Felland, C. M. 1998. Grower use of site-specific weather and pest forecasts. Pennsylvania Fruit News 78:21-24.

471

3. Felland, C. M., Travis, J. W., Kleiner, W. C., and Rajotte, E. G. 1998. Apple Orchard IPM Project: 1997 Grower feedback. Pennsylvania Fruit News 78:29-34.
4. Felland, C. M., Travis, J. W., and Kleiner, W. C. 1999. Apple Orchard IPM Project: Educational materials. Pennsylvania Fruit News 78:29-34.
5. Gleason, M., Koehler, K., and Ryan, K. 1997. SkyBit accuracy evaluation - 1997. Pages 20-21 in: Proceedings of 13th Annual Tomato Disease Workshop. Indianapolis, IN.
6. Gleason, M. L., Parker, S. K., Pitbaldo, R. E., Latin, R. X., Speranzini, D., Hazzard, R. V., Maletta, M. J., Cougill, W. P., and Biederstedt, D. L. 1997. Validation of a commercial system for remote estimation of wetness duration. Plant Disease 81:825-829.
7. Jasinski, J. 1997. Using forecasted weather data for tomato disease management: a three-year study. Pages 25-29 in: Proceedings of 13th Annual Tomato Disease Workshop. Indianapolis, IN.
8. MacNab, A. A. 1997a. Fungicide schedules and early blight levels associated with FAST forecasts generated from three sources of environmental data. Pages 15-19 in: Proceedings of 13th Annual Tomato Disease Workshop. Indianapolis, IN.
9. MacNab, A. A. 1997b. Late blight forecasts resulting from three sources of environmental data applied to the BLITECAST forecaster. Pages 36-38 in: Proceedings of 13th Annual Tomato Disease Workshop. Indianapolis, IN.
10. Royer, M. H., Russo, J. M., and Kelley J. G. W. 1989. Plant disease production using a mesoscale weather forecasting technique. Plant Disease 73:619-624.
11. Russo, J. M., Zack, J. W., and Waight, K. T. III. 1996a. E-Weather: an innovative electronic weather service. Page 262 in: Preprint of 12th Inter. Conf. Interac. Process. Sys. for Meteor. Amer. Meteor. Soc. Boston.
12. Russo, J. M., Travis, J. W., and Truxall, D. L. 1996b. Simulated mesoscale weather data as input into an apple expert system. Pages 334-335 in: Preprint of 22nd Conf. Agr. Meteor. Amer. Meteor. Soc. Boston.
13. Russo, J. M., and Zack, J. W. 1997. Commercialization of National Weather Service data for agribusiness. Pages 50-51 in:

Preprint of 13th Inter. Conf. Interac. Process. Sys. for Meteor. Amer. Meteor. Soc. Boston.

14. Seem, R. C., and Russo, J. M. 1984. Simple decision aids for practical control of pests. Plant Disease 68:656-660.

Information Management: Past, Present, and Future

R. E. Stinner

Department of Entomology and NSF Center for Integrated Pest
Management, North Carolina State University
Raleigh, NC 27695 USA

The past 5 years have seen the introduction of information
dissemination technologies that will have an impact on educational
and communication surpassing even that of television and the
telephone. The Internet and World Wide Web provide the capacity
for almost instantaneous global sharing of text, graphics, video, and
sound information, with the added capability of searching for the
needed information (see the Internet Society, **www.isoc.org/
internet-history/**). In keeping with the rapid pace of Internet
development, this paper will provide mostly links as references,
rather than paper citations.

Looking back even 5-10 years, it is important for us to
recognize some of the inefficiencies and problems with our "work-
as-usual" extension efforts. Anyone who has worked for or with the
Cooperative Extension Service in any state is familiar with the space
required for storage of the stacks of publications that for one reason
or another were not used by the public. I do not know of a single
extension specialist who does not have at least one such stack.
Conversely, there have always been those publications for which
demand was not foreseen, and clients' needs for specific publications
went unmet. This has been particularly true for time-sensitive
information, and in recent years has been exacerbated by decreasing
budgets and staff. Again, I do not know a single extension specialist
who has not had to tell an agent or grower that a specific document
was no longer available. In 1997, for example, North Carolina State
University ran out printed copies of the *1997 North Carolina
Agricultural Chemicals* manual in mid-summer, but the electronic

474

version remained available (**ipmwww.ncsu.edu/agchem**) until it was replaced by the 1998 manual.

Likewise, the telephone has proved to be both a boon and a curse to both clients and extension. During times of major pest problems, it is almost impossible for a county agent or grower to reach the necessary extension specialist or for that extension specialist to accomplish anything other than answer the telephone, typically to respond to the same questions with the same answers. Again, as budgets and personnel decrease, this problem has been exacerbated.

The development of electronic communications capabilities has had an unimaginable impact on our ability to provide permanent solutions to these information delivery problems. The solutions started with recorded phone teletips and automated faxback capabilities in the 1980s. The advent of less expensive personal computers in the late 1980s and early 1990s saw CD-ROM information become available. Unfortunately, while the teletips and faxback could be updated to provide current information, they were not interactive to the extent needed. CD-ROMs are highly interactive but cannot be used to provide current information (e.g., timely pest alerts).

Starting publicly at the end of November 1993, the advent of the World Wide Web (**www.isoc.org/internet-history/brief.html# Timeline**) changed our basic concepts of information delivery and dissemination. Although the Internet and e-mail had been available for some time, it was difficult to use, was restricted to text, and required expensive machine-to-machine connections. The acceptance and remarkable growth of the World Wide Web, I believe, is based on the confluence and co-dependence of four major factors: inexpensive personal computers with concurrent dial-up Internet access, the new availability of an easy-to-use graphical interface (the World Wide Web itself), a healthy economy allowing for the purchase of computers for home use, and the ever-increasing demand for current information.

The ability to provide text, graphics, sound, and now even limited video in an interactive environment has created an information dissemination potential that was only dreamed of a few years ago. In agriculture, almost every state now provides, via the Web, 24-hour access to weather and pest alerts, pest identification guides, scouting and production manuals, marketing and economic information, contact lists and links, and all types of software, from complex decision support models to detailed bookkeeping for safety, regulatory, and tax purposes.

This information bonanza is being provided with smaller budgets and fewer people during a time when many traditional disciplinary departments are merging. This is critical for those of us in agriculture. We are in an increasingly urban society and must accept the fact that there will continue to be erosion of public support for agricultural research and extension. There is, and will continue to be, more and more pressure for regionalization of all agricultural support, including federal and land grant university programs.

Such regionalization requires electronic information dissemination. The amount of information on the Web is increasing exponentially. But this growth comes with a price. There is already far too much information from many sources (both reliable and otherwise), and it is largely unorganized. How much frustration is created by attempting to find needed information? It is not unusual for a Web search at Yahoo or Alta-Vista to return thousands of links to information – far more than anyone could examine in a reasonable time. Although there is much discussion in the popular press about correcting bandwidth insufficiencies (the "World Wide Wait"), future technology will deal as much with how to locate the desired information as with bandwidth issues.

To understand the potential for Internet-driven information sharing, we should first examine the growth and development of the Web. The technology on which the Web is based went public in November 1993. In December 1993, there were 623 Web servers. In June 1994, less than 5 years ago, there were 2,738 (440% growth in 6 months). By January 1995, there were 120,000 public servers, and by February 1999, over 43 million Web servers. The number of servers is increasing by 53,000 per month. That's an average of 1.2 new servers per minute, and that number is increasing (**www.nw. com/**).

What are the key technologies that will impact Web-based information-sharing in the next 5 years? I picked 5 years because it is at least possible to offer some information. The technology is developing and changing so rapidly that predictions beyond 5 years would be complete speculation. Therefore, I choose to give you a glimpse of what is to come, based on technologies already being evaluated.

476

Intelligent and/or Cooperative Search

John VanDyk at Iowa State University has already developed a specialized search engine for pest management and crop production information (**www.ipm.iastate.edu/ipm/ncrsearch/**). This search engine only indexes cooperating land-grant servers, those that are registered with the National IPM Network and have agreed to NIPMN standards for accuracy, peer review, and author identification. Presently, the search is only for the North Central Region (there is also a U.S. search for cotton at **www.ipm.iastate.edu/ipm/nipmn/ cotton/cottonsearch.html**), but there is the expectation of a full U.S. search by year 2000 (VanDyk, personal communication)

Database-driven Sites

The Web is changing our basic definition of databases. With the ability to freely examine and index entire Web sites (record words with associated document locations), the largest databases ever known to man are now these collections of searchable Web page indices.

Web sites can also be database driven so that they are tailored to individual clients. By requesting individual information (e.g., state, crops grown, or other related information), Web servers can place "cookies" (a small identification file) on the user's PC (for software examples, see the Cold Fusion Web site **www.allaire.com**). When the client returns to that site, he or she can be automatically provided with the desired information or sent to the appropriate part of the Web site. For an example of this technology in practice, try **my.netscape.com**.

Even without "cookies," databases can be used to make Web sites easier to maintain and navigate. With this approach, all information is maintained in databases, where it is easy to update and search. We are experimenting with this in the National IPM Network and hope to have a database-driven site in place this year (see **www.reeusda.gov/nipmn/**).

Multiple, Concurrent Database Search

The Pest Management Information Decision Support System

being developed by USDA/CSREES contains a tool for searching pest management databases online (**ipmwww.ncsu.edu/PMAP**) for the Pest Management Alternatives Program. There is also a test system that concurrently searches the OPMP Crop Profiles, a Pipeline database, and Leonard Gianessi's Pesticide Use Database. A similar effort is being developed for invasive species under the auspices of the Invasive Species Council and involves the Departments of the Interior (USGS), Agriculture (Forest Service and APHIS), and others.

Decision Support

As we move to more complex IPM systems and regulations, decision support and/or reporting has already become mandatory for some states. As an example, CDMS (**www.cdms.net**) in California now has software that writes pesticide prescriptions, checks them against the current label, and maintains records of pounds on the ground. For consultants, this is a real time-saver; for growers, their use records are automated; and for the state of California, it is possible to work directly with individual companies representing many growers and consultants. CDMS also provides, online and free of charge, labels and MSDS from the more than 75 companies they represent.

New Technologies

Business-to-business communication is driving much of the Internet's new technology development. The cost savings in being able to share inventory and other information appropriately with partners is at the heart of the move toward enterprise-wide, intelligent communication using XML, CORBA, JAVA (**www.java. co.uk/**, but also search **www.yahoo.com** for hundreds of sites) and other standards in software. Most Internet-aware individuals have heard about Java, a platform-independent, programming language.

Perhaps even more important is XML (**www.xml.com/**), the next-generation Web language that provides structure to Web documents. This is like making every Web page part of a true database, where you assign distinct identifiers to specific sections of the page (e.g., pest name, location, author). XML Web pages will be part of an immense database, globally distributed. More important, XML allows specific assignment of variable names mutually agreed upon by a group desiring to share specific information among cooperating parties. This will allow easy search and access for any collaborating organizations, not just businesses.

CORBA (**www.omg.org/**) is a software layer that allows different machines on the Internet to communicate with each other, to run programs, and to trade results, regardless of the operating system used. Databases on Unix machines can be queried by Web-driven forms on a Windows NT server and the results combined with other database searches, then organized and returned to the user via his or her Web browser.

Most of us think of games and entertainment when we hear of Virtual Reality, but the potential for educational uses is vast. Imagine being able to take a class inside a grasshopper to view its internal anatomy up close and enlarged, or to fly with a bee from flower to flower, seeing as a bee sees. This and more is already possible at VPI, where Tim Mack and Alexei Sharov have been working with a VR laboratory to develop these teaching tools (**everest.ento. vt.edu/~sharov/3d/3dinsect.html**). They are also helping with a PC simulation of these virtual realities.

Bandwidth and Access

Many universities are already a part of the next-generation Internet, Internet2 (**www.internet2.org/**), with dramatically higher bandwidths that can even handle streaming video with ease. North Carolina State University is part of a gigabyte loop that includes the University of North Carolina at Chapel Hill and Duke University. With higher and higher transmission speeds on the horizon, full-motion video may soon become a reality.

Perhaps the real impact may well be felt in the area of Internet access. There are obviously many individuals, even in the United States, who cannot afford a computer and Internet access. What about these disenfranchised individuals? There are already efforts underway to provide free PCs and Internet access in exchange for accepting advertising, just as television has done. In a recent campaign, **www.free-pc.com** made such an offer. They expected to receive no more than 10,000 applications the first month, because these applications requested fairly personal information. They received almost 500,000 requests.

TV/Web Integration

Here, I am not referring to Microsoft's WebTV®, but to the true integration of television and the Web, where the PC uses the television as its monitor, and clicking on a part of the screen during a news broadcast will provide the full details of that story (or product). A glimpse of this can be seen in Microsoft toys that react to the TV screen, the development of on-demand movies through cable TV, DVD players that work with both PCs and televisions, and Internet "channels" that broadcast to your computer while you have an Internet connection. This technology will close the gap between television and computers and allow full integration of the spoken and printed media.

These examples provide only a small insight into where information management is headed now and in the near future. There is so much information being developed even now that I can provide only a small number of "portals" to the agriculturally related Web world.

Major Pest Management-related Web Sites

Biological Control Virtual Information Center
  (ipmwww.ncsu.edu/biocontrol/)
CGIAR SP-IPM
  (www.cgiar.org/spipm/)
CSIRO Pest Management
  (www.ento.csiro.au/research/pestmgmt/pestmgmt.html)
Directory of IPM Resources
  (www.ippc.orst.edu/cicp/)
FAO IPM Program
  (www.fao.org/WAICENT/FAOINFO/AGRICULT/AGP/ AGPP/IPM/Default.htm)
IPM Africa
  (www.icipe.org/IPMAfrica)

IPM China
   (**www.ipmchina.cn.net/**)
IPM-Europe
   (**www.nri.org/IPMEurope/homepage.htm**)
IPMnet
   (**www.IPMnet.org**)
National IPM Network
   (**www.reeusda.gov/nipmn/**)
NSF Center for IPM
   (**ipmwww.ncsu.edu/cipm**)
USAID's Africa IPM Link
   (**www.info.usaid.gov/regions/afr/alnk/**)
Virtual Library for Agriculture
   (**ipmwww.ncsu.edu/cernag**)
Virtual Library for Entomology
   (**www.colostate.edu/Depts/Entomology/WWWVL-
   Entomology.html**)

Section 8

Progress and Challenges

In the decades since the IPM concept was first formulated, many advances have been made in both the theory and knowledge that provide the foundation for IPM, as well as in the technologies used in IPM. An infrastructure to support the development and implementation of IPM has been developed, and the IPM paradigm for crop protection has been widely accepted and widely implemented. Implementation of IPM is far from universal, and in most instances has fallen short of the level envisaged by many. In this section, Peter Goodell and Frank Zalom provide a brief history of the development of IPM knowledge and its delivery to the end-user. They then discuss progress that is being made toward mitigating many of the constraints that have limited the implementation of IPM to date. In the next chapter, Kathleen Merrigan reminds us that progress in the development and implementation of IPM on a wide scale requires much more than expanding our knowledge of pests and production systems and developing powerful new technologies. It requires addressing, on a continuing basis, the institutional, political, and policy issues that determine not only funding for IPM research, development, and infrastructure, but also the regulatory environment for IPM and the incentives for and obstacles to implementation of IPM. In the final chapter, Barry Jacobsen provides an interpretive summary of the conference on which these Proceedings were based.

Delivering IPM: Progress and Challenges

P. B. Goodell[1] and F. G. Zalom[2]
UC Statewide IPM Project
Cooperative Extension, University of California, [1]Kearney Agricultural
Center, Parlier, CA 93648, and [2]Davis, CA 95616 USA

This International Conference on Emerging Technologies has two major themes, a primary one of presenting and discussing emerging technologies, and a secondary one of reviewing progress since the 1970 Conference on Pest Management. Herein we review progress made since 1970 in delivering IPM to the end-users and overcoming challenges facing IPM delivery. These challenges have changed as new knowledge, resources, and methodologies have become available during the past 30 years for delivery to a broader range of clientele.

Well before the 1970 conference, Dr. Robert van den Bosch expressed concerns about the slow rate of IPM adoption by farmers (27). He lamented the fact that good science was available to manage pests but that the knowledge was not being transferred to the field. In his summary of the 1970 Pest Management Conference, Huffaker (16) commented that the development and delivery of farm-level pest management programs had not been adequately addressed. Instead, conference speakers concentrated on the development of knowledge and theory. He also suggested that formal programs would be required to train qualified, independent practitioners, which were lacking at the time of the Conference.

We suggest three elements that are required for successful IPM transfer and adoption, including:

- The *development* of technology, tactics, and knowledge, usually the realm of researchers
- The *delivery* of technology, tactics, and knowledge, usually through extension programs

- The *integration* of technology, tactics, and knowledge into farming systems through private practitioners and farmers.

We believe that these key elements represent a progression of IPM development. They can be viewed narrowly around single sites and pests or much more broadly to incorporate ecosystem approaches. The latter represents maturity in an IPM program and is characterized by its level of complexity. As complexity increases, new challenges are identified and new obstacles must be overcome. It is our observation that the relationships of researchers, extensionists, and practitioners between these elements and their roles within any element have changed considerably through time.

Constraints to IPM Adoption

What prevents IPM knowledge and practices from being widely accepted? Diffusion theory (23) states that any new technology diffuses through a social system in a predictable manner. Initially, only a few members of the social system will adopt a technology. The number of individuals adopting the technology increases through time until the social system is saturated.

In agriculture, the rate at which the adoption process occurs and the ultimate level of adoption that is attained may be affected by many factors. Many efforts have been made to identify constraints to IPM adoption during the past 30 years. As might be anticipated, priorities among the constraints were often dependent upon which group assembled the list and the time frame within which the lists were developed. Wearing (29) organized constraints into five groups: technical, educational, financial, organizational (institutional [30]), and social. Constraints, obstacles, or barriers are any factors that prevent a practice or idea from spreading through a wide community, and the terms are often used interchangeably in the literature. These constraints may be real (e.g., no knowledge of pest biology) or perceived (e.g., IPM is too difficult).

TECHNICAL OBSTACLES

Technical obstacles primarily affect the development of IPM strategies, tactics, and knowledge. In many instances, the basic biology of pests, beneficial organisms, and their interactions in agricultural ecosystems is not understood. Similarly, the application of this knowledge to the management of pests in cropping systems through tactics such as monitoring guidelines, control action thresholds, biological controls, cultural controls, and host plant resistance is frequently lacking. The most common obstacles noted were the absence of pest sampling and decision support tools. An active research program in all these areas is essential for the continued development of IPM. Thus, lack of sufficient funding of basic and applied research often appears on lists of important constraints to IPM adoption (1,25,26).

EDUCATIONAL CONSTRAINTS

Educational constraints to the implementation of complex innovations like IPM include a lack of appropriate training methodologies, a lack of extension personnel for continued training of private-sector practitioners, insufficient educational resources, and a lack of formal degree programs to bring new researchers, extensionists, and practitioners into this field.

INSTITUTIONAL CONSTRAINTS

Institutional constraints can develop from interactions within, between, and among all private and public institutions involved in the development, delivery, and practice of IPM. Perhaps the most conspicuous constraint is the lack of coordination and communication among organizations, personnel, and disciplines.

FINANCIAL CONSTRAINTS

Financial constraints are a primary concern of growers. IPM often increases net profits and reduces risk for adopters. However, there is still a perception by growers that IPM does not offer short-term economic advantages compared to conventional control, particularly because of additional labor costs from sampling and monitoring (22). The lack of incentives to reward users of IPM is often stated as an important constraint (25). Private pest management consultants are constrained from adopting more time-consuming IPM technologies in order to provide their services at the lowest possible cost (30).

Wearing (29) found that social and marketing obstacles rated above all others for their impact on IPM implementation. Wearing (29) used the term "marketing" to refer to financial gain or risk. A broader sense of the word can also refer to the concept of marketing IPM to stakeholders beyond pest managers and growers, such as environmental or legislative communities. In this sense, barriers to communication and understanding can act as major obstacles to IPM implementation.

Development of IPM Knowledge

The 1970 Pest Management Conference piqued national interest in applying some of the principles identified to real-world problems. The 1970s and 1980s saw a major increase in IPM funding. The resulting activities focused primarily on research and development of IPM tactics and knowledge. The Huffaker Project (1972-1978) channeled $13 million through 18 universities and enabled 300 researchers to develop ecological principles that would become the foundation of insect and mite pest management (20). This project focused mainly upon insect problems in alfalfa, pome and stone fruits, citrus, cotton, pine forests, and soybean (17). It developed many of the basic concepts that improved our understanding of the dynamics of crop and pest development. This project resulted in advances in insect monitoring, efficacious use of insecticides, biological control, and evaluation of environmental and economic impacts of IPM (11).

The Consortium for IPM (CIPM) operated between 1979 and 1985. Building on the principles established by the Huffaker Project, CIPM more fully incorporated efforts to include pathogens, nematodes, weeds, and vertebrates. CIPM brought together 15 universities and continued the focus on alfalfa, apples, cotton, and soybean. A systems approach to total crop production was taken, with an emphasis on computer modeling. CIPM continued the effort to overcome institutional barriers and developed teams that integrated across disciplines.

The USDA National IPM Program was initiated in 1982 and continues into the present. Stressing the importance of multi-state and institutional collaboration through an interdisciplinary approach, it provides funding for regional IPM projects that have led to advancements in IPM knowledge, tools, and practices.

Many technical constraints were overcome, due in no small part to the contributions of these research programs. For example, simplified monitoring and decision guidelines were identified as important in overcoming obstacles to IPM adoption. Wearing (29) found that private pest management consultants and extensionists perceived these to be more important technical constraints than did researchers, probably reflecting their closer relationship to practical application. Wearing's (29) results were confirmed by Flint and Klonsky's (10) study of California pest control advisers (PCA), who rated research on monitoring and thresholds as their greatest technical need, followed, in order, by cultural control methods, biological control methods, economic analysis of controls, pesticide resistance studies, crop and pest interaction studies, and computer simulation models.

Through IPM projects in the 1970s and 1980s, breakthroughs occurred that simplified sampling and decision support methods. For example, the general introduction of binomial sampling for mites and insects has provided practitioners with rapid, reliable sampling methods in such crops as cotton, almonds, and tomato. Immuno-assays have improved and shortened the amount of time required for pathogen identification. Development of risk-assessment models, together with improved and affordable remote weather sensing equipment, has advanced disease forecasting to the point where strategic decisions can be made for pathogen management instead of relying solely upon scheduled prophylactic fungicide treatments.

Delivery of IPM Knowledge and Practices

During the same time period, efforts were undertaken to overcome educational constraints. Beginning in 1972, federal funding for IPM implementation was provided to the states on a formula-funding basis to build extension IPM infrastructure. State and federal IPM demonstration projects emphasized the value of scouting and using economic thresholds. From 1971 to 1975, 39 federally funded IPM pilot projects were conducted in 29 states (25). These projects refined implementation techniques, such as demon-stration plots, workshops, and on-site field meetings. Many also incorporated scouting programs, which eventually led to more private-sector IPM consultants. Resource materials (10) were developed to provide basic reference support to IPM practitioners in extension and the private sector. These included color manuals, color keys, published guidelines, leaflets, newsletters, and pamphlets, as well as innovative delivery approaches such as slide sets, videotapes, and beneficial insect reference collections. One key technology transfer method was the farmer-to-farmer demonstration. The involvement in field meetings of end-users who discussed their IPM success was important to convince other producers to adopt IPM practices (12). By the mid-1980s, electronic dissemination began to come into its own through dial-up computer sites. Thus, delivery of computer simulations (24) and computer-aided support systems (21) developed through the Huffaker Project and CIPM could be accomplished.

In 1972, California became the first state to require all individuals who recommend the use of any pest control method, including pesticides, to be licensed as PCAs. To fill the need for training professional pest management practitioners, several land grant universities established pest management degree programs that provided curricula in all pest disciplines.

Progress in the 1990s

As the 1980s drew to a close, interest increased in assessing adoption of IPM in U.S. agriculture and in decreasing the reliance on chemical control. Following the 1989 National Extension IPM Conference held in Washington, D.C., a joint USDA/EPA analysis was initiated in 1990. Work-groups were formed around corn and soybean, cotton, fruits, and vegetables. The work-groups issued reports that discussed the potential for IPM research to resolve pest-related problems in agriculture (14). They identified constraints to IPM adoption within their crops, as well. This report was published along with discussion papers in preparation for the National Forum on IPM. This forum was designed to create a broad consensus for accelerating the development of IPM for the end-user in agriculture (26).

The 1992 National Forum on IPM reviewed factors that limited IPM adoption. The Forum focused attention on IPM constraints in four major areas, characterized as institutional, policy, regulatory, and research, probably reflecting the interest of the organizers (EPA and USDA) in policy issues rather than those surrounding IPM development and delivery. The top 10 constraints identified by the participants of the National IPM Forum follow. For convenience, each has been placed into one of Wearing's (29) categories.

1. Lack of national commitment (institutional)
2. Insufficient funding and support for IPM implementation, demonstration, and fundamental infrastructure (institutional)
3. An EPA pesticide regulatory process that is burdensome, expensive, time-consuming, and unclear (institutional)
4. Lack of funding for applied research, lack of regulatory personnel to expedite product registration, and lack of education/promotion for growers (institutional, technical, social)
5. Shortage of independent, trained IPM practitioners (educational)
6. Inability of current USDA and EPA structures to effectively address cross-cutting agricultural and environmental concerns (institutional)
7. Agricultural policies developed without considering IPM (institutional)
8. Lack of common and specific goals for IPM (technical, social)
9. Lack of emphasis on urban IPM programs as a means to help educate the general public (educational, social)
10. Lack of coordination among federal, state, and private sectors in research planning and implementation (social, institutional)

Thus, five constraints were associated with institutional issues, one with educational issues, and four crossed two or more categories. When possible resolutions were identified to address these constraints, six of 10 were concerned with increased funding to address IPM technical and educational issues, three addressed institutional issues, and one was too vague to categorize. It is interesting to note that institutional issues were predominant, while financial issues affecting growers and practitioners did not emerge. This may reflect relatively low representation at the Forum by farmers, ranchers, and IPM practitioners (26).

In 1993, the National Foundation for IPM Education conducted a series of five Regional Workshops to enable these stakeholders to participate in the process that had begun at the Forum. These facilitated workshops sought stakeholder opinions on what they viewed as constraints to IPM adoption.

While no single set of issues emerged from these Regional Workshops, the most commonly mentioned issues and their association with the constraints identified at the national Forum included:

- Lack of financial incentives or benefits for conducting IPM programs
- Differing agendas and conflicting messages from governmental agencies (national constraint #10)
- Loss of applied funding (national constraint #4)
- Lack of funding for IPM education and research programs (national constraint #2)
- Slowness of EPA registration system (national constraint # 3)
- Lack of availability of pesticides (national constraint # 3)
- Government commodity programs discouraging crop rotations (national constraint # 7)
- Need to market IPM more effectively
- Problem with Delaney Clause restrictions (national constraint # 6)
- Lack of understanding by producers about total production system approaches to IPM

When the regional results are compared to the list of national constraints, there were six constraints in common (numbers 2, 3, 4, 6, 7, and 10). The producers and practitioners did not mention numbers 1, 5, 8, or 9 as major concerns, but rather identified other financial, educational, and social constraints (26). It is interesting to note that producers and practitioners did not perceive a shortage of trained IPM personnel to exist except in one location (Illinois-corn/soybean). This agrees with Wearing's (29) observation but is in contrast to the national concerns (constraint #5). Perhaps this reflects confusion in distinguishing between private practitioners and affiliated crop advisers.

IPM has always stressed multiple control approaches, but the introduction of the bio-intensive IPM concept (13,19) raised the level of management complexity. In pushing IPM toward increased biological reliance, Frisbie (13) recognized that increased participation by all segments of society would be required to achieve the higher levels of complexity. Larger groups, including non-traditional (and likely non-agricultural) participants, would have to work together in coordinating IPM in regional landscapes and in accepting new roles for farmers, practitioners, researchers, and extensionists.

490

The National Coalition on IPM, formed in 1988, represents a broad constituency of growers, farmer organizations, commodity groups, agricultural consultants, food processors, food marketing organizations, pest control operators, lawn care operators, and environmental groups sharing a common desire to work closely with federal and state agencies to advance IPM implementation (28).

The 1990s have been characterized by continued progress in reducing institutional and social constraints. Closer interagency collaboration has occurred between EPA and USDA and with their clientele groups. Changes in the 1995 Farm Bill removed price supports and increased farmers' rotational choices. The Food Quality Protection Act of 1996 (FQPA) replaced the Delaney Clause and mandated shorter registration times for reduced-risk pesticides. Perhaps the greatest progress was seen in the success of the IPM Forum and Regional IPM Workshops, which increased participation of stakeholders in identifying needs and proposing problem resolutions.

This progress was also reflected in the 1994 workshop jointly sponsored by the University of California, California Department of Pesticide Regulation (CDPR), and California Department of Food and Agriculture (1). Its recommendations emphasized strong grower and PCA participation in planning, development, and resource allocation. The subsequent initiatives by U.S. EPA and CDPR sought to increase participation by farmers, PCAs, environmental groups, agencies, universities, the food industry, and others. Agency-sponsored programs in California, such as Biologically Integrated Farming Systems and Pest Management Alliance, provided support in developing community-based participatory research and outreach programs. These programs provided short-term funding to encourage the development of on-farm demonstrations and the implementation of biologically based pest management tactics and strategies.

U.S. EPA provided another example of broad-based stakeholder groups (2) by showing how food companies have worked with growers to produce food while promoting environmental steward-ship, including IPM. It listed 40 food companies that have worked with growers to influence their choice of agricultural practices. They concluded that the main motivation for such programs was economic, and noted that the likelihood of adoption of any practice is greater if the practice is flexible, possesses incentives for use, or is more cost effective. The primary barriers to wider adoption of these practices did not differ much from the list identified at the National

IPM Forum. Frequently reported problems were lack of reliable and consistent information, insufficient funds to support changes in practices, and lack of financial incentives to adopt environmental practices.

IPM insurance to defer risk of changing to IPM practices was suggested in 1989 (25) and is under development (3,5,7). Under such programs, a grower could insure losses to crops caused by the failure of a new practice.

Certification of growers using IPM is another approach to overcoming constraints (25). Labeling of food products that have used IPM for their production is currently underway (15), but has not been without problems (4). A variety of food labels have been introduced, including IPM labels, "green" labels, and eco-labels. Third-party certification and education about IPM are keys to consumer acceptance of IPM labels.

Interest in measuring IPM progress has received much attention through the 1990s. One approach was the concept of a continuum of IPM practices (6) that presents the idea of movement toward a more biologically reliant pest management system. Kogan (18) presented the concept as an increase in complexity and systems integration. This concept provides a framework for measuring movement to increasingly more complex levels of IPM, rather than approaching adoption as the crossing of some static threshold level. It also provides an approach for quantification of IPM for labeling programs. No standard of evaluation is yet established, but available approaches have been reviewed (9).

Summary and Conclusions

A major transformation of IPM has occurred in the 30 years since the 1970 Pest Management Conference, which has resulted in many of the challenges facing the development, delivery, and integration of technology, tactics, and knowledge having been overcome. Many critical issues of that time have been addressed, including increased basic knowledge of crops and the pest complexes that attack them, vast improvements in delivery infrastructure, and the development of a private practitioner profession to implement IPM. However, these issues have only been addressed on a limited number of systems and then in only limited regions. In addition, changes in cropping systems, production technologies, and pest complexes necessitate constant fine-tuning of even well-established

IPM systems (8). Because IPM systems are not static, there will always be the need for development of new knowledge and the delivery, adaptation, and integration of that knowledge into usable systems for farms, homes, and other target sites.

Another major shift that has occurred is that development, delivery, and integration no longer remain the domain of any single private- or public-sector organization. This has been especially true regarding the traditional role of the land grant universities. Considerable basic technological advances are now occurring in the private sector, as is the integration of technologies and tactics into IPM systems. The land grant universities continue to play an especially important role in the discovery of basic knowledge and its adaptation for and delivery to end-users. However, this is increasingly done as a part of coalitions of different interests, and the land grant universities are in an especially good position to offer coordination and guidance for these alliances. Unfortunately, in this time of increased complexity, the erosion of support for the land grant university infrastructure, which is needed to develop, deliver, and support integration of IPM, presents a serious challenge to future progress. Efforts by coalition groups to obtain increased funding at the national level for IPM research and delivery have not been successful. For example, the National IPM Initiative of 1995 focused upon regional and national IPM planning based on stakeholders' input and participation, but failed to receive congressional funding.

The message of the 1970 Pest Management Conference still remains true: IPM is the management of ecosystems. Its success will reside in the cooperative efforts of society in developing, delivering, and integrating knowledge, practices, and policies in environmentally and economically satisfying ways.

Literature Cited

1. Anonymous. 1994. Integrated Pest Management Workshop: Developing a California Strategy. December 8, 1994. California Department of Pesticide Regulation.
2. Anonymous. 1998. Food Production and Environmental Stewardship. Examples of How Food Companies Work with Growers. US EPA Document 231-R98-001.
3. Anonymous. 1998. IPM insurance for corn rootworm and potato late blight. Pennsylvania IPM News 1:3.

4. Anonymous. 1999. IPM apple juice pulled off the market. Page 2 in: Gempler's IPM solutions. March 1999.
5. Anonymous. 1999. Promoting conservation innovation in agriculture through crop insurance. Agricultural Conservation Innovation Center. **http://www.agconserv.com/toppage11.htm**
6. Benbrook, C. 1996. Pest Management at the Crossroads. Consumers Union. 272 pp.
7. Bolkan, H. A., and Reinert, W. R. 1994. Developing and implementing IPM strategies to assist farmers: an industry approach. Plant Dis. 78:545-550.
8. Croft, B. A. 1985. Integrated Pest Management: The agricultural and environmental rationale. Pages 712-728 in: CIPM. IPM on Major Agricultural Crops. R. E. Frisbie and P. L. Adkisson, eds. Texas Agricultural Experiment Station, MP-1616.
9. Day, E. 1998. IPM Measurements System Workshop. Chicago, June 12-14, 1998. http://farm.fic.niu.edu/cae/wp/sp98-1/index.htm
10. Flint, M. L., and Klonsky, K. 1989. IPM information delivery to pest control advisors. California Agriculture 43:18-20.
11. Frisbie, R. E. 1985. Consortium for Integrated Pest Management. Organization and Administration. Pages 1-9 in: CIPM. IPM on Major Agricultural Crops. R. E. Frisbie and P. L. Adkisson, eds. Texas Agricultural Experiment Station, MP-1616.
12. Frisbie, R. E. 1985. Regional implementation of cotton IPM. Pages 638-651 in: CIPM. IPM on Major Agricultural Crops. R. E. Frisbie and P. L. Adkisson, eds. Texas Agricultural Experiment Station, MP-1616.
13. Frisbie, R. E. 1989. The future of IPM and the Cooperative Extension Service. Pages 94-99 in: Proceedings of the National Extension IPM Conference. IPM Addressing Environmental Challenges in the 90's. Washington, D.C.
14. Fry, W. E. 1992. Foreward. Pages v-vi in: Zalom, F. G. and Fry, W. E. 1992. Food, Crop Pests, and the Environment. APS Press. St. Paul, MN.
15. Hollingsworth, C. S., Coli, W. M., and Van Zee, V. 1992. Massachusetts grower attitudes toward a certification program for integrated pest management. Fruit Notes Cooperative Extension Service, Univ. Mass. 57:3-6.
16. Huffaker, C. B. 1970. Summary of a pest management conference – a critique. Pages 227-242 in: Concepts of Pest

Management. R. L. Rabb and F. E. Guthrie, eds. North Carolina State University, Raleigh, NC.

17. Huffaker, C. B., and Smith, R. F. 1980. Rationale, organization, and development of a national integrated pest management project. Pages 1-24 in: New Technology of Pest Control. C. B. Huffaker, ed. John Wiley and Sons, NY.

18. Kogan, M. 1998. Integrated pest management: Historical perspectives and contemporary developments. Ann. Rev. of Entomol. 44:243-270.

19. Madden, P. J. 1992. Beyond Pesticides. Biological Approaches to Pest Management in California. DANR Publication 3354. University of California, Division of Agricultural and Natural Resources.

20. Perkins, J. H. 1982. Insects, Experts, and the Insecticide Crisis. Plenum Press, NY.

21. Plant, R. E., and Stone, N. D. 1991. Knowledge-based Systems in Agriculture. Biological Resource Management Series. McGraw-Hill, NY.

22. Poe, S. L. 1981. An overview of integrated pest management. HortSci. 16:501-506.

23. Rogers, E. M. 1983. The Diffusion of Innovations. New York: Free Press.

24. Shoemaker, C. A. 1980. The role of systems analysis in integrated pest management. Pages 25-49 in: New Technology of Pest Control. C.B. Huffaker, ed. John Wiley and Sons, NY.

25. Sorensen, A. A. 1989. Implementation of IPM: The farmer's view. Pages 7-23 in: Proceedings of the National Extension IPM Conference. IPM Addressing Environmental Challenges in the 90's. Washington, D.C.

26. Sorensen, A. A. 1994. Proceedings of the National Integrated Pest Management Forum. American Farm Land Trust.

27. van den Bosch, R. 1964. Practical application of the integrated control concept in California. Pages 595-597 in: Proceedings of the 12th International Congress of Entomology, London.

28. Wallace, M. 1993. The National Coalition on IPM: working for safer food, cleaner water, and wildlife conservation through expanded implementation of IPM. Page 193 in: Successful Implementation of IPM for Agricultural Crops. A. R. Leslie and G. W. Cuperus, eds. Lewis Publishers. Boca Raton, FL.

29. Wearing, C. H. 1988. Evaluating the IPM implementation process. Ann. Rev. of Entomol. 33:17-38.

30. Zalom, F. G. 1993. Reorganizing to facilitate the development and use of integrated pest management. Agricultural Ecosystems and Environment 46:245-256.

Politics, Policy, and IPM

Kathleen A. Merrigan
Henry A. Wallace Institute for Alternative Agriculture
Washington, DC 20002 USA

I don't care what your computer experts tell you; I am certain that everyone in this room has a Y2K problem. How can I be so sure? Because the Y2K problem I'm referring to is our collective failure to meet the goal announced by President Clinton in 1993 to have 75% of all the agricultural land in the United States under IPM management by the year 2000. I suppose that if you stretch the definition of IPM, you could come up with some absurd calculations to suggest that this goal has been reached. But if we use the popular continuum approach of measuring IPM adoption that is advanced by Consumers Union and World Wildlife Fund (1), it becomes obvious that too few producers and other pesticide users are shifting toward biointensive IPM systems. My guess is that, no matter what your particular view on the continuum methodology may be, we can agree that we are not where we had hoped to be when the IPM Initiative was announced 6 years ago.

Our Y2K problem has many architects: the current administration and also previous administrations, all of which undervalued IPM; the U.S. Congress for failing to appropriate needed funds; our university system for failing to put environmental, systems-based research high on the agenda; the broader scientific community for clinging to old agendas or myopically pursuing biotechnology; and the environmental movement for weighing in late and with a light touch. But I won't dwell on the blame game. My goal is to stimulate a discussion of the policy hurdles that limit success in IPM. To do this, I pose six tough questions that must be answered if we have any hope of making IPM the dominant means of production in America.

Question #1: Why can't we succeed in obtaining significant budget increases for IPM?

We need more money for IPM research and education. Currently, the IPM research and education piece of the USDA budget pie is slim no matter how you slice it. Even with calculations based upon a generous definition of IPM, the amount represents just 7% of USDA Research & Development. Accounting for the whole federal budget, including components like the EPA, less than 13% of amounts spent on pesticides or pest management R&D is spent on biointensive IPM (1).

Since the Clinton administration's IPM announcement, there have been gallant attempts to increase the IPM budget. Despite the fine leadership of Deputy Secretary Richard Rominger, several USDA staff members, and a handful of non-governmental organizations (NGOs), including the Wisconsin potato growers, Consumers Union, the World Wildlife Fund, and the National Campaign for Sustainable Agriculture, the battle to increase the IPM budget has not met with great success.

Once again, we are in the midst of the annual budget battle. The administration's FY00 budget proposal requests significant new monies for IPM. Under the Cooperative States Research, Extension, and Education Service (CSREES), the request is $54.3 million, a big increase considering that CREES IPM programs at present total $28.8 million. The budget proposal would also change how the money is spent. IPM resources would be managed under Section 406 of the Research Title, along with water quality, gleaning, and an assortment of other issues. The budget proposal would also convert the IPM extension program (3d money) into a competitive grants program. And new IPM accounts are proposed: Research to "identify and implement replacements for methyl bromide," "biointensive research and extension," and "research and extension to provide transitions for crops or cropping systems" are outlined in the budget.

A new agenda and new allies will help the IPM budget battle. The proposed new accounts are where IPM needs to be in the future, and, if implemented, will undoubtedly increase support for IPM among environmental organizations. In the meantime, however, IPM leaders will need to do all they can to convince potential allies that IPM is changing and is no longer dominated by the "spray and count" kind of thinking that, truthfully, lurks in some parts of this community.

Yet, even with a change in IPM direction and new allies, obtaining budget increases will remain difficult. Unfortunately, IPM funding problems are symptomatic of the overall disinterest in agricultural research. For the first time in years, significant new monies are proposed for agricultural research, but we have to wait and see whether such increases win congressional support.

Question #2: What role will the FQPA play in ongoing IPM efforts and vice versa?

As many of you know, the Food Quality Protection Act of 1996 is in the final implementation stage. While the dust has not settled, it is nevertheless clear that this statute will lead to extensive pesticide cancellations. Now more than ever, IPM is needed in order to help farmers survive the upcoming regulatory regime where organophosphate and carbamate insecticides will be banned, along with many other chemicals. The administration's budget proposal uses FQPA as part of the rationale for expanded IPM funding. Truthfully, I find the administration's case compelling, and in my own visits to Congress, I, too, stress the heightened importance of IPM due to FQPA implementation. But please understand; I have encountered two compelling retorts.

First, IPM is being used by certain parties as the rationale for delaying and softening FQPA implementation. Frankly, it is not in the interest of some groups to find non-chemical, systems-based solutions to pest management problems. The most powerful argument the chemical industry musters against FQPA implementation is that it will derail IPM systems because chemicals essential under IPM management will be eliminated, rendering IPM useless and leaving producers with no choice but to use more chemicals. While it is true that some IPM systems are based on the limited use of existing chemicals (e.g., certain beneficials are reared to resist organophosphates), this is an overblown argument. Yet I know that some IPM leaders have been tempted to join the debate to maintain the chemical status quo because of the short-term IPM dilemmas and in exchange for commitments from the chemical industry for political support of IPM. Beware of the "bargain." It will not lead to increased IPM support in the long run, nor is it an environmentally responsible act. My advice to the IPM community--avoid being lured into the battle to defend FQPA delays.

Take the example of methyl bromide. While the USDA funds limited research efforts to find alternatives to this soil fumigant, the search has not been wholehearted. Research is capped at a level that allows the argument to be made that we are trying to find alternatives while at the same time ensuring that viable alternatives remain illusional. Why? Because industry leaders are confident that, if necessary, the lack of alternatives will ensure their success in blocking a ban on methyl bromide. They were proven correct last fall when a last-minute change to the Senate appropriations bill effectively eliminated the phase-out of this environmentally problematic chemical. Is the IPM budget being held at a similar threshold?

Second, it is not clear that alternatives to chemical pesticides will be fully considered within the FQPA debate. There is a lack of knowledge about non-chemical production at EPA. EPA is struggling to obtain information about alternatives in order to make better-informed pesticide registration decisions, since few of the experiences of organic and sustainable farmers have been catalogued and studied by our research establishment. EPA staffers are also struggling to meet their statutory imperative to identify reduced-risk pesticides and biopesticides. But without adequate guidance from the IPM community, such declarations will be based on limited information, most likely a comparison of chemical A versus chemical B, rather than a systems assessment. The Wallace Institute has strived to alert EPA staff and the environmental community to the opportunities that now exist in sustainable agriculture, and our comments on the various notices of EPA pesticide rule-making can be found on our Web site (**www.hawiaa.org**). But this task is much greater than our one little organization can handle.

Question #3: Do we really know how to stimulate and manage systems research?

Pest resistance problems provide stark evidence of our failure to contemplate the broad, systems-wide effects of our management decisions. It is estimated that 500 species of insects and mites and close to 200 different species of plant pathogens and weeds are now resistant to at least one class of synthetic pesticides (1). Bt resistance has received great attention lately, since it appears that this tool, so important to organic and sustainable growers, will soon become ineffective. Other than Fred Gould, have the leaders of the IPM community engaged in the Bt-resistance debate?

The failure to contemplate system-wide effects is also evident in the lack of non-chemical IPM weed control efforts. Let's face it: IPM is too bug-oriented. And yet, most of the crop protection chemicals used in this country are herbicides. The Organic Farming Research Foundation's 1995 survey of organic farmers found that weed control was, by far, their number-one research priority.

Many experts argue that IPM needs to be more systems based. A 1996 National Research Council report (2) criticized IPM for its almost exclusive focus on insect control and failure to address weed and disease management. This critique is echoed in the push for "biointensive IPM," defined by Consumers Union in their 1996 landmark report *Pest Management at the Crossroads* as "a systems approach to pest management" (1).

Yet, systems research is difficult, both to conceptualize and to accomplish. Institutional barriers exist, among other problems. The Wallace Institute publishes the *American Journal of Alternative Agriculture* in conjunction with Washington State University. This journal was begun more than a decade ago to provide a forum for systems-based research, the kind where, for example, an entomologist teams up with a weed scientist, an agronomist, an agricultural economist, and a farmer to solve a production problem. While having the *Journal* helps, problems remain. For example, academic promotions are based on lead authorship, making team research undesirable. The USDA Sustainable Agriculture Research and Education (SARE) Program has recognized the difficulties in generating systems research and has a special category of grants just for this purpose, although even SARE's enlightened governing council admits that systems research remains difficult to define and promote.

Question #4: Are there better ways to extend IPM knowledge than those currently underway?

The short answer has to be "yes," since our adoption rate does not come close to the Y2K goal. I spent last summer at the Food and Agriculture Organization (FAO) of the United Nations in Rome. I had the opportunity to work with the staff of the FAO IPM Global Facility. Rather than describe their work in any detail, I would just like to emphasize its importance. I am convinced that the FAO IPM "farmer field" schools are phenomenally successful because farmers are not regarded as end-users of technology, but rather are engaged as fellow investigators and educators. Is there a way we can bring the "farmer field" schools into our extension programs?

The Natural Resources Defense Council (NRDC) published a wonderful report last summer, *Fields of Change*, which included 22 profiles of farmers who have found alternatives to pesticides. Much of the report discusses ways to better extend IPM knowledge. Among other things, NRDC cites the lack of on-farm research and education programs as a major hindrance to the adoption of alternatives to pesticides. Among the recommendations, NRDC argues that USDA should increase technical and financial cost-share assistance and incentive programs for farmers seeking to change practices. Programs such as the Environmental Quality Incentives Program (EQIP) and crop insurance programs are ways to extend IPM knowledge and practice.

Question #5: Are the environmental and consumer communities really committed to IPM?

There is a cadre of environmental and consumer groups currently engaged in activities to support IPM. The roster includes NRDC, the World Wildlife Fund, the National Campaign for Sustainable Agriculture, The Nature Conservancy, the Rural Advancement Fund International, Mothers and Others for a Livable Planet, the Sustainable Agriculture Coalition, Consumers Union, and my own organization.

But truthfully, the linkage is weak, and many more environmental and consumer allies are needed. While IPM is recognized as an environmental "win," its cause does not compete with issues like global warming or endangered species. How do we position IPM so that it becomes a central concern and agenda item for the environmental and consumer lobby? Environmentalists, in some ways, have more political clout than the scientific community. The good news is that the environmental community has developed greater knowledge and sensitivities about agricultural production, much of it engendered by the good work of sustainable and organic farmers, which has paved the way for IPM partnerships.

Question #6: Can the marketplace do what the government cannot?

Marketplace incentives for alternative agricultural practices are increasing in number. Most notably, numerous food and clothing labels have been developed to indicate to consumers that products are produced in environmentally responsible ways. Several IPM labels have been introduced into the marketplace. In New York State, Wegman's supermarket has had an IPM label since 1996. Massachusetts also has an IPM label. The Wisconsin potato growers, showcased in the 1998 budget submission to Congress as an example of IPM success, have brought together scientists, environmentalists, and farmers in a team to develop a label that indicates potatoes grown under an IPM system.

But caution is necessary. Organic farmers have had labels in the marketplace for more than 20 years, yet we are currently struggling to develop national standards to guide this booming industry. Our experience with the organic industry suggests that it is not easy to devise good standards or to enforce them. But this work must be undertaken. Credible standards are necessary to maintain consumer confidence.

Labeling isn't the only marketplace opportunity for IPM. For example, Consumers Union suggests that the government use its purchasing power to assist sustainable farmers by purchasing only organically grown cotton for military uniforms. Marketplace disincentives must also be eliminated. For example, mortgage-lending institutions must be convinced to eliminate requirements for termite treatments when there is no evidence of infestation.

Tackling the IPM of IPM

Amidst the valuable scientific discussions that understandably dominate this conference, we must also turn our attention to what I call the IPM of IPM. By this I mean the *I*nterest groups, *P*olitics, and *M*oney that make the world go 'round. As we struggle to answer the six questions I have posed, I would like to conclude with three thoughts.

First, IPM needs more interest-group backing. Everyone in this audience should join an interest group and work at the grassroots level to make IPM a priority issue for the organization.

Second, IPM needs more politics. Politics is not a dirty business, although some in the scientific community treat it as such. The IPM community needs to pool private money and hire lobbyists to follow the federal budget process, and everyone in this room should be writing to his or her Congressional representatives. While the IPM staff members in Washington, D.C., are terrific, they are barred from lobbying, so the only IPM voices within the Washington Beltway are those of your few environmental and consumer allies.

Finally, IPM needs to counter the monied interests that now dominate pesticide policy-making. While this community cannot compete with the resources of the chemical industry, it is critical that you bring the IPM message to consumers and enlist their support. In the organic debate, we succeeded in rewriting the power equation in Washington when 275,708 individuals wrote letters to the Secretary of Agriculture demanding an organic rule different from that supported by large corporate interests.

Literature Cited

1. Consumers Union. 1996. Pest Management at the Crossroads. Consumers Union, Yonkers, New York.
2. National Research Council. 1996. Ecologically Based Pest Management: New Solutions for a New Century. National Academy Press, Washington, D.C.
3. Natural Resources Defense Council. 1998. Fields of Change: A New Crop of American Farmers Finds Alternatives to Pesticides, NRDC, INC., New York, New York.

Summary: International Conference on Emerging Technologies for Integrated Pest Management

Barry J. Jacobsen
Department of Plant Sciences, Montana State University,
Bozeman, MT USA

This conference is the second national/international pest management conference hosted by North Carolina State University, and only time will answer the question: Will this conference be a similarly seminal event in the development and implementation of IPM that the 1970 *Concepts of Pest Management* conference has proven to be? This conference has enriched attendees from all crop protection disciplines (entomology, plant pathology, weed science, economics, crop consultants, regulators, industry, and engineering) with documentation on the impact and implementation of IPM on contemporary and emerging pest management technologies and on new opportunities for future implementation of IPM technologies. The broad representation of disciplines at this conference gives this author great optimism that the full potential of IPM will be realized in the coming decade.

Unlike the 1970 conference that focused primarily on insect pest management, this conference has embraced the multiple-discipline IPM concepts that have evolved from the basic concepts and principles developed by entomologists in the 1960s, and this conference has clearly demonstrated that IPM is the key contemporary crop protection paradigm. The impact of IPM on agriculture and the factors affecting its adoption were addressed in excellent presentations by Ann Sorensen, Pete Goodell, and J. R. Bradley. These presentations documented that IPM has contributed to reductions in agriculture's negative environmental impacts, reductions in human risks from pesticides, and increases in profitability, and has done so by encouraging a richer context to crop protection strategies. This was presented in the context that ever-greater interdisciplinary involvement, including social sciences, will be necessary to address future changes in agriculture if we are to address IPM on an agroecosystem basis. Numerous speakers emphasized that greater progress in IPM implementation will require more attention to applied aspects of ecology. However, it is critical to understand that implementing ecologically based pest management, as proposed by the National Research Council, will require substantially increased investments to address and understand multiple trophic layers in agroecosystems. Furthermore, implementing biological controls utilizing microbes will require basic research in microecology of all plant surfaces.

These and other speakers emphasized that the rate of change and the challenges for crop producers are increasing. Factors in this increasingly dynamic, economic, and societal flux include: changes in federal farm policy; a reduced economic safety net available to producers; increased farm size; greater involvement of consumers; the involvement of nongovernmental groups and other nontraditional groups in farm and pesticide policy; the importance of the global marketplace; the critical importance of reliable information services; biotechnology; and the public disinvestment in research and extension education from the land grant university system. While these factors are relatively global in nature, speakers have emphasized that IPM concepts are general but applications are site specific and that producers need solutions and reliable, unbiased information services applicable to their local situations. Clearly, the successful establishment of a crop consultant industry throughout the USA is a major advance in IPM implementation since the 1970 conference.

Risk management is currently a major emphasis of farm policy discussions. Changes in federal farm policy with the enactment of the 1995 Farm Bill fundamentally gave producers "freedom to farm." In exchange for loss of price supports, farmers could grow almost any crop they desired (there are some limitations for fruits and vegetables), and in return, the federal government would aggressively work in the international marketplace to assure free and fair market access. World surplus production in feed grains, cotton, and oilseeds, heavily subsidized European Union production, lack of free and fair market access, and economic turmoil in Asia are major factors contributing to the lowest commodity prices in recent history. The thin-to-non-existent profit margins have increased the importance of risk management for producers, especially in light of availability of crop insurance. The crop rotation and pest management opportunities made possible by the 1995 Farm Bill have been reduced or even eliminated for farmers who do not have a

506

production history for new crops, since insurance companies require production histories. Another problem is that for some growers, the level of insurance currently available (50% of expected yield) provides inadequate risk management. Insurers are uncertain about providing insurance products covering 65% to 85% of expected yields. IPM certification may make the sale of such insurance products possible for private insurers. Crop insurance for IPM is likely to be a major issue in the near future and, if resolved, may increase IPM implementation by producers in the future.

The impact of biotechnology on agriculture, and on IPM in particular, was highlighted by many conference speakers. Significant progress was documented, not only in genetically engineered crops, but also in development of highly sensitive diagnostic technologies for detection or identification of various pests, pesticide-resistant insects or pathogens, and insecticide residues. These new technologies will play an ever more important role in IPM implementation and regulatory issues in the future.

The use of transgenic pest- or pesticide-resistant crops has had a major impact on agriculture and IPM. The land area on which this technology has been implemented has grown logarithmically in the past 5 years, and environmental impacts have been considerable. For example, cotton and corn plants expressing one or more insecticidal endotoxin proteins from *Bacillus thuringiensis* (Bt), or crops modified for herbicide tolerance, have resulted in dramatic reductions in pesticide use when compared to use of non-transgenic technologies. For example, it is estimated that the use of Bt transgenic cotton has resulted in more than 1 million fewer liters of insecticide used in 1997. Stuart Sherrick, of Monsanto Inc., indicated that herbicide-tolerant crops and transgenic plants expressing plant protection genes were only the first wave of transgenic crops; plants expressing quality factors such as enhanced protein, oil, starch, or processing characteristics would be the next wave of genetically modified crops, followed by plants that would be factories for products such as vaccines or specific proteins. David Bridges' comment that 71% of USDA permits for genetically modified organisms issued to date deal with herbicide-tolerant crops or insect resistance clearly puts in perspective the current biotechnology development for IPM.

The impact of genetically modified crops on IPM has not been without real and perceived costs. Real costs have been a reduction in scouting by growers, a decrease in multiple tactics used, and the occurrence of new problems associated with a changed agroecosystem. Reduced scouting has been noted by growers employing Bt transgenic cotton varieties. As growers employ more genetic resistance, they often perceive it to be a "silver bullet" and ignore potential problems associated with single-tactic pest management strategies. The use of herbicide-tolerant cotton cultivars has resulted in increased aphid activity associated with early weed flushes and their subsequent sudden elimination of the habitat.

Perceived problems, such as the development of widespread pest resistance to a keystone insecticide like Bt, weeds becoming resistant to glyphosate, and an increasingly narrower germplasm base for crops, are based on sound science and cannot be ignored. Resistance management strategies were addressed by several speakers in the session on Genetic Engineering for IPM, and there was clear evidence that this was a focal point for industry, government regulatory agencies, extension educators, and researchers. An excellent example of approaches to resistance management not referred to in the conference was the North Central Region Publication Number 602. This document summarizes the work of the North Central Regional Research Committee on Bt corn and European corn borer resistance management. (See the following Web site: **http://www.extension.umn.edu/Documents/D/C/DC7055.html**.) An interesting concept relating to this was identified by Gary Fitt of Australia's CSRIO. He noted that, at first view, transgenic crops were clearly more economical for growers, but that IPM approaches were cheaper when environmental costs are charged to the transgenic strategy. Similar concepts were addressed for herbicide-tolerant crops, which entail risks associated with the spread of resistance genes to other plants and the risk of growers relying on only post-emergent weed control tactics. Roger Beachy addressed the development of disease-resistant transgenic plants and noted that the slow progress in this area was more related to lack of perceived value in the marketplace than to regulatory problems. While this may be true, the success with traditional disease-resistance breeding has been so great that the perceived need for transgenic solutions is low, particularly for fungal and bacterial pathogens, where success with transgenic plants has not been as great in the field as in the laboratory. Highlights in this area are transgenic plants with virus resistance, particularly coat protein transgenic papaya for control of papaya ringspot virus, resistance to several potato viruses, and resistance to zucchini yellow mosaic and watermelon viruses in several cucurbit crops. The true value of virus-resistant transgenic potato

508

cultivars is uncertain, because potato disease-free certification programs must still be conducted for other viruses and seedborne bacterial and fungal diseases.

As a pathologist, I am concerned with an ever greater narrowing of the germplasm base utilized by our major crops. The lessons of the Southern corn leaf blight epidemics of 1970-71 will surely be painfully relearned if we do not retain a significant genetic diversity in crop plants, particularly those planted over broad geographic areas. It is estimated that transgenic corn is now planted on more than 32 million hectares worldwide. This and the aforementioned concerns must be addressed if transgenic crops are to be part of a sustainable IPM paradigm. Clearly, research and implementation of multiple-tactic IPM strategies must be part of the future, because reliance on only transgenic resistance will be no more sustainable than reliance on the old-fashioned traditional resistance breeding.

Several speakers (Federici; Ferro) discussed progress with genetic engineering of insect pathogens, including Bt, nuclear polyhedrosis virus, fungi, and protozoa. Fred Gould discussed the fascinating area of genetic manipulation of insect pests to introduce genes for conditional lethal traits or lack of vector competence. Research in this area will be critically important in that fewer new insecticides are being developed due to the emphasis that is being placed on controlling pests in major crops using transgenic technology, the increasingly difficult regulatory and fiscally risky environment for development of new pesticides, and the need for low-cost solutions (for the producer) to pest problems in minor crops. IPM in minor crops represents an area in which public research will clearly need to take the lead, since the potential for financial returns to private business is low.

Discussions of biological control ranged from classical biological control and conservation or enhancement of native biological controls to augmentative strategies and the challenges associated with the commercialization of biological control technologies. Several particularly interesting concepts to promote biological control were presented by speakers in this session. Marjorie Hoy raised a most important issue, that of the need for more public funding for both interdisciplinary research and for extension education. The first point is clearly in line with calls for a more ecological approach to the development of pest management systems, and then later acknowledges the greater complexity of biologically based pest management compared to systems that are dependent upon pesticides. She presented the interesting idea that

509

pesticide registration should take into account situations where biological control is effective, and that in these situations, pesticide use should be by prescription. This approach would help to conserve effective biological control systems. The national database she proposed would be of use to both regulatory agencies and persons responsible for crop protection decisions. However, the details of how this database would be developed and how objective criteria for scoring the efficacy of biological control at the local level would be developed was not elucidated.

Joyce Loper focused on biological control of plant pathogens. Her presentation demonstrated the progress that has been made as a result of investments in fundamental microbial ecology, genetic manipulation of microbes, improved fermentation systems, and in the understanding of pathosystems and the interactions between biological control microbes and plants. Like Marjorie Hoy, she emphasized the need to incorporate biological control strategies with a multitude of other tactics. The importance of a strong research base was evidenced by R. Charudattan's comment that, on a worldwide basis, biological control of weeds is relatively unimportant, except in more developed countries, where he annotated several successes. Douglas Landis presented evidence that habitat modification can enhance biological control of weeds by native seed predators in traditional midwestern cropping systems. Jim Ligon's description of genetic modification of *Pseudomonas fluorescens* to enhance biological control activity by increasing antibiosis activity provides a model for the future of development of proprietary microbial biological control agents. Hopefully, this strategy will be extended to other biological control agents and other modes of action for biological control agents, such as competition and induced systemic resistance.

In his presentation on the commercialization of biological control products, Jan Menely identified both the advantages and limitations of biological control products in the marketplace. His perspective, based on more than 25 years of experience in this arena, provides reason for optimism for increased development. Reasons include: his documentation of products reaching the marketplace; the lower costs of bringing products to the marketplace when compared to traditional synthetic pesticides; improved acceptance by traditional growers; and an improved regulatory environment for reduced-risk products. The importance of government support in development and registration of biological control agents through SBIR and IR-4 cannot be underestimated. While the regulatory environment for registration has improved, there is a clear need for federal/state regulatory action dealing with efficacy and quality control issues.

Realistic analysis of contemporary and near-future IPM systems demonstrates that there is and will continue to be a need for pesticide interventions by persons responsible for crop protection. Presentations by Paul Jepson, Beth Carroll, and Frank Hall provided evidence of new directions in pesticide science. Of particular interest was Beth Carroll's presentation on behalf of Novartis. She discussed the reinvention of this major crop protection company in light of regulatory emphasis on reduced-risk pesticides and the clear need for integrated solutions to address problems like resistance management. This company is supporting IPM and the need to improve agricultural efficiency in response to the increasing human population and reductions in productive land available for agriculture. Examples include: identification of the most effective isomers of synthetic molecules to reduce application rates; marketing very narrow-spectrum materials to fit IPM programs; and increasing the use of biorational concepts that use nature as a model. Eric Tedford, of Zeneca Ag Products, echoed these concepts and emphasized that new products will often be based on models from nature; they will be reduced-risk and will require integrated programs to manage pest resistance. Tim Dennehy's presentation provided evidence that more attention needs to be focused on resistance management research and extension education. That Novartis will not be just a pesticide-based crop protection company was evidenced by the announcement of a "Packaged Crop Solutions" unit that will furnish a complete set of crop protection solutions (featuring both Novartis and non-Novartis products), scouting, and other pest management approaches. The initial system targeted by Novartis will be citrus in Florida. The Optimized Banana Production system described by Douglas Marin, of Del Monte, Inc., provides a model of a successful program. It will be interesting to see how a large, multinational company addresses the myriad of local situations that we know are critical to successful IPM implementation.

Optimization of application technologies will play an important role in providing predictable residues and efficacy. Variable-rate technology, GIS/GPS, remote sensing, computer-based pest decision models, and other technologies will play a major role in the future of IPM. An area not well addressed was the potentially unique needs for the delivery of biological agents.

Of the many important and timely topics discussed at this conference, perhaps none was more important than the ability to share and extend reliable information in a timely fashion. Ron Stinner's presentation indicated that the Internet is an excellent vehicle now, and that new advances in computer-based technologies will make it ever more valuable in the future. The ability to share weather, computer simulation programs, and product information on a global basis will create greater opportunities for IPM practitioners of the future.

This conference reinforced my belief that IPM has a bright future. Passage of the Food Quality Protection Act, embraced by both the pesticide industry and environmental groups, demonstrated that the IPM paradigm provides a common path forward for crop protection. It is important to emphasize that the current level of public investment will not allow for the rate of progress that is needed to provide the systems research and extension education necessary to achieve the full potential of IPM. While the Clinton IPM Initiative to achieve 75% IPM implementation by the year 2000 was applauded by IPM practitioners, researchers, and extension educators, it did not receive significant funding from Congress. Why was this not funded, despite the modest budget request and a balanced executive budget from USDA that, for the most part, reallocated funds rather that asking for new dollars? Kathleen Merrigan's comments provide a partial answer and a challenge to those of us who want to see the full potential of the IPM paradigm achieved. She indicated that, while most will agree that more funding is justified, neither the Clinton administration nor Congress sees this as a priority, and that a broad coalition of farm groups, environmental groups, and agribusiness that see IPM as a top priority has not been formed. In addition, the broader land grant university community feels that the agricultural research and education budget is already too small and that IPM is already being funded at an appropriate level. Finally, Merrigan indicated that, while we have been very successful in IPM research and implementation, our requests for more IPM funding generally seem to be for more of the same. Perhaps this is so, although we have initiated several new thrusts, including greater stakeholder involvement in priority setting and project development, more emphasis on systems research, more emphasis on assessment, greater integration of biological controls, more emphasis on resistance management, and an overall greater emphasis on development of crop protection strategies that are multidisciplinary and that address harmony with the agroecosystem.

It is clear that Congress and interest groups respond to crisis. Is FQPA the crisis that will build the needed coalition? Certainly, commonly held perceptions of EPA's implementation plans for FQPA will create a significant crisis, and IPM-based solutions appear to be "the only game in town." As already mentioned, the crop insurance problem may be another opportunity to garner support for IPM funding and implementation. Another more distant hope is an IPM label for food and other commodities. However, based on experience with the USDA's "organic" label, an IPM label is likely to be many years in the future.

List of Participants

Ed Adams
Washington State University
Spokane, WA 99202-1662

Mahmood Ahmad
Department Crop Protection/Univ
of Adelaide
Glen Osmond
South Australia SA 5064
AUSTRALIA

Miguel Ahumada
Bear Creek Orchards
Medford, OR 97501

Jeffrey Aldrich
Insect Chemical Ecology Lab
ARS, USDA
Beltsville, MD 20705

Bader Al-Hamar
Department of Agriculture
- Botany
University of Reading
Reading, Berks RG6 2AL1
UK

Osama Anas
Department of Plant Pathology
North Carolina State University
Raleigh, NC 27695-7616

Janet Anderson
US-EPA
Biopesticides & Pollution
Prevention Division
Office of Pesticide Programs
Washington, DC 20460

Thomas Anderson
BASF Corp
RTP, NC 27709-3528

Douglas Anspaugh
Department of Entomology
North Carolina State University
Raleigh, NC 27695

J. L. Apple
Department of Plant Pathology
North Carolina State University
Raleigh, NC 27695-7616

Jorge Aragon
National Institute of Agric.
Techn.Inta, C.C. 21
Marcos Juarez
Cordoba 2580
ARGENTINA

Felipe Arauz
University of Costa Rica
Escuela de Fitotecnia
U. de Costa Rica
San Pedro, San Jose
COSTA RICA

Virginia Aris
Department of Plant Pathology
North Carolina State University
Raleigh, NC 27695-7616

Richard Axtell
Department of Entomology
North Carolina State University
Raleigh, NC 27695-7613

514

John Ayers
Department of Plant Pathology
Penn State University
University Park, PA 16802

Karen Baer
Gemplers
Belleville, WI 53508

Jack Bailey
Department of Plant Pathology
North Carolina State University
Raleigh, NC 27695-7616

Woodward Bailey
Department of Entomology
North Carolina State University
Raleigh, NC 27695-7647

Waheed Bajwa
Oregon State University
Corvallis, OR 97331

Kenneth Barker
Department of Plant Pathology
North Carolina State University
Raleigh, NC 27695-7616

O.W. Barnett
Department of Plant Pathology
North Carolina State University
Raleigh, NC 27695-7616

Roger Beachy
Donald Danforth Plant Science
Center
St. Louis, MO 63105

Edward Bechinski
University of Idaho
Moscow, ID 83843
Anastasia Becker
University of Missouri
Columbia, MO 65211

Antonio Beltran
Rohm and Haas Company
Spring House, PA 19477

Mike Benson
Department of Plant Pathology
North Carolina State University
Raleigh, NC 27695-7616

Lee Bentz
Pennsylvania Dept. of
Agriculture-Plant Industry
Harrisburg, PA 17110

Richard Berberet
Department of Entomology and
Plant Pathology
Oklahoma State University
Stillwater, OK 74078

Armando Bergamin-Filho
University of Sao Paulo
ESALQ - Fitopatologia
Piracicaba, SP 13418-900
BRAZIL

David Boethel
LSU Agricultural Center
Entomology
Louisiana State University
Baton Rouge, LA 70803

Daniel Borchert
Department of Entomology
North Carolina State University
Raleigh, NC 27695-7613

Jim Bordick
USDA National Agricultural
Statistics Service
Washington, DC 20250-2009

515

Miguel Borges
Insect Chemical Ecology Lab.
USDA
Beltsville, MD 20705

J. R. Bradley
Department of Entomology
North Carolina State University
Raleigh, NC 27695-7630

Sheila Branch
Department of Biology
UNC - Greensboro
Greensboro, NC 27402
David Bridges
University of Georgia, Crop &
Soil Sciences Department
Griffin, GA 30223-1797

Wayne Brooks
Department of Entomology
North Carolina State University
Raleigh, NC 27695-7613

Richard Brown
Zeneca Ag Products
Richmond, CA 94804

William Brown
Colorado State University
Fort Collins, CO 80523

Dean Bushey
Rhone Poulenc Ag Co
RTP, NC 27709

Susan Butkewich
Ocean Spray Cranberries, Inc.
Lakeville, MA 02349

Juan Cabrera
Department of Entomology
North Carolina State University
Raleigh, NC 27695

Gavin Campbell
Monsanto Company
San Ramon, CA 94583

Beth Carroll
Novartis Crop Protection
Greensboro, NC 27419-8300

Marco Castro
Dole Fresh Fruit International
Colonia El Naranjal
La Ceiba
HONDURAS

Margaret Caswell
USDA-Economic Research
  Service
Washington, DC 20036

Dave Cerf
Maxygen
Santa Clara, CA 95051

Alison Chalmers
Rhone-Poulenc Ag Company
RTP, NC 27709

Laurence Chandler
USDA, ARS, Northern Grain
Insects Research Lab
Brookings, SD 57006

Raghavan Charudattan
Department of Plant Pathology
University of Florida
Gainesville, FL 32611-0680

Rachael Cheverie
PEI Dept of Agriculture
  and Forestry
Charlottetown, PEI C1A 7N3
CANADA

Rodrigo Cifuentes
Fruit Development Foundation
Sargento Aldea 305
Buin
CHILE

Sharon Clay
South Dakota State University
Department of Plant Science
Brookings, SD 57007-2141

Kieran Clements
Department of Entomology
North Carolina State University
Raleigh, NC 27695-7647

Alyssa Collins
Dept. of Biological Sciences
Rutgers University
New Brunswick, NJ 08901

Thomas Currier
AgraSol
Chapel Hill, NC 27514

Catherine Daniels
Washington State University
Richland, WA 99352-1671

Timothy Dennehy
Department of Entomology
University of Arizona
Tucson, AZ 85721

Jerry DeWitt
Iowa State University
Ames, IA 50011-1010

James Dill
University of Maine
  Cooperative Extension
  Pest Management
Orono, ME 04473-1295

James Dwyer
University of Maine
Cooperative Extension
Presque Isle, MA

Michael Ellsbury
USDA, ARS
Northern Grain Insects
Research Lab
Brookings, SD 57006

Annie Engebretson
American Society of Agronomy
Madison, WI 53711

Brian Federici
Department of Entomology
University of California
Riverside, CA 92521

David Ferro
Department of Entomology
University of Massachusetts
Amherst, MA 01003

Gary Fitt
CSIRO, Division of
Entomology & Plant Industry
Cotton Research Unit
Narrabri, NSW 2390
AUSTRALIA

Mike Fitzner
USDA-CSREES
Washington, DC 20250-2220

Shelby Fleischer
Department of Entomology
Penn State University
University Park, PA 16803

517

Richard Ford
Consortium for International
Crop Protection
University of Illinois
Urbana, IL 61801-4798

Soraya Franca
BRAZIL

James Frazier
Pennsylvania State University
University Park, PA 16802-3508

Thomas Fuchs
Texas A & M University
San Angelo, TX 76901

Lyn Garling
PA IPM Program
Penn State University
University Park, PA 16802

Brhane Gebrekidan
Virginia Tech OIRD/IPM CRSP
Blacksburg, VA 24061-0334

Randy Gold
BASF Corporation
RTP, NC 27709

Eugenia Gonzalez
Department of Plant Pathology
North Carolina State University
Raleigh, NC 27695

Peter Goodell
University of California
Kearney Ag. Center
Parlier, CA 93648

Fred Gould
Department of Entomology
North Carolina State University
Raleigh, NC 27695-7634

Jennifer Grant
Cornell University
Geneva, NY 14527

Thomas Green
IPM Institute of North America
Inc.
Madison, WI 53711

Randall Griffin
Clemson University
Department of Entomology
Clemson, SC 29634-0365

Judy Grundler
Missouri Department of
Agriculture
Jefferson City, MO 65102

Nanda Gudderra
North Carolina State University
Raleigh, NC 27695-7613

Jeffrey Gunsolus
University of Minnesota
St. Paul, MN 55108

Subhash Gupta
USDA/APHIS
Riverdale, MD 20737

David Guyer
Novartis Seeds
Durham, NC 27709

Franklin Hall
OARDC-Entomology
Department
Wooster, OH 44691

George Hamilton
Rutgers University
New Brunswick, NJ 08901-8524

518

Bruce Hammock
Department of Entomology
University of California
Davis, CA 95616

Abner Hammond
Louisiana State University
Baton Rouge, LA

David Handley
University of Maine
Monmouth, ME 04259

Lawrence Hanks
Department of Entomology
University of Illinois
Urbana, IL 61801

Dale Hardin
Agricultural Information Tech
Ft. Myers, FL 33907

James Harper
Department of Entomology
North Carolina State University
Raleigh, NC 27695-7613

Robert Henkins
AndCare, Inc.
RTP, NC 27709

Linda Herbst
WR PIAP Program
University of California
Davis, CA 95616

Peter Hertl
Department of Entomology
North Carolina State University
Raleigh, NC 27695-7613

Dan Hess
Affymax Research Institute
Palo Alto, CA 94304

Carlyle Holen
University of Minnesota
Extension Service
Crookston, MN 56716

Robert Holm
Schwab One
Belle Mead, NJ 08502

Thomas Holtzer
Department of Bioagricultural
Sciences & Pest Management
Colorado State University
Fort Collins, CO 80523-1177

Eric Honeycutt
Department of Plant Pathology
North Carolina State University
Raleigh, NC 27695-7616

Marjorie Hoy
Department of Entomology
  & Nematology
University of Florida
Gainesville, FL 32608

Randy Huckaba
Dow Agrosciences
Ferrum, VA 24088

Penny Hunst
AgrEvo USA Company
Wilmington, DE 19808

Jae-Soon Hwang
Department of Plant Pathology
North Carolina State University
Raleigh, NC 27695-7616

Jimo Ibrahim
NC A&T University
Greensboro, NC 27420

Hiroyuki Ieki
National Institute of Fruit
  Tree Science
2-1 Fujimoto
Tsukuba-shi
Ibaragi-ken 305-8605
JAPAN

Barry Jacobsen
Montana State University
Bozeman, MT 59717-3140

Douglas Jardine
Department of Plant Pathology
Kansas State University
Manhattan, KS 66506

Paul Jepson
Department of Entomology
Oregon State University
Corvallis, OR 97331-2907

Xixuan Jin
EcoScience Corporation
East Brunswick, NJ 08816

Peter Johnson
EnviroLogix, Inc.
Portland, ME 04103

Dan Jones
USDA-CSREES
Washington, DC 20250

Tom Juba
Department of Entomology
North Carolina State University
Raleigh, NC 27695-7613

Joseph M. Kemble
Department of Horticulture
Auburn University
Auburn, AL 36849

David Kendra
Novartis Crop Protection
Stanton, MN 55018-4308

George Kennedy
Department of Entomology
North Carolina State University
Raleigh, NC 27695-7630

Kathleen Kidd
NCDA and CS
  Beneficial Insect Lab
Raleigh, NC 27611

Esther Kioko
Department of Invertebrate
Zoology
National Museum of Kenya
Nairobi
KENYA

Michael Klein
Wooster, OH 44691

Joseph Kovach
Cornell University
IPM Program, NYSAES
Geneva, NY 14456

Jennifer Kuzma
National Research
Council/Board on Biology
Washington, DC 20418

Jose Laborde
State of Guanajuato Growers
  Industry
Guanajuato
MEXICO

Doug Landis
Department of Entomology
Michigan State University
East Lansing, MI 48824-1115

Joy Landis
Michigan State University
  IPM Program
East Lansing, MI 48824

Marlo Lascono
Universidad Católica del
Ecuador
Quito
ECUADOR

James Lashomb
Rutgers University
New Brunswick, NJ 08903

Jesusa Legaspi
Texas Agricultural Experiment
Station
Weslaco, TX 78596

Ana Legrand
Department of Plant Science
University of Connecticut
Hartford, CT 06105

Gary Leibee
CFREC/University of Florida
Sanford, FL 32771

Mary Jane Letaw
National Research Council
Falls Church, VA 22046

James Ligon
Novartis Biotechnology
Research
Durham, NC 27709-2257

Mike Linker
Department of Crop Science
North Carolina State University
Raleigh, NC 27695

Joyce Loper
USDA-ARS
Horticulture Crops Lab
Corvallis, OR 97330

Gus Lorenz
University of Arkansas
Cooperative Extension Service
Little Rock, AR 72203

Gregory Luther
OIRD, Virginia Tech
Blacksburg, VA 24061-0334

Ian MacRae
Department of Entomology
University of Minnesota
St. Paul, MN 55108

Richard Maples
Total Ag Services
Wee Waa, NSW 2388
AUSTRALIA

Douglas Marin
Del Monte Corp.
P.O. Box 4084-1000
San Jose
COSTA RICA

Frank Marks
Agriculture & Agri-food Canada
London, Ontario N5w 6b3
CANADA

Ronald Mau
University of Hawaii
Honolulu, HI 96822

Paul McDonald
Uniroyal Chemical
Bethany, CT 06524-3402

521

Heike Meissner
Department of Entomology
North Carolina State University
Raleigh, NC 27695

Charles Mellinger
Glades Crop Care, Inc.
Jupiter, FL 33458

Rick Melnicoe
Department of Environmental
  Toxicology
University of California
Davis, CA 95616

Tom Melton
Department of Plant Pathology
North Carolina State University
Raleigh, NC 27695-7616

Jan Meneley
Ag Bio Development, Inc.
Westminster, CO 82030

Kathleen Merrigan
Henry A. Wallace Institute for
  Alternative Agriculture
Washington, DC 20002

Leah Millar
Department of Entomology
North Carolina State University
Raleigh, NC 27695-7613

David Miller
EcoScience Corp.
Amherst, MA 01002

Robert Miller
Clemson University
Department of Plant Pathology
& Physiology
Clemson, SC 29634-0377

Jorge Miranda
Chiquita Brands International
Edificio D, Centro Empresarial
Forum
San Jose
COSTA RICA

Russell Mizell, III
University of Florida
Monticello, FL 32344

William Morrison
Department of Entomology
Texas A & M University
College Station, TX 77843

Vincent Morton
VIVA, Inc.
Reidsville, NC 27320

Brian Moser
Potlatch Corporation
Hybrid Poplar Program
Boardman, OR 97818

Cary Munkrold
Novartis Crop Protection
Stanton, MN 55018-4308

John Murphy
Department of Plant Pathology
Auburn University
Auburn, AL 36849

Gerald Musick
Department of Entomology
University of Arkansas
Fayetteville, AR 72701

Yuki Nakano
K-I Chemical U.S.A. Inc.
Leland, MS 38756

Godfrey Nalyanya
Department of Entomology
North Carolina State University
Raleigh, NC 27695-7613

Lawrence Nea
Seminis Vegetable Seeds, Inc.
Woodland, CA 95695

Nolan Newton
Public Health Pest Management
Division/Division of
Environmental Health
Raleigh, NC 27626-0593

Patricia O'Leary
Cotton, Inc.
Raleigh, NC 27612

Sara Oppenheim
Department of Entomology
North Carolina State University
Raleigh, NC 27695-7634

Eldon Ortman
Purdue University
West Lafayette, IN 47907

Lawrence Osborne
University of Nebraska
Lincoln, NE 68583

Peter Oudemans
Rutgers University
Chatsworth, NJ 08019

Sarote Panasahatham
Entomology Department
Oregon State Univ.
Corvallis, OR 97331

Curtis Petzolof
Cornell University
IPM NYSAES
Geneva, NY 14456

Daniel Pitts
Monsanto Company
Lexington, SC 29072

Sridhar Polavarapu
Rutgers University
Chatsworth, NJ 08019

Donald Prostak
American IPM Inc.
Glen Gardner, NJ 08826

Eduardo Rangel
State of Guanajuato Growers
  Industry
Guanajuato
MEXICO

Marjorie Rayburn
North Carolina
  Cooperative Extension Service
Edenton, NC 27932

Gene Reagan
Department of Entomology
Louisiana State University
Baton Rouge, LA 70803

Michael Reding
Utah State University
Logan, UT 84322-5305

Dennis Ring
Louisiana Cooperative
Extension Service
Baton Rouge, LA

David Ritchie
Department of Plant Pathology
North Carolina State University
Raleigh, NC 27695-7616

Daniel Robison
Department of Forestry
North Carolina State University
Raleigh, NC 27695

Michael Roe
Department of Entomology
North Carolina State University
Raleigh, NC 27695-7647

Ana Romero
Department of Plant Pathology
North Carolina State University
Raleigh, NC 27695-7616

Ronald Romero
Chiquita Brands
Oficentro FORUM, Santa Ana
San Jose
COSTA RICA

Tom Royer
Oklahoma State University
Stillwater, OK 74078-3033

Mark Rumph
IPM Alabama Program -
Auburn University
Auburn University, AL 36849

Joseph Russo
Skybit, Inc.
Boalsburg, PA 16827-0010

Rosa Sanhueza
EMBRAPA-CNPU
Rur Livramento, 515
Bento Gonsalves, RS
BRAZIL

Sandra Sardanelli
Department of Entomology
University of Maryland
College Park, MD 20742

Harald Scherm
Department of Plant Pathology
University of Georgia
Athens, GA 30602

Daniel Schiffhauer
Ocean Spray Cranberries
Chatsworth, NJ 08019

Katherine Schmidt
University of Wisconsin-
    Madison, UWEX
Madison, WI 53706

Greta Schuster
West Texas A & M University
Division of Agriculture
Canyon, TX 79016-0001

Veeresh Sevala
Paradigm Genetics
RTP, NC 27709-4528

Venkata Sevala
Department of Entomology
North Carolina State University
Raleigh, NC 27695-7647

Gary Sewell
University of Maine
Aroostook Farm
Presque Isle, Maine 04769

Amy Sheck
Department of Entomology
North Carolina State University
Raleigh, NC 27695

Stewart Sherrick
Monsanto
Youngsville, NC 27596

Edward Sikora
Auburn University
Auburn , AL 36849-5624

George Smith
University of Missouri
Columbia, MO 65211

Ann Sorensen
American Farmland Trust/CAE
Dekalb, IL 60115

Sandy Stewart
Immunovation
RTP, NC 27709

Ronald Stinner
NSF Center for IPM
North Carolina State University
Raleigh, NC 27606

Robert Stoltz
University of Idaho
Twin Falls, ID 83303-1827

Nicholas Storer
Department of Entomology
North Carolina State University
Raleigh, NC 27695-7630

Joyce Strand
University of California
Statewide IPM Project
Davis, CA 95616-8621

Glenn Strasburg
USDA NASS
Independence Ave.
Washington, DC 20250

Turner Sutton
Department of Plant Pathology
North Carolina State University
Raleigh, NC 27695-7616

Chester Sutula
Agdia, Inc.
Elkhart, IN 46514

Narayan Talekar
Asian Vegetable Research and
Development Center
Shanhua, Tainan 741
TAIWAN

Deirdre Tate
Zeneca Plant Science
Wilmington, DE 19850

Mary Thurman
Department of Crop Science
North Carolina State University
Raleigh, NC 27695-7620

Elizabeth Tomlin
Department of Biology
UNC - Greensboro
Greensboro, NC 27402

Stephen Toth
Department of Entomology
North Carolina State University
Raleigh, NC 27695-7613

Neal Van Alfen
Department of Plant Pathology
& Microbiology
Texas A&M University
College Station, TX 77843

John Van Duyn
Department of Entomology
North Carolina State University
Raleigh, NC 27695

James Van Kirk
Northeast Region IPM
Auburn, NY 13021-1495

Keith Waldron
Department of Entomology
Cornell University
Ithaca, NY 14853-0901

525

James Walgenbach
Department of Entomology
North Carolina State University
Fletcher, NC 28732

Wes Watson
Department of Entomology
North Carolina State University
Raleigh, NC 27695

Kerry Watts
Total Ag Services
Wee Waa
NSW 2388
AUSTRALIA

Donald Weber
Ocean Spray Cranberries
Lakeville, MA 02349

John Westbrook
USDA, ARS, Areawide Pest
Management Research Unit
College Station, TX 77845

Susan Whitney
Department of Entomology &
Applied Ecology
University of Delaware
Newark, DE 19717-1303

Phillip Wilcox
Mycogen Corp.
Cordova, TN 38018

Stephen Wilhite
University Park, MD 20782

Michael Williams
Mississippi State
  Extension Service
Mississippi State, MS 39762

Sharon Williamson
Department of Plant Pathology
North Carolina State University
Raleigh, NC 27695-7616

Daniel Woods
Developmental Biology Center
University of California
Irvine, CA 92697

Ken-ichi Yamaguchi
Mitsui Life Sciences Research
  Center
1144 Togo
Mobara, Chiba 297-0017
JAPAN

John Yaninek
Falls Church, VA 22043

Geoffrey Zehnder
Department of Entomology
Auburn University
Auburn, AL, 36849-5629

Paul Zorner
Mycogen Corporation
San Diego, CA 92121